Model-based Parameter Estimation in Computational Electromagnetics

Other related titles:

You may also like

- SBEW559 | Vipiana and Peng| Integral Equations for Real-Life Multiscale Electromagnetic Problems | 2023
- SBEW567 | Miller | Charge Acceleration and the Spatial Distribution of Radiation Emitted by Antennas and Scatterers | 2022
- SBEW549 | Gennarelli, D'Agostino, Ferrara, and Guerriero | Non-Redundant Near-Field to Far-Field Transformation Techniques | 2022

We also publish a wide range of books on the following topics:
Computing and Networks
Control, Robotics and Sensors
Electrical Regulations
Electromagnetics and Radar
Energy Engineering
Healthcare Technologies
History and Management of Technology
IET Codes and Guidance
Materials, Circuits and Devices
Model Forms
Nanomaterials and Nanotechnologies
Optics, Photonics and Lasers
Production, Design and Manufacturing
Security
Telecommunications
Transportation

All books are available in print via https://shop.theiet.org or as eBooks via our Digital Library https://digital-library.theiet.org.

Electromagnetic Waves 577

Model-based Parameter Estimation in Computational Electromagnetics

Edmund K. Miller

The Institution of Engineering and Technology

About the IET

This book is published by the Institution of Engineering and Technology (The IET).

We inspire, inform and influence the global engineering community to engineer a better world. As a diverse home across engineering and technology, we share knowledge that helps make better sense of the world, to accelerate innovation and solve the global challenges that matter.

The IET is a not-for-profit organisation. The surplus we make from our books is used to support activities and products for the engineering community and promote the positive role of science, engineering and technology in the world. This includes education resources and outreach, scholarships and awards, events and courses, publications, professional development and mentoring, and advocacy to governments.

To discover more about the IET, please visit https://www.theiet.org/.

About IET books

The IET publishes books across many engineering and technology disciplines. Our authors and editors offer fresh perspectives from universities and industry. Within our subject areas, we have several book series steered by editorial boards made up of leading subject experts.

We peer review each book at the proposal stage to ensure the quality and relevance of our publications.

Get involved

If you are interested in becoming an author, editor, series advisor, or peer reviewer please visit https://www.theiet.org/publishing/publishing-with-iet-books/ or contact author_support@theiet.org.

Discovering our electronic content

All of our books are available online via the IET's Digital Library. Our Digital Library is the home of technical documents, eBooks, conference publications, real-life case studies and journal articles. To find out more, please visit https://digital-library.theiet.org.

In collaboration with the United Nations and the International Publishers Association, the IET is a Signatory member of the SDG Publishers Compact. The Compact aims to accelerate progress to achieve the Sustainable Development Goals (SDGs) by 2030. Signatories aspire to develop sustainable practices and act as champions of the SDGs during the Decade of Action (2020–30), publishing books and journals that will help inform, develop, and inspire action in that direction.

In line with our sustainable goals, our UK printing partner has FSC accreditation, which is reducing our environmental impact to the planet. We use a print-on-demand model to further reduce our carbon footprint.

Published by SciTech Publishing an imprint of The Institution of Engineering and Technology, London, United Kingdom

The Institution of Engineering and Technology (the "**Publisher**") is registered as a Charity in England & Wales (no. 211014) and Scotland (no. SC038698).

First published: 2025

The Institution of Engineering and Technology
Futures Place,
Kings Way,
Stevenage,
Herts, SG1 2UA, United Kingdom

www.theiet.org

British Library Cataloguing in Publication Data

A catalogue record for this product is available from the British Library

ISBN 978-1-83724-537-6 (hardback)
ISBN 978-1-83724-538-3 (PDF)

Typeset in India by MPS Limited

Cover Image: Image shows geotomography, a method that uses electromagnetic waves to find tunnels. Used with permission from R. J. Lytle and D. L. Lager, Lawrence Livermore National Laboratory, USA.

Dedication

This book is dedicated to my wife of 65 years, Patricia Denn Miller. We met at the University of Michigan in 1957 and were married in 1958. Pat was a life partner and active participant in my professional life. This began when she rewrote 300+ pages of my dissertation since it was doubtful it could otherwise have been deciphered for typing given my terrible handwriting.

She also uncomplainingly accepted my 8 job changes and the various moves that they entailed. And she accommodated the travel that these jobs could involve, often accompanying me when circumstances permitted. As our children grew up she returned to school to earn a certificate in dental assisting, working thereafter in chairside surgical and hospital procedures. The picture was taken at our 50th wedding anniversary.

Contents

Preface[1]

There is rarely a human creative endeavor that does not involve at some point the discrete sampling and collection of quantitative data and its subsequent analysis. The latter activity most often involves looking for underlying relationships that reveal patterns and connections in the data that yield information about its source. Such relationships typically can be used to derive mathematical models that serve various applications.

One is extrapolation where a model of the past provides a way to predict future behavior for time-dependent phenomena. A second is interpolation where the model is used to join the discrete data samples to thus reveal accurately phenomenological behavior that has not been collected. For either extrapolation or interpolation, a model based on the available data is required, to be called in the following a fitting model (FM). Whatever its mathematical form, the FM will invariably incorporate parameters whose numerical values are determined by the sampled data.

System identification (SID) and model-based parameter estimation (MBPE) both use sampled data to develop FM parameters while also being significantly different. SID uses experimental measurements or observational data to identify the best fit to the data being processed from a set of candidate FMs. In other words, SID begins with data for which the FM itself is unknown. The identification step is accomplished by finding the FM that best fits the given data while at the same time determining that model's parameters.

MBPE, however, begins with a known FM that comes from a first-principles model (FPM) analytical description of whatever physical phenomenon is being investigated. For electromagnetic (EM) applications, the FPM is Maxwell's equations (ME). In this case, the FM does not itself come directly from ME but from fields, sources, and other observables that result from solutions produced from ME. But MBPE can also be used for experimental measurements to which an ME-generated FM is also applicable.

Chapter 1 highlights some issues associated with SID and the closely related mathematical tool of MBPE. Observe that SID always seems to begin with measured and/or observed data for which a plausible mathematical description is sought. However, MBPE involves an analytically pre-determined model for which the associated parameters are sought. In both cases, the goal is to develop a

[1]Bracketed references in the Preface and numbered equations are located in the related chapters of the book.

quantitative mathematical model that will be useful for interpolatory or predictive purposes. For the most part, this book is addressed to MBPE applications associated with wave-equation problems, as exemplified by ME. While EM SID is also feasible, most applications will involve MBPE used for analytical or computational data coming from MEs or experimental measurements.

Chapter 1 introduces these concepts in more detail and the particular applications of MBPE in EM computer modeling. The author's personal initial motivation for MBPE evolved from research at the University of Michigan Radiation Laboratory on Tropospheric Scatter Communication in early 1960s [3]. This was followed by Radar Cross Section (RCS) modeling in the late 1960s at MBAssociates in Danville, CA [4] and subsequently at Lawrence Livermore National Laboratory in connection with nuclear electromagnetic pulse (EMP) research.

In the early 1970s, time-domain modeling in electromagnetics began to receive more attention [7,8] together with the introduction of the singularity expansion method [5]. This, in turn, inspired interest in a rediscovered technique for processing of transient data called Prony's method (PM) [10] originally published in 1795. (An editor once changed this date to 1975, thinking that the numbers were transposed!) Much of the material throughout the remainder of this book employs applications of PM to not only transient data but for other kinds of exponential observables as well. Several examples of these applications complete the chapter.

Chapter 2 begins with a more detailed discussion of SID and includes several references to this topic. Note that when SID or MBPE or parameter estimation is included in the reference titles, their listing are displayed in boldface to highlight them. Several references are included for a variety of disciplines along with some excerpted examples from some of them. The discipline areas listed are:

- **System identification**
- **Acoustics**
- **Aerodynamics and aerospace**
- **Astronomy and cosmology**
- **Biological**
- **Chemistry and chemical engineering**
- **Electrical engineering**
- **Electromagnetics**
- **General and miscellaneous**
- **Matrix mathematics**
- **Mechanical and structural**
- **Medicine and medical**
- **Physics**
- **Power and devices**
- **Signal processing**
- **Thermodynamics**

Note that the choice of these categories is somewhat arbitrary. An ongoing internet search for publications that included some SID, MBPE, or related content produced 1,000+ references. Two-thirds to three-quarters of these references fall into the electromagnetics category.

Chapter 3 introduces the mathematical background for MBPE from the perspective of EM applications. The chapter begins with a discussion of FMs for waveform and spectral data in the time domain and frequency domain, respectively. A time-domain signal that is given by a series of complex exponentials whose exponents vary sequentially in uniform steps is a candidate for PM. This is the kind of waveform for which PM was developed. After a few mathematical steps, a polynomial, or characteristic, equation results whose roots contain its exponents, or poles, as shown in (3b). Similar exponential series occur for other variables such as angle, space, and frequency as discussed in Chapters 7, 8, 12, and 13. An interesting and useful extension of waveform sampling is the sampling of waveform derivatives [6], briefly outlined in Chapter 3 and discussed in detail in Chapter 14.

Also discussed in Chapter 3 are the spectral counterparts of PM where the FM is a rational function. Several examples of spectral-derivative sampling are presented, including some involving up to fourth-derivative sampling. The chapter concludes with a brief discussion of angle sampling of radiation and scattering patterns, of the possibility of using MBPE for phase recovery of magnitude-only data, and a summary of Appendix A where an extensive variety of other angle-sampling EM applications are summarized.

Because PM became an important tool in using MBPE in EM, Chapter 4 is devoted to discussing sampling strategies for its effective implementation. A factor in motivating this discussion had been the skepticism in the signal-processing community during our original exploration of its utility for EM applications. This chapter summarizes our experimentation in seeking the most effective approach to produce acceptable results

The results summarized in Chapter 4 led to the following conclusions and recommendations:

1. The characteristic-equation (polynomial) coefficients are the least accurate parameters obtained, although the accuracy can be correlated closely to that of the real pole components.
2. Use of the moving-window procedure and overdetermined data improve the accuracy of extracted parameters.
3. **Averaging the coefficients can produce pole accuracies comparable to averaging the poles directly.**
4. **Pole accuracy is approximately equal to the peak-signal-to-peak-noise ratio of input data.**

Chapter 5 continues the discussion of PM from the perspective of its performance as an information-preserving process, that is, by estimating the information content of the data presented to PM and the resulting information content of its output poles and residues. Information content is “counted” by expressing an item’s normalized accuracy in terms of the number of correct digits, *D*, it contains. Thus, if a quantity has normalized accuracy N_A of 0.001, then

$$D = -\log_{10}(N_A) = -\log_{10}(0.001) = -\log_{10}\left(10^{-3}\right) = 3$$

Note that D will most often not be an integer since N_A would rarely have a negative-integer exponent.

A variety of computer experiments are described to establish whether PM does conserve information in transforming time-domain exponential series to a pole-residue series. The input and output digit counts were compared as the processing parameters were systematically varied. The major conclusion reached was that PM can preserve the information content of data that is appropriately sampled and processed.

Chapter 6 considers the possibility of minimizing the number of frequency samples needed to represent an EM frequency response using adaptive sampling. The approach developed and tested was the possibility of sampling a spectrum using a series of overlapping FMs or windows. The sampled data may originate from a computational model or experimental measurement and is described as a generating model (GM). The GM provides the data samples from which a spectrum is to be estimated to some prescribed accuracy.

An appropriate FM for an EM frequency response is provided by a rational function, a ratio of two polynomials. Such an FM naturally includes the kinds of resonances included in the PM as the roots of the denominator polynomial. Equations (6.1) and (6.2) illustrate the connection between the time or waveform domain and the spectral or frequency domain. The non-pole terms in these FMs are included to handle the influence of effects that lie outside the FM window.

After an initial GM sampling is done over the overall frequency range of interest, the overlapping FMs are compared to determine the maximum difference or mismatch between all pairs. If that difference exceeds the specified level, a new GM sample is computed there and the affected FMs updated. A new maximum difference between FMs is then determined with this process repeated until the specified accuracy is achieved. This is illustrated conceptually in Figures 6.1 and 6.2. Several examples of adaptive sampling using this approach are included. Of particular interest might be results like those shown in Figure 6.4 where the initial GM uniformly spaced sample frequencies are plotted together with the adaptively added samples.

Chapter 7 introduces the application of PM for developing discrete-source approximations (DSAs) that produce the radiation patterns of continuous source distributions or prescribed radiation patterns, the classic pattern-synthesis problem. PM is shown to produce the standard Dolph–Chebyshev (DC), equal-sidelobe pattern, and a variation that has uniform, but different-level sidelobes on either side of the main beam. One unique feature offered by Prony pattern synthesis is that it produces a non-uniform element spacing for patterns that naturally require this. Also demonstrated is the application of singular-value decomposition to quantify the complexity of a pattern. Several other kinds of source distributions are explored.

Chapter 8 continues the theme of pattern synthesis but with a new concept, the integer exponentiation of a given pattern. For the example of a DC pattern raised to the 2nd, 3rd power, the sidelobes are decreased in value from their original –X dB, to −2X dB, −3X dB, etc. However, the number of sidelobes remains as it was for the original pattern, while the main lobe decreases successively in width and the array width and the number of elements increase in proportion to X. For a 10 element, 7.5 wavelength DC array, the −3deg main lobe decreases from 7.5 to 4 dB, while the array width increases in proportion to X. The SVD spectra are also found to extend in proportion to X, while their minimum values fall off exponentially as do the current strengths. These results indicate that exponentiation may be a practical design tool only for patterns having relatively modest initial sidelobe values.

The use of MBPE for reducing the cost of computing radiation and scattering patterns for large objects is discussed in Chapter 9. But in contrast with the previous two chapters, PM is not the final choice for this problem. As an introductory observation, it is mentioned that a measure of the complexity of a given pattern is its rank or the number of degrees of freedom (DoF) it has. This is represented by the number of numerically significant singular values the pattern has. A good approximation for the DoF of an array L wavelength long is $2L + 1$, a relevant quantity in establishing how many samples are needed to develop a continuous estimate of a pattern to an acceptable accuracy or uncertainty.

After discussing various options for FMs for pattern sampling and error estimation, and an approach using overlapping FM windows similar to what has been previously described for transfer functions, three different approaches based on DSAs are initially studied for pattern estimation. These can be described as Residue (RFM), Fourier and Prony FMs, where the RFM is a hybrid PM except that the source locations are specified and only the residues are computed. Of these three, the RFM was found to need the fewest pattern samples for effective performance.

The chapter continues with the application of RFM to several kinds of prescribed sources and adaptively developed patterns together with scatter plots that illustrate the progression of the sampling process. Figure 9.23 illustrates that the difference between the overlapping FMs converges to about 0.1 dB as the number of pattern samples is increased to about 2 samples per wavelength of aperture. This convergence rate is essentially independent of the aperture size varying from 10 to 25 wavelengths.

One question that arises in the pattern-synthesis exercise is whether the array whose geometry has been derived is realizable physically? This question is investigated in Chapter 10 by computer modeling several of the various arrays developed using PM and another procedure [22,23]. The NEC computer model was used for this investigation.

This involves two steps: (1) using NEC to model the particular array geometry to obtain the mutual interactions among its elements; and (2) determining the voltages required for the elements of the array that would produce the individual element currents determined from the synthesis procedure. The first two cases shown are nine-element DC arrays with −20 dB sidelobes, one a uniform array and the other a randomized array. Tables 10.1 and 10.2 show the voltages from NEC and corresponding currents along with the synthesized and NEC-model patterns, with both being graphically indistinguishable.

For the remaining examples, the tables include only the array currents, with the synthesized and NEC patterns while plots of the array geometries are used only when they are nonuniform or nonlinear. The two patterns generally continue to agree similarly to that seen in the first two examples, except for the exponentiated array in Figure 10.11 where the NEC cubed-array pattern becomes quite different. This is probably due to the limited precision of the NEC result compared with the 24-digit precision used for the synthesis results. These results overall provide mutual validation of the synthesis procedure and the NEC models.

An unanticipated bonus of MBPE is demonstrated in Chapter 11 where a way to estimate the numerical accuracy of spectral-domain or frequency-response data is provided. This opportunity is made feasible by the overlapping FMs that are involved in adaptive sampling of each new frequency, as previously described in Chapter 6. This capability exploits the best-fit agreement that can be achieved by two FMs in their region of overlap.

The rational for this is that the accuracy of the GM samples themselves is limited; however, they are obtained. For example, a classical boundary-value solution involves an analytical series that is necessarily truncated as it converges with a specified limit. This means that the accuracy of any FM based on those GM samples is subject to the same limit. It follows that the agreement between overlapping FMs is limited in the same way.

A variety of computer experiments are discussed in Chapter 11 that demonstrate this. Perhaps the clearest example in this respect is Figure 11.6. The plots there exhibit the agreement in digits between overlapping FMs sampled at frequencies other than those used to compute the FM parameters. The straight lines are the best fits to their respective convergence limits, L, used for the cylindrical functions in the analytical solutions for the cylinder current. Figure 11.7 summarizes these results where the average FM accuracy or agreement is plotted versus L, showing a 0.93 slope with a data match of 0.997.

Another way of estimating the GM error, or uncertainty, is also discussed in Section 6.4. There, the GM accuracy is revealed by adding noise of varying levels to the GM samples used for the FM evaluation. This is done by adding random noise averaging X% of the maximum GM level to the GM samples used to compute the FM parameters. The FMs will exhibit this noise in the average match of their GM estimates, $FMFM_{AM}(f)$, in their overlapping frequency band. If the initial value of X is 10%, then $FMFM_{AM}(f)$ will be about 10% or 1 digit, assuming that the error in the GM values is less than 10%.

This process is continued using successively smaller values of X%, for example 1%, 0.1%, etc. until $FMFM_{AM}(f)$ no longer decreases. When this happens, the X% value will provide an estimate of the GM digit accuracy. An example of this approach is illustrated in Figures 11.15 and 11.16. Alternatively, more GM frequency samples could be developed and those not incorporated for the FM computation could be used instead to obtain $FMGM_{AM}(f)$. These two results are plotted in Figure 11.16 where the $FMFM_{AM}(f)$ match is slightly higher than the $FMGM_{AM}$

(*f*). Alternatively, the extra GM data can also be processed as an overdetermined system, as demonstrated in Figure 11.17.

The concluding results of Chapter 11 exhibit the application of the MBPE approach to experimental data. The example included here shows, as might be expected, that processing complex data may be more challenging than magnitude-only data.

Chapter 12 deals with the application of PM to image wires of varying length and radius from their far fields. The image consists of spatial poles that represent point sources and whose residues represent their strengths. Motivation for this possibility was provided [2] where microwave holography was explored as a way to image a monopole antenna from its near field. Extending this idea to the far field seemed like a reasonable possibility for PM. A variety of computer experiments were subsequently conducted beginning with a discrete source distribution and the sinusoidal current filament (SCF). The SCF served as a special case with the 1/sin (θ) removed from the denominator of its far field to yield a three-point source as an analytical test of PM. To be consistent, the fields of the other currents analyzed using PM were also multiplied by sin(θ) to obtain results consistent with the reference SCF. The chapter concludes with several examples of currents on straight wires excited as antennas and scatterers and a discussion of their implications for radiation.

Another version of the inverse problem is determining the electrical properties of a layered medium using the EM field reflected from its surface by an incident plane wave, which is the topic of Chapter 13. This can be done by sampling the far field of a linear array as a function of cos(θ). The latter is necessary for the field to be sampled in uniform steps in the exponent of the pattern for the PM application. The resulting poles are determined by the physical location of the array elements and their directivity. This technique is also applied to the far field of the sinusoidal current filament and to various straight-wire antennas and scatterers. The wire results provide an indication about where the radiated field originates for such objects.

The concluding Chapter 14 expands on using GM derivative sampling briefly discussed in Chapter 3. The initial example of using what came to be called MBPE in EM was described as frequency-derivative interpolation [11]. This was originated by G. J. Burke for modeling the RCS frequency-dependence of radar decoys. The incorporation of MBPE and its implementation into a moment-method, integral-equation model such as NEC is further described in Chapter 14. The leading sequence of Figures 14.1–14.6 is included to demonstrate and explore the capability of derivative sampling. The applications that follow provide more examples of using frequency derivatives in the estimation of spectral-domain EM responses.

Appendix A includes a detailed summary of a wide variety of EM field observables that are candidates for processing using PM. Appendix B contains a list of notations and abbreviations used in the main text. Finally, Appendix C provides a representative sampling of publications that feature a variety of MBPE applications.

About the author

Edmund K. Miller earned a Ph.D. in electrical engineering from the University of Michigan in 1965 with an emphasis on computational electromagnetics. His working career has been varied, including employment at four universities (Michigan Technological University, University of Michigan, Kansas University and Ohio University), three companies (MB Associates, Rockwell International Science Center, and General Research Corporation, all in California) and two national laboratories (Lawrence Livermore and Los Alamos). He was the first president of ACES. He is a life fellow of IEEE and an ACES fellow.

Chapter 1

System identification and model-based parameter estimation[1]

Abstract

System identification (SID) and model-based parameter estimation (MBPE) are two related computational tools that are employed in almost any discipline and field of study that involves some sort of quantitative research and analysis. This includes not only obvious subjects such as science and engineering but such varied topics as economics, epidemiology and medicine, statistics, agriculture, and social studies. Literally, anything that can be measured or quantified is a candidate for SID and/or MBPE. Both employ models that involve parameters whose numerical values are estimated from data. For SID, on the one hand, the data comes from measurements or observations that are used via numerical analysis to determine the best-fit model and its parameters. For MBPE, on the other hand, the model is derived from known analytical equations that provide the data for estimating its parameters.

Then discussed is a brief account of the author's introduction and subsequent involvement in the evolution of MBPE in computational electromagnetics (CEM) and the role of Prony's method (PM) in determining the parameters of time-series transient data. The chapter concludes with several examples of different ways MBPE and PM have been used in CEM.

1.1 Introduction

This introductory chapter highlights some issues associated with system identification (SID) and a closely related mathematical tool of model-based parameter estimation (MBPE). Observe that SID seems always to begin with measured and/or observed data for which a plausible mathematical model is sought. However, MBPE involves an analytically predetermined model for which the associated parameters are sought. In both cases, the goal is to develop a quantitative mathematical model that will be useful for interpolatory or predictive purposes. For the most part, this book is addressed to MBPE applications associated with wave-equation problems, as

[1]The only prior reference that I have found in ongoing internet searches to MBPE is Dr. Pao-Ping Chang, "Model-based Parameter Estimation of the Caninie Systemic Circulation System," PhD thesis, Univ. of Wisconsin-Madison, 1973.

exemplified by Maxwell's equations (ME). While electromagnetic (EM) SID is also feasible, most applications will involve MBPE used for analytical or computational data coming from MEs or experimental measurements.

1.2 Background

An almost universally encountered problem in science and engineering is the development and application of analytically derived mathematical or first-principles models (FPMs). Initially, FPMs generally originated from observation and experimentation and may be more or less complex. A famous example of a less-complex model is Newton's development of the law of gravity and Kepler's related derivation of the laws of planetary motion. An example of a more complex analytical model is the ME, derived from various laboratory experiments and their mathematical relationships. Still other examples are provided by the Navier–Sokes equations applicable to problems in fluid dynamics, or the Vlasov equation for problems involving plasma physics. An internet search for publications related to MBPE and related topics will easily result in 1000+ references.

Analytical models based on such fundamental descriptions, denoted in the following as generating models (GM), often involve differential equations that are the province of integral and differential calculus. They can appropriately be described as microscopic models since they deal with the "fine" or high-resolution behavior that such models can provide. Consider the use of ME to determine the EM fields scattered from some object such as an airplane in flight. This might be done to obtain its radar cross section (RCS). Obtaining that information about such an object can require using a GM model to find the currents excited over its surface that the integration of which will yield the fields it scatters from which its RCS can be computed.

But there are many applications where the main information being sought is not the detailed complexity of the GM solution but instead a result to be obtained from it. An example is provided by the airplane where the goal is to find the frequency and angle dependencies of the aforementioned RCS. These results can be more appropriately described as being macroscopic as they are much less detailed than finding the entire current distribution over the object. Such results can be obtained from samples of the GM using what may be called, as is done in the following, a fitting model (FM). The process of using a less-detailed FM is often referred to as being derived from "model-order reduction" (MOR). It is the use of FMs in place of the rigorous but more computationally costly GMs where MBPE becomes relevant.

Key to the advantage of MBPE in such a situation is the substantial reduction in computer time required to more finely sample the macroscopic RCS in a way that avoids first finding the current distributions using a GM. While computing an RCS might seem to be a specialized case, similar situations arise in many areas of science and engineering. Besides the reduced computation cost of the FM samples compared with the GM that MBPE makes possible, there is the advantage that some assurance is provided that fine-scale variations in the RCS frequency dependence or viewing angle will not be missed.

It should be noted that MBPE is not limited to computational or analytical sampling of a phenomenon of interest, but it can also be used in the processing of measured results. Consider that one definition of parameter estimation found on the internet is "Parameter estimation is the process of computing a model's parameter values from measured data." A question that immediately arises is "What models?" An internet search using terms like "Model-based parameter estimation" or variations of this yields a seemingly unlimited variety of examples.

Historically, perhaps the first example of MBPE can be traced back nearly 2,000 years to observations of planetary motions that led the Greek astronomer Ptolemy to propose a model for the solar system. An internet description about Ptolemy from Britannica includes the statement "In several fields his writings represent the culminating achievement of Greco-Roman science, particularly his geocentric (Earth-centered) model of the universe now known as the Ptolemaic system." As can happen when the analytical basis for a model turns out to be wrong, the earth-centered assumption in this case, obtaining more and better data revealed that a planet sometimes appeared to reverse its direction of motion, or retrograde. This required revising the model and making it more complicated by superimposing smaller circular orbits or "epicycles" to the original geocentric circular orbit. Finally, many centuries later with the development of Newton's law of gravity, and with further analysis by Copernicus and Kepler, the correct sun-centered, elliptical-orbit model was developed.

An extensive and continuing major use of MBPE and related methods continues to be the processing of observations and purposeful measurements to quantify models in a variety of applications. In these cases, such data can provide a way to derive an analytical model. A simple example of this possibility is the situation where the model is expected from physical considerations to be a straight line. Two points would be sufficient to estimate the slope and x or y intercept of that simple model. If more than two data points are available, a best-fit straight line to that data can be derived that would also provide a quantitative error, or uncertainty, measure.

Generally speaking, the appropriate model can be expected to be more complex, and unless there is a known, analytical source for the model, a different approach is needed to determine one. A procedure called SID uses statistical methods to build mathematical models of dynamical systems from measured data that will usually be a function of time or frequency [1]. SID involves estimating a model of the system using measured input–output data. There are several ways to describe a system and to estimate such descriptions that involve three basic ingredients:

- The input–output data.
- A set of candidate models (the model structure).
- A criterion to select a particular model in the set, based on the information in the data (the identification method).

The identification process amounts to repeatedly testing a model structure, to find out the best model in a set of candidate models, and evaluating this model's properties to see if they are satisfactory for predictions that involve these steps:

1. Designing an experiment and collecting input–output data from the process to be identified.

2. Examining the data and if necessary removing outliers and selecting useful portions of the original data while also possibly filtering it to enhance important frequency ranges.
3. Selecting and defining a model structure (a set of candidate system descriptions) within which an appropriate model might be found.
4. Determining the best model in this set according to the input–output data and a given criterion of fit.
5. Examining the model's properties.
6. Terminating this process if the selected model is found to be acceptable. If this is not the case, then return to Step 3 to try another model set. Alternatively, try other parameter-estimation methods using the same data, or obtain further input–output data.

To summarize, a criterion is used to select a particular model in the set, based on the information in the data identification process which amounts to repeatedly selecting a model structure, computing the best model in the structure, and evaluating this model's properties to see if they are satisfactory. SID also includes the optimal design of experiments for efficiently generating informative data for fitting such models as well as MOR. A common approach is to start from measurements of the behavior of the system and the external influences (inputs to the system) and try to determine a mathematical relation between them without going into many details of what is actually happening inside the system; this approach is called black box SID. Observe that once it has been determined that acceptable and appropriate data has been acquired, the remainder of the SID process becomes model-based parameter estimation. However, this term is rarely found in published work that specifically includes SID, being primarily then referred to instead simply as parameter estimation.

Alternatively, when an appropriate model is available from theoretical analysis or has been determined by other means, only the parameter-estimation step remains. This is the case for much of the MBPE applications in EMs and other wave-equation disciplines such as acoustics and structural analysis where frequency responses are well-represented by rational functions and the associated time responses by a series of complex-argument exponentials. The cost of computing EM radiation patterns can also be greatly reduced using MBPE with the FM in this case not necessarily having any corresponding connection with the problem physics [2].

The author's personal involvement with the development of what eventually became known in the field of EM as MBPE began in 1960 working in the Radiation Laboratory at the University of Michigan [3]. Later in 1969 while working at a small company called MBAssociates in Danville, CA, a colleague there, Gerald J. Burke, developed a procedure that he called "Frequency Derivative Interpolation" for expediting computing the RCS frequency dependence of a wire loop scatterer. Rather than having to sample an excessive number of frequency steps, he demonstrated that many fewer samples could be used if a low-order polynomial FM were used to approximate the RCS values between these samples. A further reduction in the number of frequencies needed can be realized if frequency derivatives of the

RCS are also available. These samples enabled the development of a simple polynomial approximation or FM that provided an accurate and continuous plot of the RCS frequency dependence. This technique was described in a report to our sponsor but nothing further was done beyond that at the time [4].

This idea was not further pursued until the late 1970s after I joined the Lawrence Radiation Laboratory [the name of which later became the Lawrence Livermore Nation Laboratory (LLNL)]. Some aspects of our CEM activities began to indirectly include various signal processing and related topics that eventually evolved into MBPE. The initial CEM activities at LLNL concerned the electromagnetic pulse (EMP) produced by a nuclear explosion above the earth's ionosphere. An EMP poses a threat to electrical and electronic civilian and military systems, the evaluation of which had become a major focus of CEM work in the USA.

Analysis and computer modeling of EMP effects had motivated a variety of research studies to assess the threat and develop ways to mitigate the potential damage that it might cause. One area of research that received protracted attention over the years involved the resonances, or complex-frequency poles, of an object subject to EM excitation. For a passive object, these poles have negative real parts and determine its overall frequency response and its time-domain ringdown when impulsively excited. While knowledge of these poles was a long-known phenomenon, their potential use for application to EMP problems was introduced and popularized by Baum [5].

The usefulness of poles and Singularity Expansion Method (SEM) turned out not to be limited to EMP applications, as a variety of other potential uses for EM poles were investigated. For example, poles are a function of an object's size and shape. The object's size determines its oscillatory frequencies that are associated with the pole's imaginary components. A pole's real components however establish the damping or loss rate, an effect caused by radiation, an effect that is primarily dependent on the object's shape. These properties seem to offer the potential for poles to be used for automatic target recognition based on an object's scattered fields or RCS. As previously mentioned, another attractive use for EM poles that subsequently developed was the possibility of substantially reducing the time required to compute or measure the object's frequency dependence for literally any kind of problem. Some representative examples include RCS, antenna impedance, gain and radiation patterns, coupling effects between antennas, etc. Various ways of exploiting EM poles have been investigated and given different names such as asymptotic waveform evaluation [6], vector fitting [7], frequency-derivative interpolation [8], and Cauchy's method [9]. Whatever such methods are called, they all fit under the rubric of MBPE while also incorporating the concept of MOR.

Concern with EM poles at LLNL where I worked from 1971 to 1985 originated with an EMP project in Electronics Engineering. This evolved naturally to include a focus on time-domain electromagnetics (TDEM), including computer modeling [10,11] and experimental measurements [12]. Historically, perhaps the first pole-based analytical tool was Prony's method (PM) [13]. But PM had become regarded somewhat skeptically because of its perceived inaccuracy and sensitivity when applied to noisy data. This prompted our study of how reliably PM could be used for computing poles from TDEM results [14]. A related investigation was subsequently performed to determine the feasibility of obtaining EM poles directly from frequency-domain (FD) data [15]. This led to the evaluation of a FD counterpart of time-domain PM (TDPM) [16] as FDPM. Additional studies demonstrated that when appropriately used TDPM and FDPM could be valuable tools for determining EM poles from computed or measured data. This initial pole-related research was initially performed primarily to show how the poles of simple wire objects depended on their geometry.

Additional EM applications for EM poles were subsequently examined further. Included among them were:

1) A series of 3 LLNL reports on testing the performance of FDPM for:
   a) Analytically specified test spectra [17];
   b) Computed EM spectra [18]; and
   c) Identification of objects from their poles [19] with the latter summarized in a brief journal article [20]. The three "ship" models are shown here with the pole-based correlation results Table 1.1.
2) An evaluation of PM performance in transforming information from time-series data to an exponential series [21]. The results obtained indicate that when the data and processing parameters are judiciously chosen, PM essentially preserves the information content of the original information. The graph here shows the average accuracy of the PM poles and residues in digits on the vertical axis compared with the digit accuracy of the input time-domain data (Figure 1.2).
3) EM poles have been found to be useful in radiation-pattern synthesis and analysis. [22]. When used for pattern analysis, PM provides the source locations and strengths in a linear array. When a pattern is specified instead as shown below on the next page, the PM determines the array geometry and excitations needed to produce the pattern (Figure 1.3).
4) An overview of EM poles, the variety of ways that they occur, and the insight they provide for the analysis of EM fields. Especially influential is their control

Table 1.1 Tabulation showing the correlation results for the three ship models of Figure 1.1

X	*i*		
	1	**2**	**3**
1	1.000	0.329	0.637
2	0.316	1.000	0.701
3	0.523	0.536	1.000

The library poles were first extracted from the backscattered spectrum of each ship for a 60° incident wave. The FD correlation procedure was then applied to each of these spectra using the above poles.

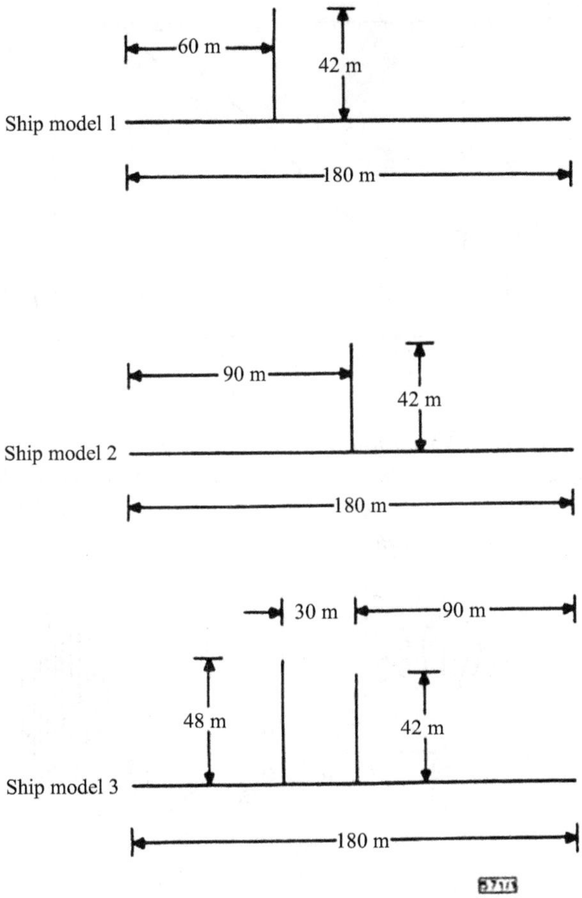

Figure 1.1 Stylised ship models used to test identification routines

of EM frequency transfer functions and the possibility this offers for MOR and more efficient descriptions of FMs [23]. The odd-numbered poles of the center-loaded dipole have a larger negative real component due to current maxima there and consequent resistive loss (Figure 1.4). The opposite situation occurs at the right-angle center bend case due to charge maxima there accelerating around the bend (Figure 1.5).

5) Continuing investigation using PM for inverting EM data to determine the thickness of a planar, penetrable layer and to determine the scattering centers along a straight-line source and the related problem of pattern synthesis. [24,25]. Below in Figure 1.6 is the frequency-dependent reflection coefficient for the layer and below the poles representing the periodic reflection peaks that occur with increasing frequency. The following Figure 1.7 is the radiation

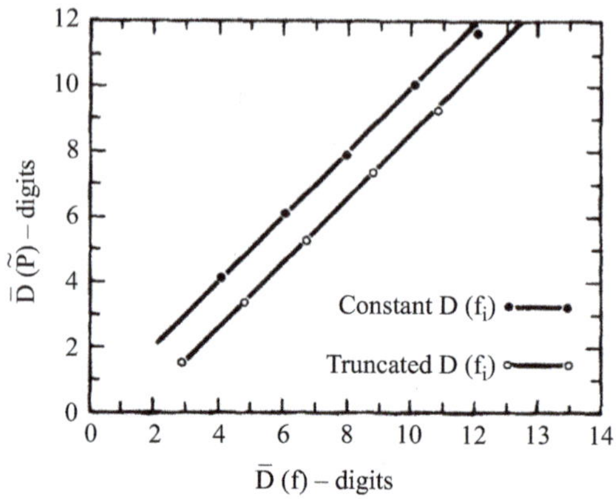

Figure 1.2 The output information remains proportional to the input information when the data are noisy, as shown here. In this case, the noise was simulated by truncation of the input at a given level relative to the peak signal amplitude.

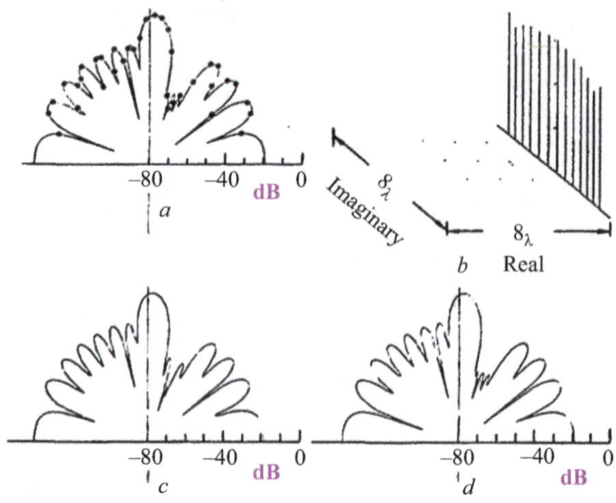

Figure 1.3 Synthesis of a polar pattern (dB scale) of a continuous 7× (700 m) aperture: (a) original pattern; (b) its derived distribution, using this technique; (c) reconstructed pattern; (d) constrained pattern

pattern of a linear array and the poles determined from PM that show the source locations (Figures 1.6 and 1.7).

6) It is important to note that MBPE is not limited to pole-based problems as it can exploit MOR in problems where other kinds of FMs are appropriate. One of the first examples of doing so was for problems having more complex Green's functions than required for an infinite-medium application. Modeling of an object

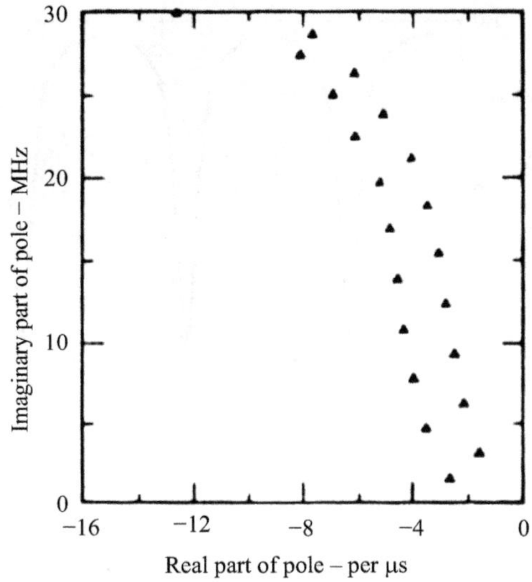

Figure 1.4 Poles of a center-loaded dipole

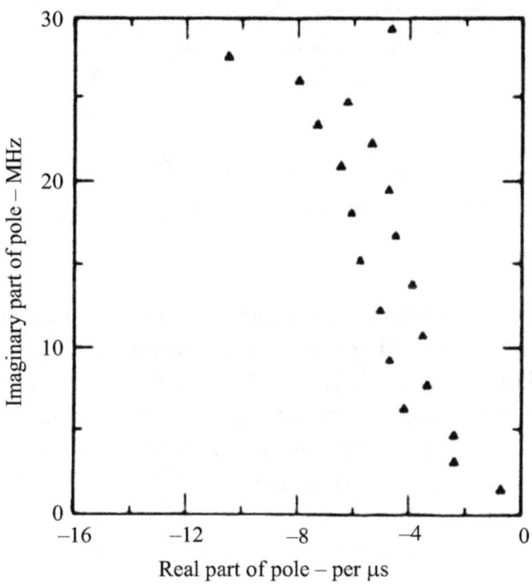

Figure 1.5 Poles of a dipole with right-angle center bend

near the planar, earth-air interface is one such example. Although the analytical treatment of this problem was developed by Sommerfeld early in the 20th century, the complexity of this formulation resisted accurate numerical treatment until 1960 or so. An approach that differed from the various analytical approximations

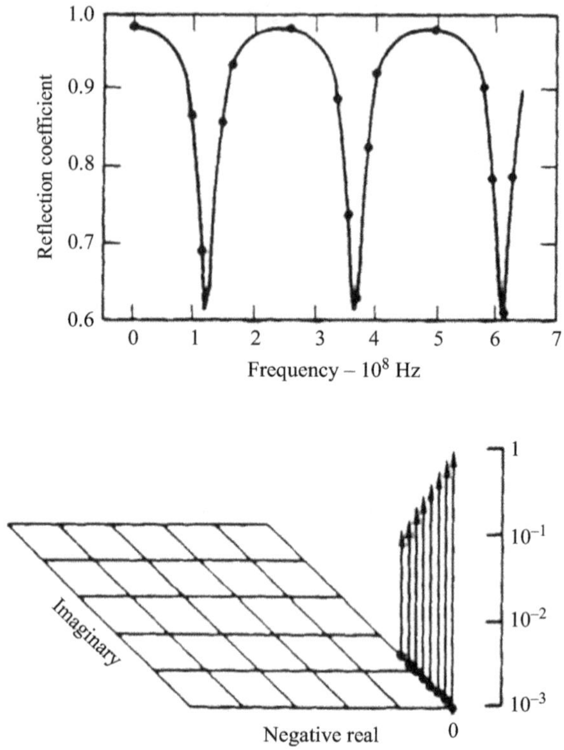

Figure 1.6 Frequency-dependent reflection coefficient and the poles due to periodic reflection maxima

derived from the Sommerfeld formulation instead used numerical interpolation to evaluate the fields of an EM source near the interface [26].

The interpolation approach is a 2-D procedure for the case where the EM source and field point are located on the same side of the interface, or the one-sided problem. In this case, the field is a function of two variables, the source-field separations along the interface and the sum of their distances perpendicular to it. A representative plot of one component of the field is shown below in Figure 1.8. As shown in this plot, interpolation is suitable for the one-sided problem but was found to be less suitable when the source and field points are on opposite sides, a two-sided problem.

In that case, an interpolation approach is necessarily 3-D and so a hybrid procedure was developed for the 3-D problem summarized briefly as [27]:

> "Least squares approximation (parameter estimation)—For efficient interpolation the interpolated functions were made as smooth as possible, by suppressing the singularity and phase factor, to better match the linear or higher order interpolation functions. As an alternative, we might

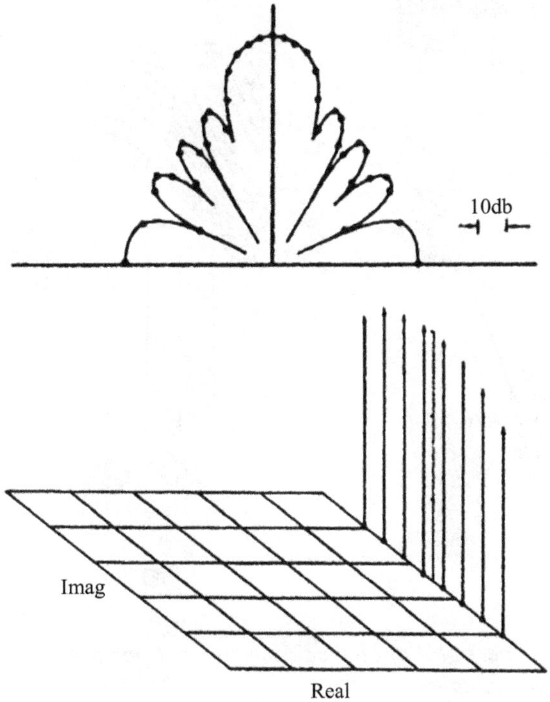

Figure 1.7 Radiation pattern of a linear array and the source locations yielded from PM

consider a model-based procedure. The basic difference is to employ interpolation functions chosen on physical or mathematical grounds to be close to the expected field behavior. The function amplitudes are obtained by enforcing a match to Sommerfeld integral values. If suitable model-based interpolation functions can be found, the possibility exists that the number of stored data values can be greatly reduced using a form suggested by appropriate asymptotic approximations . . . ."

The details of which can be found in the referenced article. Observe that while MBPE was not explicitly mentioned in this article per se, it was nevertheless used to obtain some of those results. In subsequent articles, that terminology appeared in the titles of numerous articles (Figures 1.8 and 1.9).

7) All of the preceding developments contributed to the evolutionary implementation of MBPE as a way to improve the computational efficiency of CEM. It seems that the first example to mention using MBPE as a means of reducing the sampling density required to obtain frequency response functions was mentioned by Miller [28].

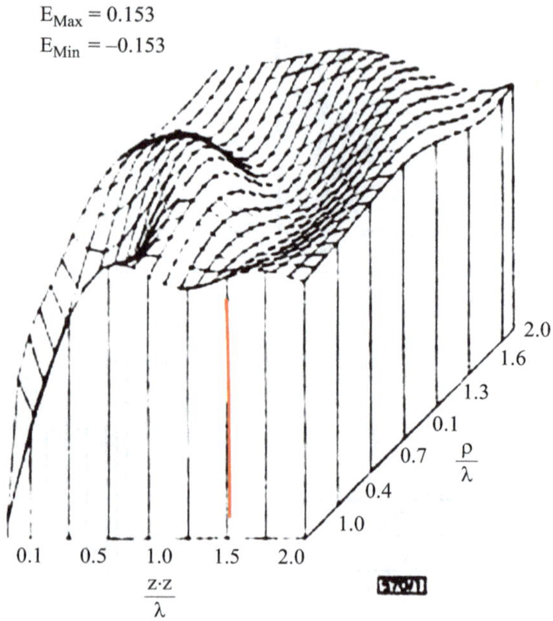

Figure 1.8 3-D aspect plot of Sommerfeld-integral portion of radial electric field for a vertical dipole

An excerpt from this publication begins

"Section B-4-b) Model-based parameter estimation: Identifying trends in data by sequentially connecting adjacent points with straight lines or using smoothed approximations thereto is an exercise familiar to any engineering student, a procedure that is typically described as curve fitting. The basic idea is to see whether curves which 'fit' the data exhibit any sort of explainable behavior which reveals some underlying physics and/or mathematics. A classical example of curve fitting is polynomial interpolation and extrapolation as previously discussed in connection with selecting basis and testing functions for using the moment method.

When the curve-fitting procedure employs an analytical description based on the physics of the process which produced the data, the exercise may be more accurately described as one of model-based parameter estimation (MBPE). The analytical description provides the model whose parameters are adjustable constants needing numerical determination before the model is quantifiable. These constants are estimated by fitting the model to the data, usually in some least-mean-square sense. In electromagnetics, and in the context of our previous discussion that source-field relationships establish a transfer-function representation, we observe that MBPE is a tool for solving the inverse problem. One example is that

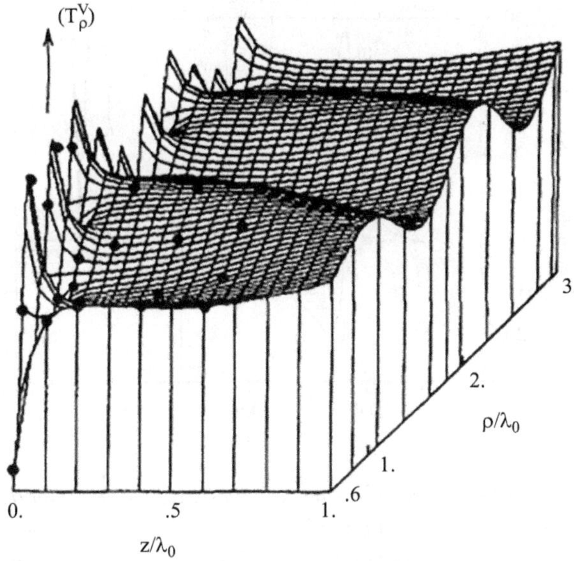

Figure 1.9 *Transmitted field, shown for T_p^y component with $\in^- = 16\,j0$ and source depth of z' $-0.1\lambda_0$, as modeled by least squares approximation. A total of 88 points are fit (36 in the plane of this figure with visible ones shown) to cover the 3-D region $0.6 \leq \rho/\lambda_0 \leq 3.0$, $0 \leq z/\lambda_0 \leq 2.0$, and $0 \leq |z'|/\lambda_0 \leq 0.25$ with less than 4% error.*

> of finding the complex resonances or poles of objects from their temporal or spectral responses using an exponential-or pole-series model."

A more detailed discussion of the roles of exponential FMs in CEM and MBPE was included in a 1987 NATO Advanced Study Institute [29] that was not published until 1991. In subsequent chapters, frequency sampling and ways of reducing other kinds of sampling densities in CEM will be discussed in more detail, some examples of which are described in [30] and [31]. One example from [29] is shown below where samples of a forked-dipole admittance and four derivatives are used at only two frequencies to obtain graphically indistinguishable results from 0.2 to 1.0 in units of h/wavelength (Figure 1.10).

Besides the interpolation and MBPE interface treatments considered above, another kind of Green's function for which MBPE has been found useful are the series forms that occur for 1-D, 2-D, and 3-D boundaries of planar, rectangular, and box-like enclosures [32]. The graph here shows the current on a half-wave dipole between two infinite, PEC planes obtained using a brute-force approach and our MBPE version. As mentioned in [32],

> As far as computation time is concerned. the MBPE the of Green's Function is most efficient, especially for larger numbers of segments on a given scatterer. According to our timings on a VAX 8650, the computation

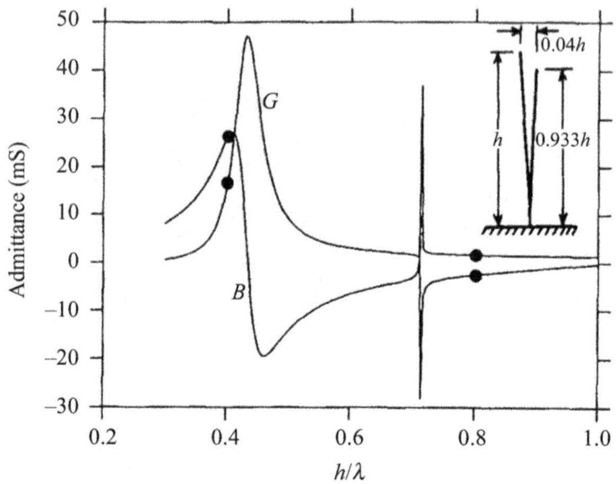

Figure 1.10 Input admittance of a forked monopole on a perfectly conducting ground. Directly computed values (————————) are compared with a rational function model (– – – –) with n = 5 and d = 4, matched to the value and four derivatives at h/λ = 0.4 and 0.8.

of the current on a half-wave wire with 6 segments was 20 times faster with our MOM model incorporating MBPE as compared to the brute force Green's function. Significantly greater time savings would be expected for 2- and 3-D enclosures since the MBPE numerical grid would be relatively unchanged, whereas the 'brute force' Green's functions contain double, and triple infinite sums, respectively (Figure 1.11).

8) EM radiation patterns are another problem that can require excessive computational cost as problem sizes become larger and more complex for which MBPE can substantially reduce the sampling requirements [33,34]. An example of using three kinds of FMs in sampling the radiation pattern of a 20-wavelength uniform current filament are included in [34]. A result for a discrete-source approximation FM is shown here using 13 overlapping FMs and 14 initial GM samples. A total of 21 additional samples were needed to obtain a final pattern that satisfied the specified error criterion for a total of 35 to derive a 20-lobe pattern, the cosine angles locations of which are shown in a graph on the next page (Figures 1.12 and 1.13).

$$\mathrm{FME}(\theta) = \mathrm{A}_1 + \mathrm{A}_2\left[\left|\mathrm{GM}(\theta_{\max}) - \frac{1}{M}\sum_{\alpha=1}^{M}\mathrm{FM}_{\alpha}(\theta)\right|\right] \leq \mathrm{FME}_{\max},$$

The permitted FM error varies with the observation angle, so it varies in proportion to the decrease in the FM values relative to the maximum GM. For the

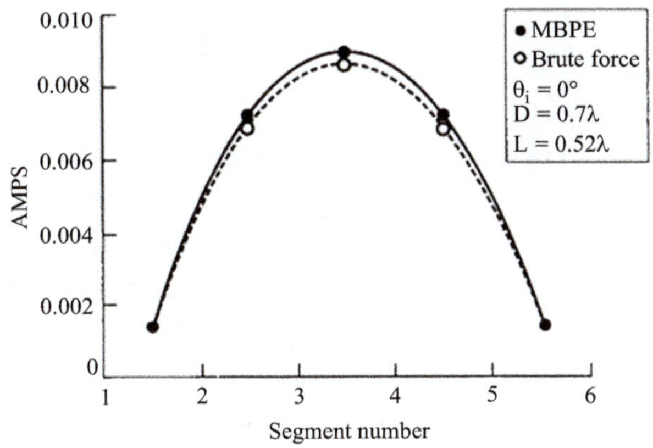

Figure 1.11 Current distribution on a 1/2 dipole using brute force and MBPE

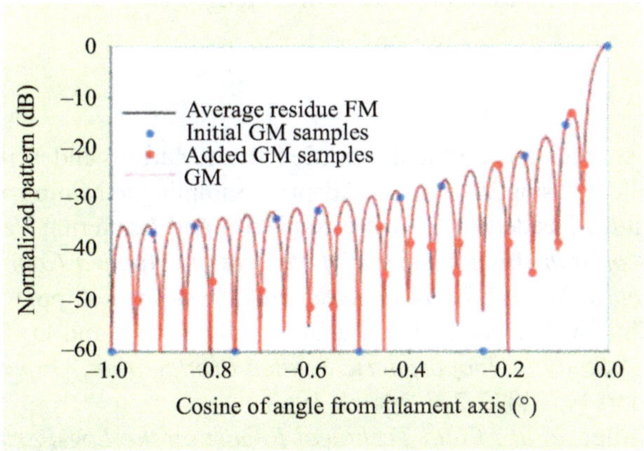

Figure 1.12 The normalized actual field of the 20-wavelength UCF compared with the average of the 13 residue FMs as a function of the cosine of the observation angle measured from the filament axis. The circles indicate 34 of the 35 GM samples computed from (6) (one is below −60 dB) that were used in the estimation process. Because the cosine angle values used to obtain the average and actual patterns do not always precisely match those used to obtain the original GM samples, some of the latter do not coincide with the pattern plots

results shown here, the nominal values for A1 and A2 were 0.1 and 0.05 dB, respectively, with the maximum FME = 3 dB. Other possible variations that might be used to set the acceptable error or uncertainty could be to relax the desired accuracy in certain angular regions or increase it in other regions consistent with the requirements of a particular application.

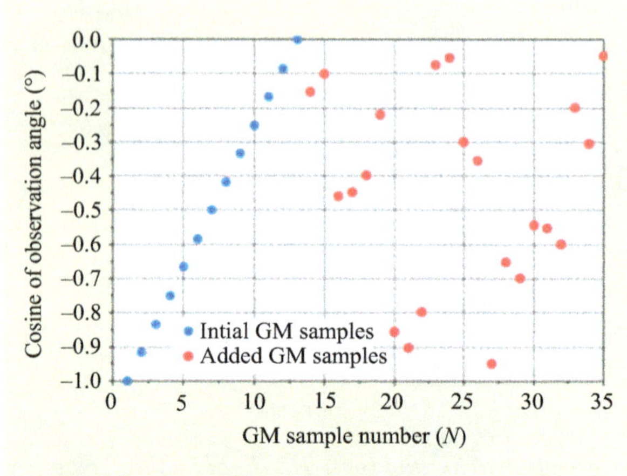

Figure 1.13 Sequence of GM samples showing their locations in the cosine of the observation angle for a 20-wavelength UCF using 13 FMs

References

[1] Internet sources adapted and modified from MatLab and Wikipedia, 2024.

[2] E. K. Miller, "Using Windowed, Adaptive Sampling to Minimize the Number of Field Values Needed to Estimate Radiation and Scattering Patterns," in *Proceedings of 14th Annual Review of Progress in Applied Computational Electromagnetics, Naval Postgraduate School, Monterey, CA*, pp. 958–963, 1998.

[3] C. M. Chu, A. W. Wren, J. J. LaRue, E. K. Miller, and D. T. Politis, *Final Report: A Study of Tropospheric Scatter Propagation*, University of Michigan, Report No. 2872-2-F, March 1960.

[4] E. K. Mliller, *et al.*, *Final Technical Report on the Log-Periodic Scattering Army Program*, MBAssociates. Report No. HB-R-70/105, Contract No. F04701-68-C-0188, 1970.

[5] Carl E. Baum, "The Singularity Expansion Method (SEM): Background and Developments," *IEEE Antenna and Propagation Society Newsletter*, Vol. 18, pp. 15–23, 1986.

[6] C. R. Cockrell, and F. B. Beck, "Asymptotic Waveform Evaluation (AWE) Technique for Frequency Domain Electromagnetic Analysis," *NASA Technical Memorandum*, Vol. 110292, 1996.

[7] B. Gustavsen, and A. Semlyen, "Rational Approximation of Frequency Domain Responses by Vector Fitting," *IEEE Transactions on Power Delivery*, Vol. 14, No. 3, pp. 1052–1061, 1999. DOI:10.1109/61.772353.

[8] E. K. Miller, G. J. Burke, and S. Chakrabarti, "An Extended Frequency-Derivative Model for Adaptive Sampling of Electromagnetic Transfer Functions", National Radio Science Meeting, Boulder, CO, January 4–6, p. 164, 1989.

[9] Raviraj S. Adve, *et al.*, "Application of the Cauchy Method for Extrapolating/Interpolating Narrow-Band, System Responses," *IEEE Transactions on Microwave Theory and Techniques*, Vol. 45, No. 5, pp. 837–845, 1997.

[10] E. K. Miller, A. J. Poggio, and G. J. Burke, "An Integro-Differential Equation Technique for the Time-Domain Analysis of Thin-Wire Structures, Part I: The Numerical Method," *Journal of Computational Physics*, Vol. 12, pp. 24–48, 1973.

[11] A. J. Poggio, E. K. Miller, and G. J. Burke, "An Integro-Differential Equation Technique for the Time-Domain Analysis of Thin-Wire Structures. Part II: Numerical Results," *Journal of Computational Physics*, Vol. 12, pp. 210–233, 1973.

[12] R. M. Bevensee, and E. K. Miller, *The LLNL Electromagnetic Transient Facility: Design, Test, and Applications of a Time-Domain Range, Chapter 7 in Time-Domain Measurements in Electromagnetics*, Van Nostrand, New York, 1986.

[13] C. Prony, "Essai expérimental et analytique sur les lois de la dilabilité des fluides élastiques, et sur celles de la force expansive de la vapeur de l'eau et de la vapeur de l'alkool, à différentes températures," *Journal de l'École polytechnique*, Vol. 2, pp. 24–77, 1795.

[14] A. J. Poggio, M. L. VanBlaricum, E. K. Miller, and R. Mittra, "Evaluation of a Processing Technique for Transient Data," *IEEE Transactions on Antennas and Propagation, AP*, Vol. 26, pp. 165–173, 1978.

[15] J. N. Brittingham, E. K. Miller, and J. L. Willows, "Pole Extraction from Real-Frequency Information," *Proceedings of the IEEE*, Vol. 68, pp. 263–273, 1980.

[16] S. Chakrabarti, K. R. Demarest, and E. K. Miller, "An Extended Frequency-Domain Prony's Method for Transfer Function Parameter Estimation," *International Journal of Numerical Modelling*, Vol. 6, No. 4, pp. 269–282, 1993.

[17] J. N. Brlttingham, E. K. Miller, and J. I,. Willows, *The Derivation Of Simple Poles in a Transfer Function from Real-Frequency Information*, UCRL-52050, April, 1976.

[18] J. N. Brittingham, E. K. Miller, and J. L. Willows, *The Derivation of Simple Poles in a Transfer Function From Real-Frequency Information, Part 2: Results from Real EM Data*, UCRL-52118, August, 1976.

[19] E. K. Miller, J. N. Brittingham, and J. L. Willows, *The Derivation of Simple Poles in a Transfer Function From Real-Frequency Information, Part 3: Object Classification and Identification*, UCRL-52211, January, 1977.

[20] E. K. Miller, J. N. Brittingham, and J. L. Willows (1977), "Identification of EM Spectrum by Known Pole Sets," *Electronics Letters*, Vol. 13, pp. 774–777.

[21] E. K. Miller, and D. L. Lager, *Information Extraction Using Prony's Method*, UCRL-52329, August, 1977.

[22] E. K. Miller, D. L. Lager, and J. T. Okada, *Imaging of Linear Source Distributions*, UCRL-52822, November 1979.

[23] E. K. Miller, *Natural Mode Methods in Frequency and Time Domain Analysis, Theoretical Methods for Determining the Interaction of Electromagnetic Waves with Structures*, Sijthoff and Noordhoff, pp. 173–212, 1979.

[24] E. K. Miller, and D. L. Lager, "Inversion of One-Dimensional Scattering Data Using Prony's Method," *Radio Science*, Vol. 17, pp. 211–217, 1982.

[25] E. K. Miller and D. L. Lager (1983), "Imaging of Linear Source Distributions," *Electromagnetics*, Vol. 3, pp. 21–40.

[26] E. K. Miller, J. N. Brittingham, and J. T. Okada, "Bivariate- Interpolation Approach for Efficiently and Accurately Modeling Antennas Near a Half Space," *Electronics Letters*, Vol. 13, pp. 690–691, 1977.

[27] G. J. Burke, and E. K. Miller, "Modeling Antennas Near to and Penetrating a Lossy Interface," *IEEE Transactions on Antennas and Propagation, AP*, Vol. 32, pp. 1040–1049, 1984.

[28] E. K. Miller, "A Selective Survey of Computational Electromagnetics," *Invited Tutorial-Review Paper, IEEE Transactions Antennas and Propagation*, Vol. 36, pp. 1281–1305, 1988.

[29] E. K. Miller, "Model-Based Parameter-Estimation Applications in Electromagnetics," in *Electromagnetic Modeling and Measurements for Analysis and Synthesis Problems*, edited by B. de Neumann (eds.), Kluwer Academic Publishers, *NATO ASI Series E: Applied Sciences*, Vol. 199, pp. 205–256, 1991.

[30] G. J. Burke, E. K. Miller, S. Chakrabarti, and K. R. Demarest, "Using Model-Based Parameter Estimation to Increase the Efficiency of Computing Electromagnetic Transfer Functions," *IEEE Transactions on Magnetics*, Vol. 25, No. 4, pp. 2807–2809, 1989.

[31] E. K. Miller, "Using Adaptive Estimation to Minimize the Number of Samples Needed to Develop a Radiation or Scattering Pattern to a Specified Uncertainty," *Applied Computational Electromagnetics Society Journal*, Vol. 17, No. 3, pp. 176–186, 2002.

[32] K. R. Demarest, E. K. Miller, K. Kalbasi, and L-K Wu, "A Computationally Efficient Method of Evaluating Green's Functions for 1-, 2-, and 3D Enclosures," *IEEE Transactions on Magnetics*, Vol. 25, No. 4, pp. 2878–2880, 1989.

[33] E. K. Miller, "Using Windowed, Adaptive Sampling to Minimize the Number of Field Values Needed to Estimate Radiation and Scattering Patterns," in *Proceedings of 14th Annual Review of Progress in Applied Computational Electromagnetics, Naval Postgraduate School, Monterey, CA*, pp. 958–963, 1998.

[34] Edmund K. Miller, "Adaptive Sparse Sampling to Estimate Radiation and Scattering Patterns to a Specified Uncertainty with Model-Based Parameter Estimation," *IEEE Antennas and Propagation Society Magazine*, Vol. 57, No. 4, pp. 103–113, 2015.

Chapter 2

A brief sampling of system identification and model-based parameter estimation applications in various disciplines

Abstract

This chapter continues with a more detailed discussion of system identification and model-based parameter estimation, their similarities and differences. A list of some subject areas where these methods have been used together with a few representative references of each and one or two example applications are given below. The list begins with System Identification, and follows in alphabetical order with:

Acoustics
Aerodynamics and aerospace
Astronomy and cosmology
Biological
Chemistry and chemical engineering
Electrical engineering
Electromagnetics
General and miscellaneous
Matrix mathematics
Mechanical and structural
Medicine and medical
Physics
Power and devices
Signal processing
Thermodynamics

2.1 Introduction

System identification (SID) began as a **methodology for building mathematical models of dynamic systems using measurements of the input and output signals of the system** [1]. The process of SID requires that the measured input and/or output signals from a system to be obtained in time or frequency domain and that a model structure or structures be selected for testing.

This is often done using *statistical* methods to build mathematical models of dynamical systems from the measured data. Possibly also included in SID is the optimal design of experiments for efficiently generating the most appropriate data for testing a candidate reduced-order fitting model (FM). A common approach is to start from measurements of the behavior of the system and the external influences (inputs to the system) and try to determine a mathematical relation between them without going into many details of what is actually happening inside the system; this approach is called black box SID and generally involves computing the parameters of candidate FMs. A dynamic mathematical model in this context is a mathematical description of the dynamic behavior of a system or process in either the time or frequency domain. Examples include:

- Physical processes such as the trajectory of a bullet affected by the influence of gravity.
- Economic processes such as commodity prices that react to external influences such as the weather and fuel prices.

One of the many possible applications of SID is in control systems. For example, it is the basis for modern data-driven control systems, in which concepts of SID are integrated into the controller design, and lay the foundations for formal controller optimality proofs. SID techniques can utilize both input and output data (e.g. eigensystem realization algorithm) or can include only the output data (e.g. frequency-domain decomposition). Typically, an input–output technique would be more accurate, but the input data is not always available. The process of testing candidate FMs involves what may be described as model-based parameter estimation (MBPE).

The variety and number of MBPE-related publications that can be found in a cursory internet survey provide evidence of just how much attention has been devoted to this general area during the last few decades. A few representative samples are listed here from several different disciplines to give an idea of some of the research and applications that can be found. The items are listed in somewhat arbitrarily designated areas in chronological order of publication.

Thus, the problem of developing the pole sets for various kinds of data belongs to the more general class of SID. A succinct summary of SID is "the determination, on the basis of input and output, of a system within a specified class of systems to which the system under test is equivalent." The origin of SID began in situations where experimental and/or observational data provided the basis for developing analytical FMs that would yield useful insight or be helpful in optimizing designs of more or less complex systems. With the ever-growing capability of computers, a new source of analytical data has become available for problems where first-principles models (FPMs) have become practically solvable.

As mentioned previously, an FPM can be described as "microscopic" as it can provide as finely detailed solution as desired for a given problem. But in many cases, what is needed from that microscopic model are aggregated results that can be appropriately described as "macroscopic." The case of determining the radar cross section of, for example, a B-52 aircraft at a wavelength of 1 GHz bandwidth centered at 10 GHz, a macroscopic result, from microscopic FPM data is a representative example that emphasizes the essential role of MBPE in SID: Does the FPM need to be solved at enough frequency samples to ensure no important details are missed or can some FM be applied to reduce the number of FPM samples to minimize the computer cost? The original and traditional role of SID has been to test candidate FMs using whatever measured or observed data is available to identify the "best" FM for the application of interest. Having FPM computed results combined with reduced-order FMs that are based on physical principles is the advantage MBPE offers and is the primary focus of this book.

Thus, SID remains a methodology to find best-fit models for measured and/or observational data while incorporating what can properly be called MBPE even though that term has rarely been employed in connection with it. Extending the application of MBPE to computational results has become an important and useful tool to make better use of such data, and for measured data as well. In the remainder of this chapter, a brief survey of some of the many disciplines in which SID and MBPE are being used will be given. Note that what distinguishes traditional SID and the incorporation of MBPE is that the former is mainly based on what might be properly curve-fitting FMs, while the latter instead incorporates physically motivated FMs, and so could be alternately referred to as "Smart Curve Fitting."

A list of a few SID articles begins this survey. This is followed by a more extensive listing of MBPE articles divided into some of the many various disciplines in which this technology can be found. Boldface fonts are used to highlight SID- and MBPE-associated content. Where appropriate excerpted results for some sources are included for illustrative purposes.

2.2 System identification

1980-M. B. Beck, Applications of **system identification** and **parameter estimation** in water quality modelling, *Prévisions hydrologiques* (*Proceedings of the Oxford Symposium*, April, Actes du Colloque d'Oxford, 1980): IAHS-AISH Publ. no. 129.

Abstract: Applications of techniques of system identification and parameter estimation in water quality modeling are surveyed. This survey of the literature covers three areas: river water quality, lake water quality, and wastewater treatment plant modeling. The applications cited are classified according to the type of algorithm used for calibration, the type of model, and the field data used. To assist the classification, a number of estimation algorithms are very briefly introduced. Although there are clearly different lines of development in each area of water quality modeling, it is possible to identify problems common to all three areas. The major problems discussed concern the availability of field data, levels of noise in the data, and model structure identification.

1989-Torsten Söderström and Petra Stoica, ***System Identification***. New York: Prentice Hall, ISBN 978-0138812362.

2009-David James Murray-Smith, Methods of System Identification, Parameter Estimation and Optimisation Applied to Problems of Modelling and Control in Engineering and Physiology, A Thesis submitted for the Degree of Doctor of Science in Engineering (DSc (Eng)) at the University of Glasgow.

2012-E. Reynders, System identification methods for (operational) modal analysis: review and comparison, *Archives of Computational Methods in Engineering*, 19(1), 51–124.

2016-Tamara Nestorovic´, Miroslav Trajkov and Matthias, Patalong, Identification of modal parameters for complex structures by experimental modal analysis approach, *Advances in Mechanical Engineering*, Vol. 8(5) 1–16.

2022-Mirna N. Kapetin *et al.*, Complex-Order Models: A System Identification Point of View, *Applied Science*, 12, 4768. https://doi.org/10.3390/app12094768.

2.3 A sampling of other subject areas

Acoustics

The science of acoustics, a wave-equation discipline that shares many similarities with electromagnetics, is one in which MBPE applications are frequently mentioned as shown in the citations below. The first two references here use time-domain models. Pade approximation and frequency derivatives are employed as ways to speed up frequency-domain modeling as well.

2010-James S. Hall and Jennifer E. Michaels, A **model-based** approach to dispersion and **parameter estimation** for ultrasonic guided waves, *J. Acoust. Soc. Am.* **127** 2, February 2010.

Abstract: A model-based algorithm is presented that adaptively estimates *in situ* ultrasonic-guided wave system parameters. Dispersion curves, propagation loss, transducer distances, transmitted signal, and mode weighting coefficients are estimated using minimal *a priori* information and assumptions. The five-part algorithm is scalable to accommodate two or more receivers and one or more propagating modes, provided that mode separation can be achieved prior to use of the algorithm (Figure 2.1).

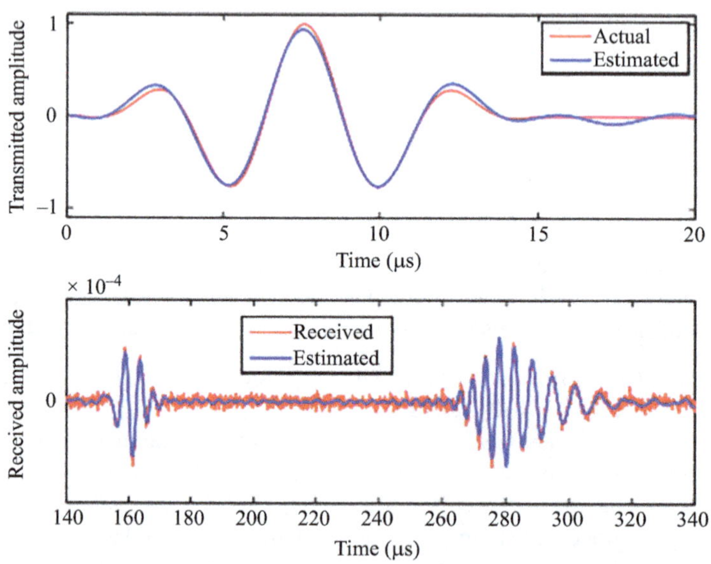

Figure 2.1 Comparison of estimated and actual transmitted signals top and estimated and actual received signals bottom for the test case. Signals in the bottom plot correspond to the receiver located at 803 mm.

Comparison of estimated and actual transmitted signals top, and estimated and actual received signals bottom for the test case. Signals in the bottom plot correspond to the receiver located at 803 mm.

2015-Huiqun Denga and Jun Yang, Modeling and estimating acoustic transfer functions of external ears with or without headphones, *J. Acoust. Soc. Am.* 138, 694–707.

2016-Yufeng Lu and Jafar Saniie, **Model-based parameter estimation** for defect characterization in ultrasonic NDE applications, *2015 IEEE International Ultrasonics Symposium Proceedings*.

2016-Sanda Lefteriu, Marcos Souza Lenzi, *et al.*, Fast frequency sweep method for indirect boundary element models arising in acoustics, *Engineering Analysis with* Boundary Elements, Volume 69, August 2016, Pages 32–45.

Abstract: This paper presents an efficient approach for computing frequency sweeps with fine increments of acoustic systems described by the variational indirect boundary element method. The matrices arising from such boundary element discretizations are fully populated and their assembly is computationally demanding as it involves the calculation of double surface integrals for each system matrix entry. Therefore, the two operations performed when computing the response at one frequency, namely assembling the system matrix and solving the associated linear system, may be of the same order of magnitude in terms of the required computational time .... **Matrices are only assembled at a few master frequencies, while for the rest, interpolation is used on scaled quantities determined at the master frequencies.** ... **Padé approximants constructed via the well-conditioned asymptotic waveform evaluation algorithm are used to extrapolate the response around expansion frequencies using derivative information** ....

2022-Giacomo Moretti1and Gianluca Rizzello, **Model-based parameter** analysis of dielectric elastomer loudspeakers, Conference Paper, July, DOI: 10.1109/AIM52237.2022.9863404.

Aerodynamics and aerospace

These various articles all mention parameter estimation as a goal while employing different approaches for different applications. The first reports on using control derivatives in the process. Data segmentation is used in the second article to divide a nonlinear problem into several local models. The last two articles deal with launch vehicles and unmanned aerial vehicles.

2018-Jitu Sanwale and Dhan Jeet Singh, Aerodynamic **Parameters Estimation** Using Radial Basis Function Neural Partial Differentiation Method, Defence Science Journal, Vol. 68, No. 3, May 2018, pp. 241–250, DOI. Figure from this article (Figure 2.2).

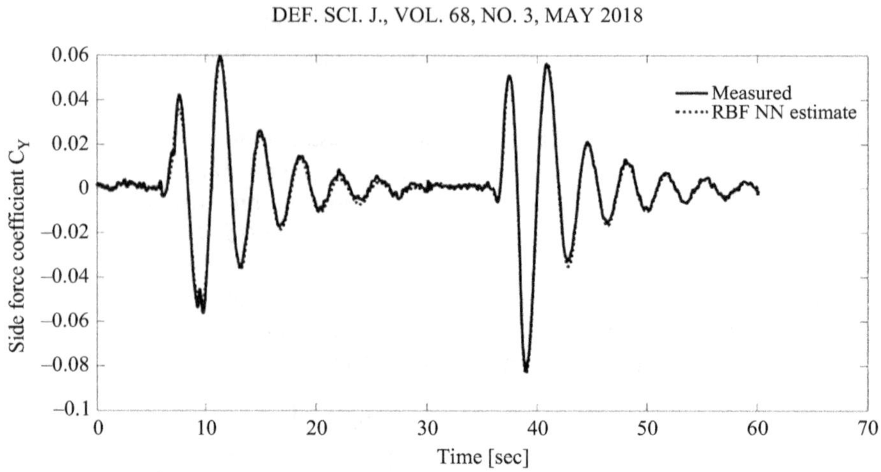

Figure 2.2 Comparison of radial basis function neural network estimate and measured side force coefficient (CY) during neural network testing phase (Validation method 1).

Comparison of radial basis function neural network estimate and measured side force coefficient (CY) during neural network testing phase (Validation method 1).

2022-Hao Liu, *et al.*, Aerodynamic Modeling Based on Segmented **Parameter Estimation** in Frequency Domain, *Proceedings of 2021 International Conference on Autonomous Unmanned Systems (ICAUS 2021).*

2022-Jian Jia, Weifeng Chen, and Zixuan Wang, Chapter 7 Aerodynamic **Parameter Estimation** for Launch Vehicles, in *Autonomous Trajectory Planning and Guidance Control for Launch Vehicles*, Zhengyu Song, Dangjun Zhao and Stephan Theil, Editers, Springer.

2024-Zhe HUI, *et al.*, Aircraft **parameter estimation** using a stacked long short-term memory network and Levenberg-Marquardt method, *Chinese Journal of Aeronautics.*

Astronomy and cosmology

Theories in this discipline rely entirely on observational data covering astronomical distances and times while having no control over the collection circumstances, only the choice of what is observed. The various cosmological hypotheses and theories that have been developed are dependent entirely on whatever relevant data are available and consequently change over time as more data become available. Combined with earth-based laboratory measurements such as the emission and absorption spectra of various elements, atmospheric models of distant stars and planets can be determined.

2006-Andrew Liddle, Pia Mukherjee and David Parkinson, **Model** Selection in Cosmology, **A&G** • August 2006 • Vol. 47 (Figure 2.3).

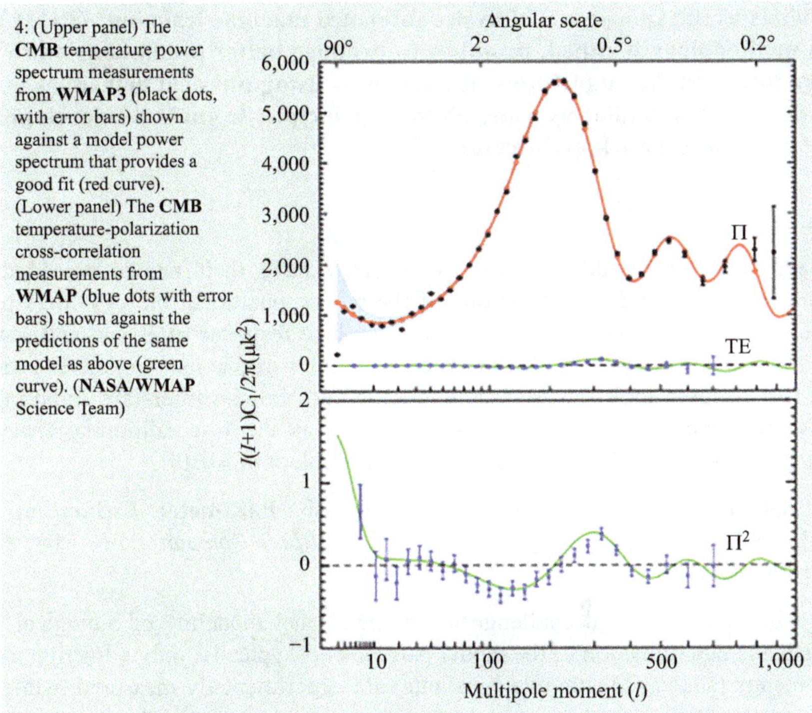

Figure 2.3 From model selection in cosmology, 2006. From 2006.

2013-S. Benitez-Herrera, *et al.* Cosmological **parameter estimation** from SN Ia data: a model-independent approach, arXiv:1308.5653.

2014-Santanu Das and Tarun Souradeep, SCoPE: An efficient method of Cosmological **Parameter Estimation**, arXiv:1403.1271v1 [astro-ph.CO].

2023-J. R. Eskilt, *et al.*, COSMOGLOBE: Towards end-to-end CMB cosmological **parameter estimation** without likelihood approximations, *A&A* 678, A169.

2024-Helen Meskhidze, Beyond Classification and Prediction: The Promise of Physics-Informed Machine Learning in Astronomy and Cosmology, Juan Dur´an and Giorgia Pozzi (Eds.): Philosophy of Science for Machine Learning: Core Issues and New Perspective, Synthese Library.

Abstract: Though the use of machine learning (ML) is ubiquitous in astrophysics and cosmology, many still see the opacity of ML algorithms as a major issue to their scientific utility. One way of addressing this opacity is through an emerging trend in ML research of "teaching" ML algorithms physical laws and domain-specific knowledge. **"Physics-informed machine learning" (PIML), as this methodology is called, promises to produce better predictions and yield more interpretable algorithms. It does so by using physical principles in the training process and/or by using physical principles to guide the development of the neural network architecture** ....

Biological

It seems that many different disciplines now have their names preceded by "Computational" and biology is one of them. Computational biology finds parameter estimation useful for such things as the heat response of *E. coli* and a wide class of biological systems. As for other disciplines in this list, parameter estimation is a featured tool. Bayesian computation and Kalman filters are some of the tools mentioned in these publications as well as the ill-conditioning that can sometimes be encountered, a not-uncommon problem in MBPE.

2010-Gabriele Lillacci and Mustafa Khammash, **Parameter Estimation** and Model Selection in Computational Biology, *PLoS Comput. Biol.*, Mar; 6(3): e1000696.

Abstract: A central challenge in computational modeling of biological systems is the determination of the model parameters. Typically, only a fraction of the parameters (such as kinetic rate constants) are experimentally measured, while the rest are often fitted. ... In this work, we present a new approach to the problem of parameter selection of biological models. **We show how one can use a dynamic recursive estimator, known as the extended Kalman filter, to arrive at estimates of the model parameters**. First, we use a variation of the Kalman filter that is particularly well suited to biological applications to obtain a first guess for the unknown parameters. Second, we employ an *a posteriori* identifiability test to check the reliability of the estimates. Finally, we solve an optimization problem to refine the first guess in case it is not accurate enough. ... We demonstrate these ideas by applying our methods to two examples, namely a model of the heat shock response in *E. coli*, and a model of a synthetic gene regulation system. The methods presented are quite general and may be applied to a wide class of biological systems where noisy measurements are used for parameter estimation or model selection.

2014-Juliane Liepe, Paul Kirk, Sarah Filippi, Tina Toni, Chris P. Barnes, and Michael P.H. Stump, A Framework for **Parameter Estimation** and Model Selection from Experimental Data in Systems Biology Using Approximate Bayesian Computation, *Nat. Protonics*. 2014 Feb; 9(2): 439–456.

2014-Choujun Zhan, Wuchao Situ, Lam Fat Yeung, Peter Wai-Ming Tsang, Genke Yang, A **Parameter Estimation** Method for Biological Systems modelled by ODE/DDE Models Using Spline Approximation and Differential Evolution Algorithm, *EEE/ACM Transactions on Comp.*, Volume, 11 Issue 6.

2015-Attila Gábor and Julio R. Banga, Robust and efficient **parameter estimation** in dynamic models of biological systems, *BMC Systems Biology*, 9:74.

2015-L. Hu, Z.G. Zhang, A. Mouraux, and G.D. Iannetti, Multiple linear regression to estimate time-frequency electrophysiological responses in single trials, *Neuroimage*, May 1; 111: 442–453.

Chemistry and chemical engineering

Again parameter estimation is a common theme in this area. Optimal model-based design of experiments is employed to yield the most informative process data to be used for the estimation of the parameters of a first-principles dynamic process model. Chemical engineers who develop fundamental models often have difficulties in estimating all model parameters due to problems with parameter identifiability and estimability. When some parameters are not estimable from the data, modelers must decide whether to conduct new experiments, change the model structure, or estimate only a subset of the parameters and leave the others at fixed values.

2001-Peter Englezos and Nicolas Kalogeraki, *Applied* ***Parameter Estimation*** *for Chemical Engineers*, Marcel Dekker Inc., New York, Basel.

2009-Federico Galvanin, Massimiliano Barolo and Fabrizio Bezzo, Online Model-Based Redesign of Experiments for **Parameter Estimation** in Dynamic Systems, *Industrial & Engineering Chemistry Research*, Vol 48, Issue 9.

2011-Kevin A. P. McLean and Kim B. McAuley, Mathematical modelling of chemical processes—obtaining the best model predictions **and parameter estimates** using identifiability and estimability procedures, *Canadian Journal of Chemical Engineering*.

2017-Ernie Che Mid and Vivek Dua, **Model-Based Parameter Estimation** for Fault Detection in Process Systems using Multiparametric Programming, *Industrial & Engineering Chemistry Research*, 56 (28) pp. 8000–8015. 10.1021/acs.iecr.7b00722.

2022-Duo Zhang, Kexin Wang, Zuhua Xu, *et al.*, Generalized **Parameter Estimation** Method for **Model-Based** Real–Time Optimization, *Chemical Engineering Science*, Volume 258, 31 August 2022, 117754.

Abstract: **Real-time optimization is the technology to find the optimal setpoint of a plant under uncertainty**. An iterative two-step approach with alternating parameter estimation and optimization is popular and intuitive in practice. However, this approach does not guarantee convergence if structural plant-model mismatches exist. Using the modifier adaptation principle, we improve the iterative two-step approach by the proposed generalized parameter estimation method. … Simulation shows that the proposed method converges to a plant optimum in the presence of structural plant-model mismatch in a relatively noise-insensitive way, and the model parameters are properly adapted during the iteration.

Electrical engineering

Electrical engineering is represented here in two categories, the other being electromagnetics. Other subspecialties might also be included but these two seem adequate to provide reasonable coverage. Of the seven articles included, six actually cite MBPE in their titles. This might have occurred since many of their publications appear in IEEE journals. The applications range from lightning, and electric motors to battery technology, photovoltaics and power electronics. The motor article was involved in controlling the load torque of the motor. MBPE use in automotive batteries and power systems particles as do others represent control in a variety of power applications.

2007-Hamid. R. Karami, Rouzbeh Moini, S. H. Hesam Sadeghi, A Model-Based Parameter Estimation of Current Distribution along Lightning Return Stroke Channel, Amirkabir University of Tech., Electrical Engineering Dept. 424, Hafez. Ave., Tehran 15914, Iran (Figure 2.4).

2012-Song Wang, Venkata Dinavahi, Jian Xiao, Multi-rate real-time **model-based parameter estimation** and state identification for induction motors, *IET Electric Power Applications.*

2015-Ryan Ahmed, *et al.*, **Model-Based Parameter Identification** of Healthy and Aged Li-ion Batteries for Electric Vehicle Applications, Copyright *SAE International* doi:10.4271/2015-01-0252 saealtpow.saejournals.org.

2015-Claudius Ortbandt, Cezary Dzienis, Robert Matussek and Horst Schulte, **Parameter Estimation** in Electrical Power Systems Using Prony's Method, *Journal of Physics: Conference Series* 659, 012013.

2016-Emerson A. Silva, Fabricio Bradaschia, Marcelo C. Cavalcanti, Aguinaldo J. Nascimento, Jr., **Parameter Estimation** Method to Improve the Accuracy of Photovoltaic Electrical Model, *IEEE Journal of Photovoltaics*, Vol. 6, No. 1, January 2016.

Abstract: This paper proposes an estimation method to identify the electrical model parameters of photovoltaic (PV) modules and makes a comparison with

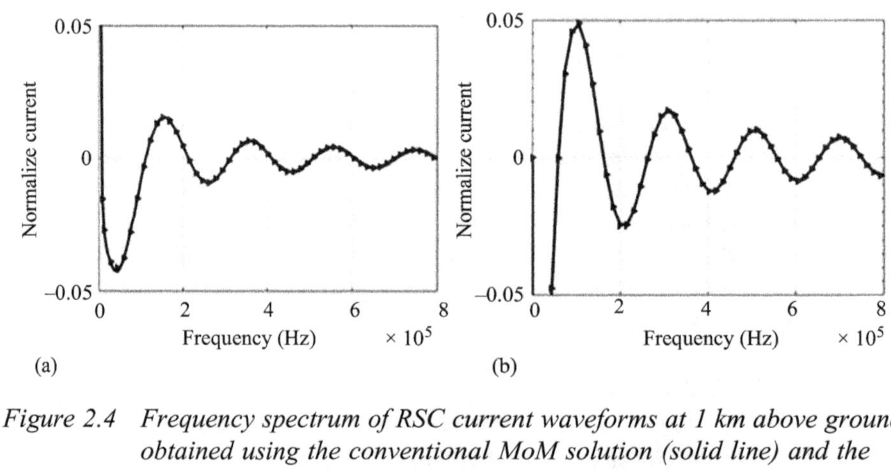

Figure 2.4 Frequency spectrum of RSC current waveforms at 1 km above ground obtained using the conventional MoM solution (solid line) and the MBPE—MoM solution (triangle line); (a) real part and (b) imaginary part. From 2007 above.

other methods already popular in the technical literature .... The first advantage of the proposed method is that the parameters are estimated using the *P–V* curves instead of *I–V* curves, since most of the applications that use the electrical model want to accurately estimate the extracted power. The second advantage is that the value ranges of each unknown parameter respect their physical meaning. In order to prove the effectiveness of the proposition, a comparison among methods is carried out using both types of *P–V* and *I–V* curves: those obtained by manufacturers' data sheets and those extracted experimentally in the laboratory.

2019-Ernie Che Mid and Vivek Dua, **Model-Based Parameter Estimation** for Fault Detection in Process Systems using Multiparametric Programming, *Conference: UCL*

2021- Alia SalahOmar Abu MoharebHans-Christian Reuss, **Model-Based** Fault Detection Using **Parameter Estimation** in Automotive EPGS Systems, *International Aegean Conference on Electrical Machines and Power Electronics (ACEMP) & 2021 International Conference on Optimization of Electrical and Electronic Equipment (OPTIM)*

Electromagnetics

Electromagnetic applications represent the initial mention of MBPE in all of the articles found in an extensive internet search. There are several distinct applications of MBPE in CEM with the two primary FMs being exponential series and rational functions. Besides reducing the sampling requirement of frequency domain transfer functions, it has been similarly effective in the numerical computation of radiation and scattering patterns. Other applications have been effective for specialized Green's functions and in the computation of temporal and spatial poles, with the latter applicable for array pattern synthesis and specialized point scatterers (Figure 2.5).

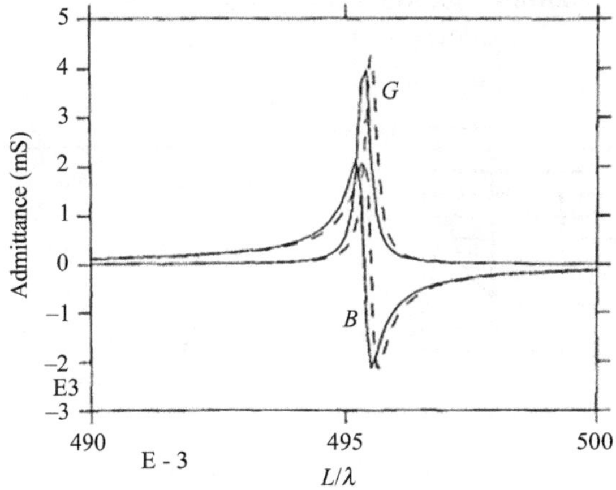

Figure 2.5 Input admittance of a dipole antenna with length L and thick ness parameter Ω-15 at height L/50 above a perfectly conducting ground. Directly computed values (————) are compared with a ration al function model (- - - -) with n = 5 and d = 4, matched to the value and four derivatives at L/λ = 0.25 and −0.25. From 1984 below.

1984-G. J. Burke and E. K. Miller, "Modeling Antennas Near to and Penetrating a Lossy Interface," *IEEE Transactions on Antennas and Propagation*, AP-32 (10), pp. 1040–1049. (Figure 2.5).

1987-E. K. Miller, "**Model-based Parameter Estimation** in Computational Electromagnetics," presented at the *NATO Advanced Study Institute Electromagnetic Modelling and Measurements for Analysis and Synthesis Problems*, Il Ciocco, Castelvecchie Pascoli, Tuscany, Italy, August 10–21, and published in 1991 in Electromagnetic Modeling and Measurements for Analysis and Synthesis Problems, edited by B. de Neumann, Kluwer Academic Publishers, NATO ASI Series E: Applied Sciences Vol. 199, pp. 205–256l.

1989-G. J. Burke, E. K. Miller, S. Chakrabarti, K. Demarest, **Using Model-based Parameter Estimation** to Increase the Efficiency of Computing Electromagnetic Transfer Functions, *IEEE Transactions on Magnetics*, Vol. 25, No. 4, July. Figure 2 below from this article.

1989-E. K. Miller and G. J. Burke, "Some Applications of **Model-Based Parameter Estimation** in Computational Electromagnetics," in *Proceedings of NATO AGARD Lecture Series No. 165*, Modern Antenna Design Using Computers and Measurement: Application to Antenna Problems of Military Interest, October, pp. 4-1 to 4-26.

1993-Paul M. Baggenstoss and Ramdas Kumaresan, On the **Estimation** of Rational Transfer Functions from Samples of the Power Spectrum, *IEEE Transactions on Signal Processing*, Vol. 41, No. 3, March.

1998-R. J. Reddy, "Application of **Model Based Parameter Estimation** for RCS Frequency Response Calculations Using Method of Moments," NASA.

2002-Manohar D. Deshpande and Robin L. Cravey, Multi-Variable **Model-based Parameter Estimation** Model for Antenna Radiation Pattern Prediction, NASA Langley Research Center, Hampton, VA. USA.

2017-Fatih Kaburcuk, Fast Wideband Solutions Obtained **Using Model-based Parameter Estimation** with Method of Moments, *Advanced Electromagnetics*, Vol. 6, No. 3, October (Figure 2.6).

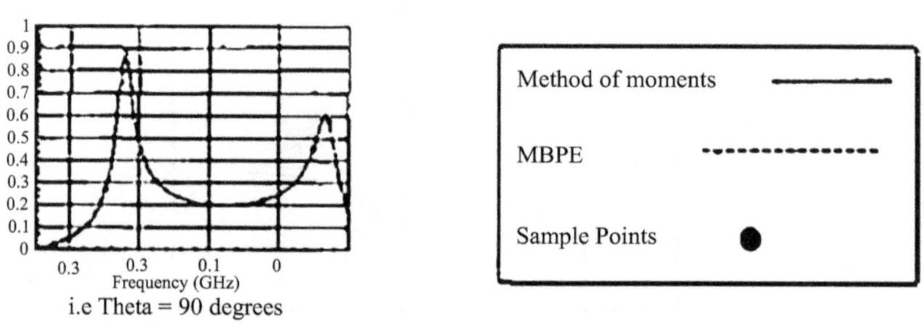

Figure 2.6 Far field intensity as a function of frequency for a dipole with length $L = 0.4836$ meters and diameter $d_0 = 00003$ meters as calculated by MOM and MBPE at theta = 10, 30, 50, 70, and 90 degrees, with the seven fitting frequencies shown as points. From 2017.

Fluid dynamics

The term "fluid dynamics" could first generate images of pipelines, rivers and wells, airplanes and rockets, but blood vessels and aneurysms might be a little unexpected. The first article here concerns estimating the probability of floods in arid and semi-arid regions. The fourth article describes an approach to stabilize the parameter estimation. The other three articles deal with modeling blood flow through human vascular systems and specific problems such as aneurysms using observational data to improve model parameters.

2019-Wei Ting, *et al.*, Probability Weighted Moments-based **Parameter Estimation** for Kinematic Diffusion and Muskingum-based Distributions, *Journal of Hydrologic Engineering* 24(12):04019054 1-11, DOI: 10.1061/(ASCE)HE.1943-5584.0001863.

2020-Sarah N Lipp *et al.*, Computational Hemodynamic **Modeling** of Arterial Aneurysms: A Mini-Review, *Frontiers in Physiology* 2020; 11: 454.

Abstract: Arterial aneurysms are pathological dilations of blood vessels, which can be of clinical concern due to thrombosis, dissection, or rupture. Aneurysms can form throughout the arterial system, including intracranial, thoracic, abdominal, visceral, peripheral, or coronary arteries. Currently, aneurysm diameter and expansion rates are the most commonly used metrics to assess rupture risk .... Since simple diameter is often insufficient to reliably determine rupture and thrombosis risk, computational hemodynamic simulations are being developed to help assess when an intervention is warranted. ... In this review, we highlight recent work that investigates aneurysm geometry, model parameter assumptions, and other specific considerations that influence computational aneurysm simulations.

2022-Amirtahà Taebi, Deep Learning for Computational Hemodynamics: A Brief Review of Recent Advances, *Fluids* 2022, 7, 197.

2022-Zheqi Shen and Youmin Tang. A two-stage inflation method in **parameter estimation** to compensate for constant parameter evolution in Community Earth System Model, *Acta Oceanologica Sinica*.

2024-Sumant R. Morab, Atul Sharma and Janani S. Murallidharan, Computational Hemodynamics in Human Vasculature: A Review on Role of Rheology, Multiphase Flow, and Fluid–Structure Interaction, *Journal of the Indian Institute of Science*.

General and miscellaneous

This category is a "catch-all" for applications that are quite variable but are related by the common goal of parameter estimation. Many include MBPE in their titles.

1976-M.E. Polis and R. E. Goodson, "**Parameter identification** in distributed system: A synthesizing overview," *Proc. IEEE*, vol. 64, pp. 45–61, January.

2003-Richardson and Brian Schwarz, Modal **Parameter Estimation** from Operating Data, *Sound and Vibration*, January.

2007-Adriaan van den Bos, ***Parameter Estimation*** *for Scientists and Engineers*, Wiley-Interscience, A John Wiley & Sons, Inc., Publication.

2013-Jie Ding, Chunxia Fan, and Jinxing Lin, Auxiliary **model based parameter estimation** for dual-rate output error systems with colored noise, *Applied Mathematical Modelling*, Volume 37, Issue 6, 15 March, Pages 4051–4058.

2019-Ernie Binti Che Mid, **Model-Based Parameter Estimation** for Fault Detection in Process Systems using Multiparametric Programming, Department of Chemical Engineering, University College London, Torrington Place, London WC1E 7JE.

2021-Damien Gueho, Manoranjan Majji, Puneet Singla, Data-Based Modeling and Control of Dynamical Systems: **Parameter Estimation**, *60th IEEE Conference on Decision and Control (CDC)* (Figure 2.7).

Parameter convergence to estimate unknown input from 2021 article.

Matrix mathematics

As the next generation of detectors come online, the work of gravitational wave physics will begin to shift from just making a detection to actually extracting information about the generating source of some detected signal. In the 1970 article, it is shown that measurements of bone and joint motion can be performed based on surface markers placed on the skin.

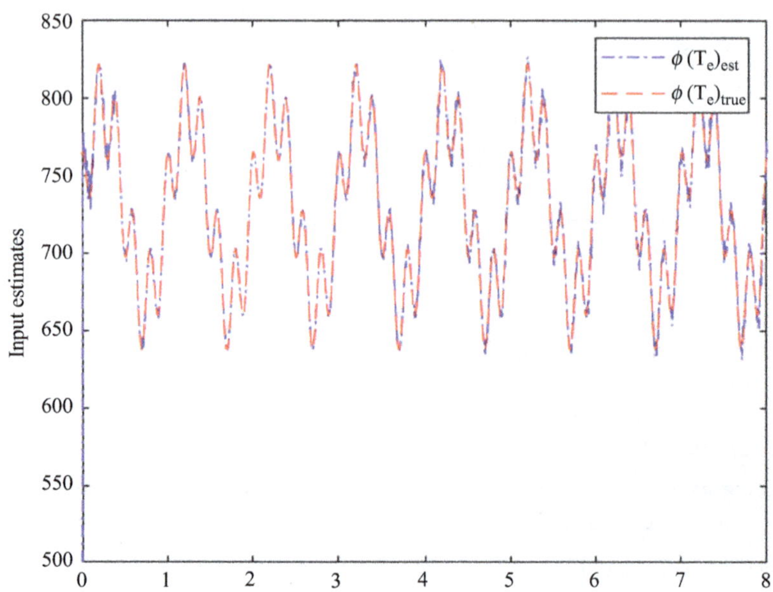

Figure 2.7 Parameter convergence to estimate unknown input. From 2021 article below.

2013-Peter Shaffery, **Parameter Estimation** consistency between MCMC (Markkov Chain and Monte Carlo) and Fisher Information Matrix, University of Florida, Seoul National University and University of Massachusetts Lowell.

1970-Brother Kenneth E. Fitzgerald, Error Estimates for the Solution of Linear Algebraic System, *Journal of Research of the National Bureau of Standards B: Mathematical Sciences* Vol. 74B, No. 4, October–December.

2021-Elena Hernandez, Octavio Manero, *et al.* Analytic Matrix Method for Frequency Response Techniques Applied to Nonlinear Dynamical Systems, *Mathematics*, 9, 3287. https://doi.org/10.3390/math9243287.

Mechanical and structural

The 2007 article shows that measurements of bone and joint motion can be performed based on surface markers placed on the skin. The other articles share the related goal of modeling and parameter estimation for various mechanical structures to monitor their conditions and/or frequency characteristics (Figure 2.8).

2007-M. Mor, A. Wolf, and O. Gottlieb, Nonlinear **Model-based Estimation** of Rigid-Body Motion via an Indirect Measurement of An Elastic Appendage, *Proceedings of the ASME 2007 International Design Engineering Technical Conferences & Computers and Information in Engineering Conference IDETC/CIE*, Las Vegas, Nevada, USA, DETC2007.

2012-Michaels, Jennifer E., Jacobs, Laurence J., Ruzzene, Massimo, A Multi-Scale Structural Health Monitoring Approach for Damage Detection, Diagnosis and Prognosis in Aerospace Structures, Air Force Office of Scientific Research, AFOSR 875 North Randolph Street, Suite 326, Arlington, VA 22203-1768.

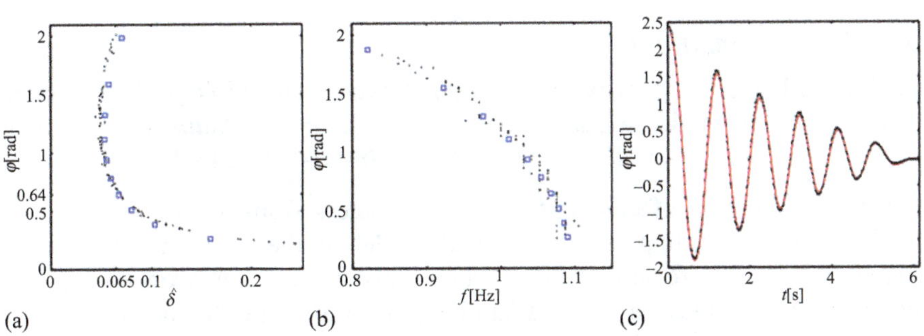

Figure 2.8 Experimental (black dots) and numerical (blue squares) damping backbone (a) and frequency backbone (b) curves. The experimental results are obtained from a superposition of ten different free vibration decays. (c): Free vibration decay: black dots: experimental result, solid red line: numerical result. For the numerical simulations: $\omega_n = 6.9$ *rad/s* $\delta = 0$, $\delta_2 = 0.028$ *rad*$^{-1}$, $\beta = 2$, $\mu_0 = 2.6$ *rad/s*^{2} *and* $\alpha = 0.77$ $(\mu_0(1+\alpha) = 4.6$ *rad/s*$^2)$. *From 2017 article below.*

2017-Giuseppe Habib. Akos Miklos, Eniko T. Enikov, Gabor Stepan and Giuseppe Rega, Nonlinear **model-based parameter estimation** and stability analysis of an aero-pendulum subject to digital delayed control, *International Journal of Dynamics and Control*, DOI: 10.1007/s40435-015-0203-0. (Figure 2.8).

2018-Tiago Silva1 and João Pereira, Coupling of structures using frequency response functions, *MATEC Web of Conferences* 211, 06005.

2020-Krzysztof Bartecki, Rational Transfer Function Model for a Double-Pipe Parallel-Flow Heat Exchanger, *Symmetry* 2020, 12, 1212, doi:10.3390/sym12081212.

Abstract: Transfer functions of typical heat exchangers, resulting from their partial differential equations, usually contain irrational functions which quite accurately describe the spatio-temporal nature of the processes occurring therein. However, such an accurate but complex mathematical representation is often not convenient from the practical point of view, and some approximation of the original model would be more useful. This paper discusses approximate rational transfer functions for a typical thick-walled double-pipe heat exchanger working in the parallel-flow configuration. ... Based on the rational transfer function representation, the frequency and steady-state responses of the approximate model are evaluated and compared with those resulting from its original irrational transfer function model. The presented results show better approximation quality for the "crossover" input–output channels where the in-domain heat conduction effects prevail as compared to the "straightforward" channels, where the transport delay associated with the heat convection dominates.

2023-XuYoua, Xinping Yana, *et al.*, A novel **model-based parameters estimation** combining local optimization and global optimization of nonlinear ship models with physical experiment dataset, Preprint submitted to Elsevier.

Medicine and medical

2009A.L. Lloyd, Sensitivity of **Model-Based** Epidemiological **Parameter Estimation** to Model Assumptions, *Mathematical and Statistical Estimation Approaches in Epidemiology*, DOI 10.1007/978-90-481-2313-1 6.

Abstract: Estimation of epidemiological parameters from disease outbreak data often proceeds by fitting a mathematical model to the data set. The resulting parameter estimates are subject to uncertainty that arises from errors (noise) in the data; standard statistical techniques can be used to estimate the magnitude of this uncertainty. ... Here we focus on the estimation of the basic reproductive number from the initial growth rate of an outbreak as this is a setting in which parameter estimation can be surprisingly sensitive to details of the time course of infection (Figure 2.9).

2012-Andrew H. Briggs, **Model Parameter Estimation** and Uncertainty: A Report of the ISPOR-SMDM Modeling Good Research Practices Task Force-6, *International Society for Pharmacoeconomics and Outcomes Research (ISPOR)*.

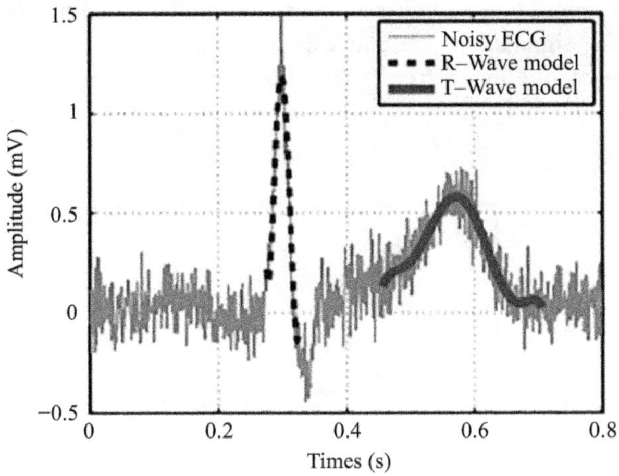

Figure 2.9 A noisy ECG beat and the two polynomial models of order M = 6 fitted over the R- and T-waves. From 2012 article citation above.

2012-Sun K. Hong, Resonance-based Techniques for Microwave Breast Cancer Applications, Dissertation submitted to the faculty of the Virginia Polytechnic Institute and State University in partial fulfillment of the requirements for the degree of Doctor of Philosophy In Electrical Engineering.

2021-Reza Sameni and Davood Fattahi, Theoretical Performance Bounds of **Model-Based** Electrocardiogram **Parameter Estimation**, researchgate.net/publication/353992606.

2022-Leah S. Wilk, Gerda J. Edelman and Maurice C. G. Aalders, Next-generation time of death estimation: combining surrogate **model-based parameter** optimization and numerical thermodynamics, https://doi.org/10.1098/rsos.220162.

Physics

2016-Sohel Rana, Jeevan Kanesan, Ahmed Wasif Reza, and Harikrishnan Ramiah, Tickhonov-ased Well-Condition Asymptotic Waveform Evaluation For Dual-Phase-Lag Heat Conduction, *Thermal Science*, Vol. 20, No. 6, pp. 1891–1902.

2024-A L Milder *et al.*, Qualitative and quantitative enhancement of parameter estimation for model-based diagnostics using automatic differentiation with an application to inertial fusion, *Mach. Learn. Sci. Technol.* 5 015026 (Figure 2.10).

2024-Javier Roulet and Tejaswi Venumadhav, Inferring Binary Properties from Gravitational Wave Signals, Annual Reviews.

Abstract: This review provides a conceptual and technical survey of methods for parameter estimation of gravitational wave signals in ground-based interferometers such as LIGO and Virgo. We introduce the framework of Bayesian inference and provide an overview of models for the generation and detection of gravitational waves from compact binary mergers, focusing on the essential features that are observable in the signals. ... We end with a discussion of recent developments in the application of likelihood-free (simulation-based) inference methods to gravitational wave data analysis.

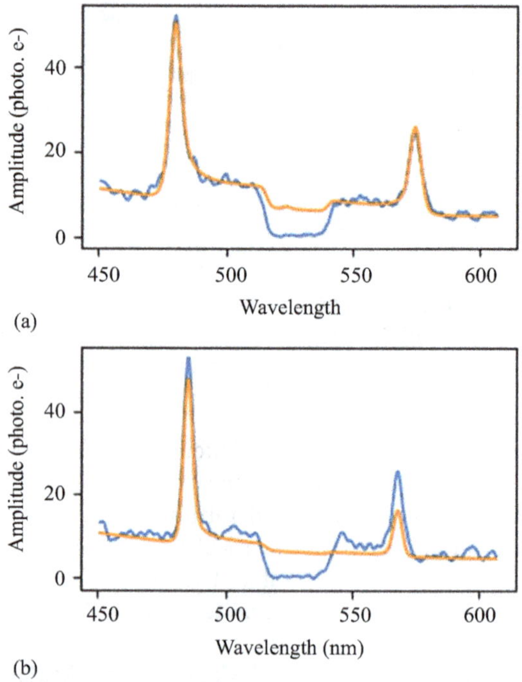

Figure 2.10 *(a) Example of a "good" and (b) "bad" fit. (a), (b) Data (blue) is shown in units of photocathode photoelectrons the minimal counting unit in the diagnostic system and compared to the codes best fit (orange). From the above article.*

Power and devices

2002-Jorge Pleite, Emilio Olias, Andres Barrado, Antonio Lazaro, Ramon Vazquez, Frequency response modeling for device analysis, *IEEE Conference Paper*, December, DOI: 10.1109/IECON.2002.1185493, Source: IEEE Xplore.

2020-Chen Fiona F., Xiang Junting, David G. Thomas, Anthony B. Murphy, **Model-based parameter** optimization for arc welding process simulation, *Applied Mathematical Modelling*, Volume 81, May, Pages 386–400.

Abstract: This paper presents a model-based parameter optimization for simulating a metal-inert gas welding process. The computational model used in this study is based on computational fluid dynamics methods and implemented using the finite volume approach on a 3D computational domain. The wire electrode, the arc plasma, and the workpiece are treated as a self-consistent system. Important welding parameters, including arc current, wire feed rate, workpiece thickness, welding speed and geometry, as well as the metal alloy types used for the wire and workpiece, were implemented as adjustable parameters. By tuning these parameters, the performance of the arc welding can be predicted, and different settings can be compared to optimize welding performance. … The work presented shows the potential of model-based parameter optimization to assist process engineers in the practical improvement of the welding process.

2021- Artun Sel, *et al.*, GLSDC (Gaussian Least Squares Differential Correction) **Based Parameter Estimation** Algorithm for a PMSM Model, *Energies* 2021, 14 (3), (https://doi.org/10.3390/en14030611).

In the graph below, the circumflex terms are the model estimates and those without are the measured values where the numbered states denote direct and quadrature axis currents, angular velocity, and angular position of the rotor, respectively (Figure 2.11).

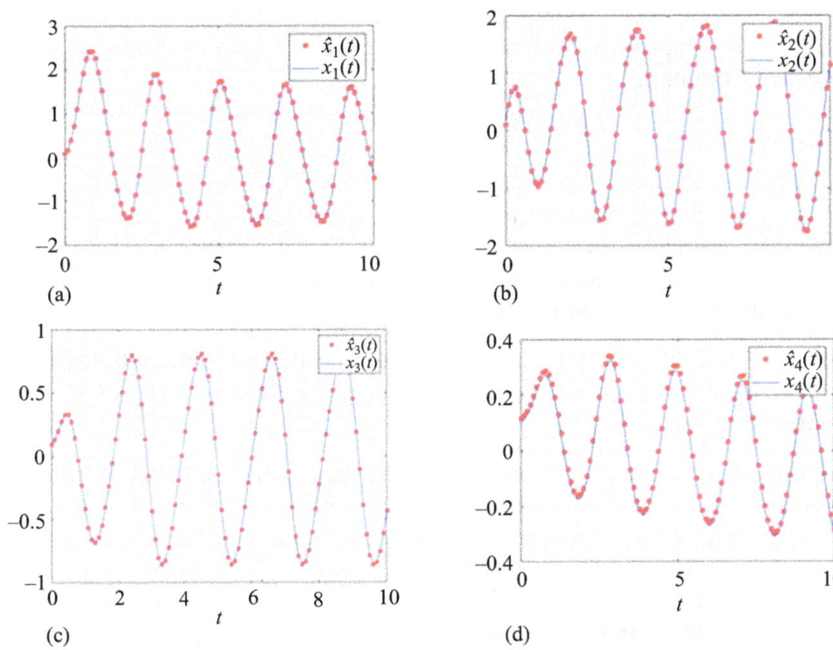

Figure 2.11 The state signals of the real and model system. listed as: ***(a)*** *$x_1(t)$ and $\widehat{x}_1(t)$.* ***(b)*** *$x_2(t)$ and $\widehat{x}_2(t)$.* ***(c)*** *$x_3(t)$ and $\widehat{x}_3(t)$.* ***(d)*** *$x_4(t)$ and $\widehat{x}_4(t)$. From 2021 article.*

2022-A. N. Ali *et al.*, **Parameter Estimation** of DC Motor using Multiparametric Programming, *J. Phys.: Conf. Ser.* 2312 012035.

Signal processing

Signal processing is generally more productive when the physical origination of the signal of interest is somehow involved in its processing. Thus, knowing that a transient signal comes from electromagnetics, acoustics or a vibrating structure naturally indicates that a damped, exponential series is an appropriate FM.

2004-Emmanuel Candes, Justin Romberg, and Terence Tao, Exact Signal Reconstruction from Highly Incomplete Frequency Information, *IEEE Transactions on Information Theory*, Vol. 52, Issue 2, February.

2014-C.M. Velte1, P. Buchhave, and W.K. George, Power Spectrum **Estimation** of Randomly Sampled LDA Signals, *17th International Symposium on Application of Laser Techniques to Fluid Mechanics*, Lisbon, Portugal.

2016-Dušan Agrež Periodic signal **parameters' estimations** from random sampling measurements under noncoherent sampling condition, *21st IMEKO TC4 International Symposium and 19th International Workshop on ADC Modelling and Testing.*

2019-C. Sidney Burrus, Prony, Pade, and Linear Prediction for Interpolation and Approximation in the Time and Frequency Domain Design of Infinite-Impulse Response Digital Filters and in **Parameter** Identification, Open license under Creative Commons.

Thermodynamics

Among the models used in the first article is a reduced-order version based on mass and energy balances in a given zone in a building that serves as the basis to estimate unmeasured internal loads with two major assumptions: The zone is well mixed. Long-wave radiation exchanges between surfaces are ignored. Other sources listed are basically summarized in their respective titles.

2010-Zheng O'Neill, Satish Narayanan, and Rohini Brahme, **Model-based** thermal load **estimation** in buildings, *Fourth National Conference of IBPSA-USA*, New York City, New York.

Abstract: Understanding and optimizing the energy performance of a building require insights into the dynamics of thermal loads at a variety of spatial scales within the building. ... The use of physics-based thermal balance models with measured data can provide more robust and accurate estimates of the loads. In this paper, a model-based approach is used to estimate zonal loads in a medium-sized commercial building and verified with measurements.

2013-López, C., *et al.* **Model-based** identifiable **parameter** determination applied to a simultaneous saccharification and fermentation process model for bio-ethanol production. *Biotechnol. Prog.* 2013, 29, 1064–1082.

2021-Yuxiang Chen, *et al.*, **Parameter estimation** of resistor–capacitor models for building thermal dynamics using the unscented Kalman filter, *Journal of Building Engineering*, Volume 34, February, 101639.

Abstract: Accurate and computationally efficient building energy models are critical to the development of online or pseudo-online control strategies and other building management activities. However, such models need to overcome the large uncertainty involved with continuously changing occupant activities and building status. The present study uses unscented Kalman filtering in the model parameter estimation for simple yet accurate resistor-capacitor models to develop reliable building energy models ....

2024-Heru Berian Pratama and Katsuaki Koike, Thermodynamic **model-based** spécification of optimal geothermal power generation system for high-temperature liquid dominated systems using flash and flashbinary cycles, *Renewable Energy*, Volume 220, January 2024, 119634.

Chapter 3
Mathematical background of MBPE

Abstract

An overview of model-based parameter estimation (MBPE) applications in computational electromagnetics and its use in a variety of other science and engineering disciplines have been explored in Chapters 1 and 2. In this chapter, a more detailed mathematical discussion of MBPE will be covered focusing on electromagnetic (EM) applications. Motivating this interest in MBPE is that the mathematical detail associated with most EM analysis can obscure the possibility of representing physical observables in simpler ways by using reduced-order models, a process that is generically called "model-order reduction" (MOR). Knowledge of such models, ideally physically based, can be helpful in ways ranging from reducing the computer cost of achieving desired solutions to developing more compact representations of observables of interest. One specific approach to MOR is to estimate unknown parameters of the reduced-order models from sampled data, the process called "model-based parameter estimation." Various applications of MBPE in EM modeling are illustrated and their resulting benefits are demonstrated here. It should be observed that the designation MBPE has evolved in EM primarily from application to integral-equation (IE) models, whereas the term MOR has primarily been associated with application to differential-equation models. While conceptually equivalent, it might be observed that MOR describes a general goal, whereas MBPE is a particular approach to achieving that goal. The topics to be considered here are: (1) fitting models for waveform (time domain) and spectral data (frequency domain); (2) function sampling and derivative sampling; (3) adaptive sampling of frequency spectra; and (4) adaptive sampling of radiation and scattering patterns.

3.1 Fitting models for waveform and spectral data

As a representative example of the kind of problem that can be typically encountered in computational electromagnetics (CEM) consider the plots below of the conductance and susceptance of a dipole antenna having a short, straight center section and vee-shaped ends whose wire lengths are slightly different. This configuration is what causes the sharp peaks in these plots and which could be easily missed if the frequency sampling of the generating model (GM) were too widely spaced. This is the kind of problem for which MBPE is well suited (Figure 3.1) [1–3].

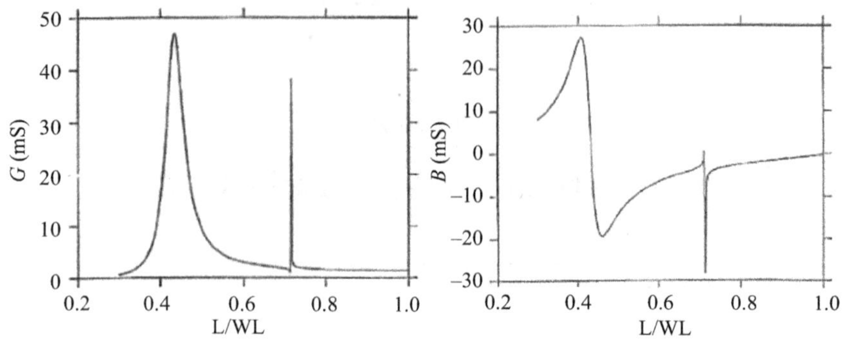

Figure 3.1 Admittance of a vee-dipole having slightly different end lengths

The salient points of MBPE can be succinctly summarized:

A **fitting model**, whose **parameters** are to be **estimated** to a specified accuracy requiring as few GM samples as possible using various signal-processing techniques, often in the presence of noise, incomplete data, and/or uncertainty.

It can be observed that MBPE is connected to digital signal processing (DSP) and classical numerical analysis (CNA). While DSP is concerned with the representation of signals by sequences of numbers or symbols and the processing of these sequences, CNA formulas such as interpolation integration and differentiation "are certainly DSP algorithms" [4].

In summary, it should be noted that straightforward implementation of MBPE involves these two basic steps:

1) Data or complex formulations are replaced by a simpler mathematical representation, the "fitting model" (FM) which is ideally physically based.
2) The model parameters are obtained from sampling the original description, a first-principles solution such as Maxwell's equations, or "generating model" (GM), i.e., the parameter estimation step.

The FM can provide a more efficient overall numerical approach by reducing the complexity and cost of the GM evaluation and/or making more efficient use of the GM samples that are computed, something that might be referred to as "smart curve fitting." Observe that MBPE is an umbrella term that covers a variety of methods that can be used for reduced-order descriptions. A representative but not exhaustive list of such methods includes Pade approximation, Prony's method, Cauchy's method, Shanks' method, Richardson extrapolation, recursion, and asymptotic waveform evaluation. Whatever specific one of such methods is employed, an exponential series FM is a common ingredient for a waveform-domain (WD) description as given by

$$f(x) = f_p(x) + f_{np}(x) = \sum_{\alpha=1}^{W} R_\alpha e^{s_\alpha x} + f_{np}(x) \tag{3.1}$$

with a corresponding pole series or spectral domain (SD) form given by

$$F(X) = F_p(X) + F_{np}(X) = \sum_{\alpha=1}^{W} \frac{R_\alpha}{(X - s_\alpha)} + F_{mp}(X) \tag{3.2}$$

where "*np*" denotes a non-pole component. In these FMs, the pole-based terms come from the resonance-dominated behavior of EM phenomena where "*x*" represents a WD variable and "*X*" its SD counterpart. For a time-frequency transform pair, "*x*" is the time-domain variable "*t*" and "*X*" is the complex frequency "*s*." The associated parameters are the "residues" or amplitudes R_α and the poles s_α which are generally complex numbers. Time-domain application of Prony's method will not be further considered, but will be demonstrated to be useful for both pattern synthesis and analysis and some inverse problems.

3.2 Solving the time-domain parameters

Waveform models can be processed using an exponential series via Prony's method [5] with sample f_i given by

$$f_i = f(x_i) = \sum_{\alpha=1}^{P} \widehat{R}_\alpha (\widehat{X}_\alpha)^i; i = 0, 1, \ldots, 2P - 1 \tag{3.3a}$$

where

$$\widehat{X}_\alpha = e^{\hat{s}_\alpha \delta}, \tag{3.3b}$$

with δ the time step with the poles $\widehat{s}_\alpha$ given by

$$\widehat{s}_\alpha = \frac{1}{\delta}\left[Ln\left(\widehat{X}_\alpha\right)\right], \tag{3.3c}$$

Introducing the characteristic equation

$$\sum_{\alpha=0}^{P} \widehat{a}_\alpha X^\alpha = 0; X = X_i, i = 0, 1\ldots, P - 1 \tag{3.4}$$

together with (3.3a) leads to a linear system for the coefficients

$$\begin{bmatrix} f_0 & f_1 & \cdots & f_{P-1} \\ f_1 & f_2 & \cdots & f_P \\ \vdots & \vdots & \vdots & \vdots \\ f_{P-1} & f_P & \cdots & f_{2P-2} \end{bmatrix} \begin{bmatrix} \widehat{a}_0 \\ \widehat{a}_1 \\ \vdots \\ \widehat{a}_{P-1} \end{bmatrix} = \begin{bmatrix} f_P \\ f_{P+1} \\ \vdots \\ f_{2P-1} \end{bmatrix} \tag{3.5}$$

where the caret denotes a quantity to be estimated and a_P has been set to one to make the system of equations inhomogeneous. Note that each sample of the right-hand side vector is given by the preceding P samples on the left to permit extrapolation beyond the sampled data which is why the system is known as a "linear predictor." An "energy" constraint could also be used, i.e.

$$\sum_{i=0}^{N} a_i^2 = 1 \tag{3.6a}$$

Still another possibility would be a "non-causal" constraint where

$$a_k = 1, k \neq P \tag{3.6b}$$

The poles can then be computed from the roots of the characteristic equation, i.e. values of X that make it zero. Time-domain Prony's method has been found useful for obtaining EM poles and computing their frequency-domain or time-domain CEM results.

Waveform derivative sampling [6] is another option where the data samples are then

$$f^{(i)} = \frac{d^i f(x)}{dx^i} = \sum_{\alpha=1}^{P} s_\alpha^i R_\alpha e^{s_\alpha x} \tag{3.7}$$

which together with the characteristic equation

$$\sum_{\alpha=0}^{P} \widehat{b}_\alpha s^\alpha = 0; s = s_i, i = 1, 2, \ldots, 2P \tag{3.8}$$

yields the linear system

$$\begin{bmatrix} f^{(0)} & f^{(1)} & \cdots & f^{(P-1)} \\ f^{(1)} & f^{(2)} & \cdots & f^{(P)} \\ \vdots & \vdots & \vdots & \vdots \\ f^{(P-1)} & f^{(P)} & \cdots & f^{(2P-2)} \end{bmatrix} \begin{bmatrix} \widehat{b}_0 \\ \widehat{b}_1 \\ \vdots \\ \widehat{b}_{P-1} \end{bmatrix} = - \begin{bmatrix} f^{(P)} \\ f^{(P+1)} \\ \vdots \\ f^{(2P-1)} \end{bmatrix} \tag{3.9}$$

3.3 Solving the spectral-domain parameters

3.3.1 Using spectral-domain samples

The SD samples are found using a rational function modified to

$$F(X)D(X) = N(X) \tag{3.10}$$

where

$$N(X) = N_0 + N_1 X^1 + N_2 X^2 + \cdots + N_n X^n \tag{3.11a}$$

and

$$D(X) = D_0 + D_1 X^1 + D_2 X^2 + \cdots + D_d X^d \tag{3.11b}$$

so that using sampled SD data

$$F_i = F(X_i) \tag{3.12a}$$

in

$$F_i \widehat{D}(X_i) = \widehat{N}(X_i) \tag{3.12b}$$

a linear system is generated for the model coefficients, where now $D_d = 1$ has been used to again produce the inhomogeneous equations that follow.

Thus there is obtained

$$\begin{bmatrix} F_0 & X_0F_0 & \cdots & X_0^{d-1}F_0 & -1 & -X_0 & \cdots & -X_0^n \\ F_1 & X_1F_1 & \cdots & X_1^{d-1}F_1 & -1 & -X_1 & \cdots & -X_1^n \\ \vdots & \vdots & \vdots & \vdots & \vdots & \vdots & \vdots & \vdots \\ \vdots & \vdots & \vdots & \vdots & \vdots & \vdots & \vdots & \vdots \\ F_{D-1} & X_{D-1}F_{D-1} & \cdots & X_{D-1}^{d-1}F_{D-1} & -1 & -X_{D-1} & \cdots & -X_{D-1}^n \end{bmatrix} \begin{bmatrix} \widehat{D}_0 \\ \vdots \\ \widehat{D}_{d-1} \\ \widehat{N}_0 \\ \vdots \\ \widehat{N}_n \end{bmatrix} = - \begin{bmatrix} X_0^dF_0 \\ X_1^dF_1 \\ \vdots \\ \vdots \\ X_{D-2}^dF_{D-1} \\ X_{D-1}^dF_{D-1} \end{bmatrix} \tag{3.13}$$

The spectral-domain FM can be explicitly written in an expanded form as

$$\sum_{\alpha=1}^{P} \frac{R_\alpha}{X - s_\alpha} = F_p(X) = \frac{N_0 + N_1X + \cdots + N_{P-1}X^{P-1}}{1 + D_1X + \cdots + D_PX^P} = \frac{N(X, P-1)}{D(X, P)} \tag{3.14}$$

where the non-pole part can generally be approximated in one of two forms

$$F_{np}(X) \cong C_1 \Rightarrow F(X) \approx \frac{N(X, P)}{D(X, P)} \tag{3.15a}$$

or

$$F_{np}(X) \cong C_1 + C_2X \Rightarrow F(X) \approx \frac{N(X, P+1)}{D(X, P)} \tag{3.15b}$$

An example of an MBPE window successively moved through sparsely sampled data for a dipole antenna is shown below. Input conductance and susceptance of a monopole antenna (a/L=0.01) versus length in wavelengths as obtained from a series of overlapping rational-function FMs using $d = n = 3$ (the solid line) based on 20 GM samples spaced 0.15 apart in L/wavelength are shown as open crosses and joined by a dashed line [Burke (1992)]. A comparison of the FM values with GM samples at other frequencies reveals a numerical agreement of 1% or better (Figure 3.2).

3.3.2 *Using spectral-domain derivative samples*

Spectral derivatives can also be used for the frequency domain using a differentiated rational-function model, which is discussed in more detail in Chapter 12.

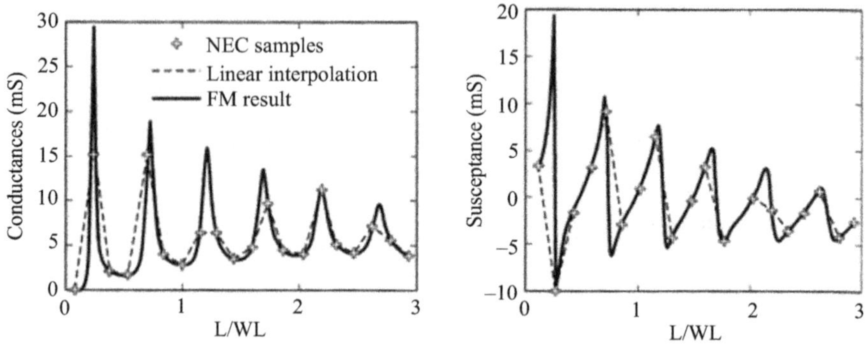

Figure 3.2 Input conductance and susceptance of a monopole antenna (a/L = 0.01) versus length in wavelengths

With samples up to the tth derivative, there is obtained

$$\begin{aligned}
&FD = N \\
&F'D + FD' = N' \\
&F''D + 2F'D' + FD'' = N'' \\
&\vdots \\
&F^{(t)}D + tF^{(t-1)}D^{(1)} + \ldots + C_{t,t-m}F^{(m)}D^{(t-m)} + \ldots + FD^{(t)} = N^{(t)}
\end{aligned} \tag{3.16}$$

where F here represents any EM field or source variable whose frequency behavior can be modeled by a rational function such as current, antenna admittance, radar cross section, and near or far fields.

A linear system for the associated numerator and denominator polynomial coefficients can then be developed as

$$\begin{bmatrix}
1 & 0 & \cdots & 0 & 0 & 0 & \cdots & 0 \\
0 & 1 & \cdots & \vdots & -F_0 & 0 & \cdots & 0 \\
0 & 0 & \cdots & 0 & -F_1 & -F_0 & \cdots & 0 \\
0 & 0 & \cdots & 1 & -F_2 & -F_1 & \vdots & \vdots \\
0 & 0 & \cdots & 0 & \vdots & -F_2 & \vdots & \vdots \\
\vdots & \vdots & \vdots & \vdots & \vdots & \vdots & \vdots & 0 \\
0 & 0 & \cdots & 0 & -F_{D-1} & F_{D-2} & \cdots & -F_{D-d}
\end{bmatrix}
\begin{bmatrix} \widehat{N}_0 \\ \widehat{N}_1 \\ \vdots \\ \widehat{N}_n \\ \widehat{D}_1 \\ \vdots \\ \widehat{D}_d \end{bmatrix}
=
\begin{bmatrix} F_0 \\ F_1 \\ F_2 \\ \vdots \\ \vdots \\ \vdots \\ F_D \end{bmatrix} \tag{3.17}$$

where

$$F_m = F^{(m)}/m! \tag{3.18}$$

Another example of an MBPE FM derived using derivative samples is also shown next for a dipole antenna located over a perfect ground plane. A GM sample is located at +0.25 wavelengths in (a) and a second, conjugate-frequency sample added at −0.25 wavelengths in (b) for which no additional GM sampling is needed since a negative L/λ sample is conjugate-related with one at the positive L/λ. In both cases, four derivatives are also included at each frequency sample. The wavelength spanned in these plots is only 2% and their respective Numerical Electromagnetics Code (NEC) and FM results are within 1% and nearly identical. The FM in (a) has a non-pole constant term and in (b) there is an added linear-frequency term which produces an even better match (Figure 3.3).

The results in Figure 3.4 use the same ten GM samples to further demonstrate the advantage of using an FM having non-pole terms. In (a), the FM has only pole terms while in (b) there are also constant and linear-frequency terms.

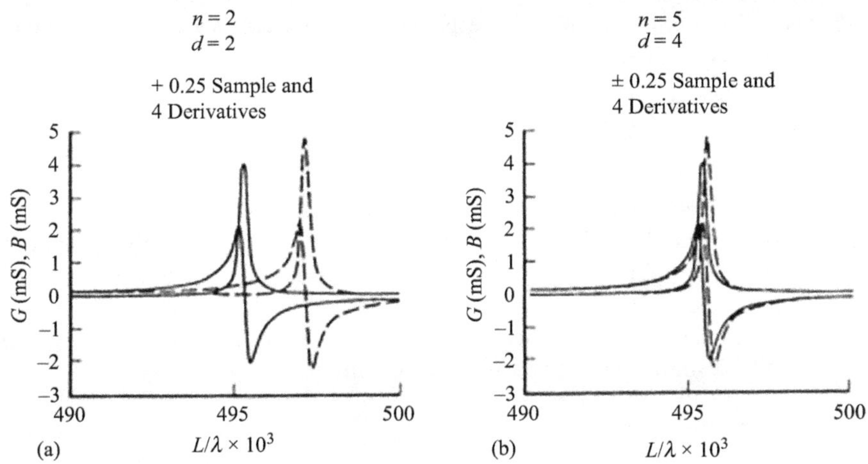

Figure 3.3 Input admittance of a dipole over perfect ground versus length in wavelengths with (a) sampled only at $L/\lambda = 0.25$ and (b) adding a conjugate sample at $L/\lambda = -0.25$ using FMs having the parameters shown

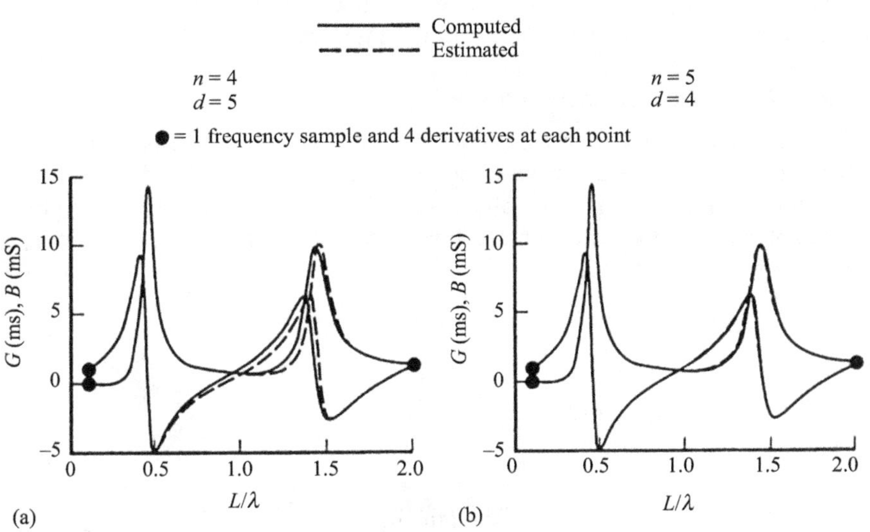

Figure 3.4 Input admittance of a dipole antenna in free space versus length in wavelengths with (a) using a FM having only pole terms and with (b) reversing the FM numerator and denominator orders to one having constant and linear terms in frequency

3.4 Some additional Prony applications

Essentially any phenomenon, physical or otherwise, that can be described by an exponential series is a candidate for analysis using Prony's method.

3.4.1 Radiation and scattering patterns

Far-field patterns of wave-equation observables are obvious candidates for Prony processing as discussed in more detail in Chapters 8 and 9 in terms of pattern synthesis and pattern estimation. The pattern of a discrete set of sources located along the z-axis is given by

$$P_{DSA}(\theta) = \sum_{\alpha=1}^{N} S_\alpha e^{kz_\alpha \cos(\theta)} \tag{3.19}$$

for N sources $S_\alpha, \alpha = 1, \ldots, N$ located at position $z_\alpha = z_{\alpha,real} + iz_{\alpha,imag}$ along the axis of a linear array lying in the complex z-plane. In complex-variability notation, these are usually referred to as residues and poles, respectively. The far-field observation angle θ is measured from the array axis. Note that $z_{\alpha,imag}$ in (3.1) corresponds to the real space location of source α, while a non-zero real part implies that the source has a directivity D_α given by

$$D_\alpha = e^{kz_{\alpha,real} \cos(\theta)} \tag{3.20}$$

The exponential series in (3.19) is a counterpart of a time series with which Prony's method has been most associated. When used for pattern synthesis, the desired pattern is specified and the source locations and strengths are to be determined. When the application is instead pattern estimation, the source locations and strengths are known, where the goal is to minimize the number of samples that are needed to determine the pattern to a specified uncertainty.

3.4.2 A 1D layered medium

The backscattered field from a structure comprising thin sheets having specified reflection coefficients and separated by variable spacings is also described by an exponential series. For the case where multiple reflections are small for product terms to be ignored, the reflection coefficient from the front surface of the structure can be approximated as

$$\begin{aligned} R(k) &= R_n + R_{n-1}e^{[i2kd_n]} + R_{n-2}e^{[i2k(d_n+d_{n-1})]} + \\ &\cdots + R_1 e^{[i2k(d_n+d_{n-1}+\cdots+d_2)]} = \sum_{\alpha=1}^{n} e^{\left[\left(i2k\sum_{\beta=n}^{\alpha+1} d_\beta\right)\right]} \end{aligned} \tag{3.21}$$

where:

R_α is the reflection coefficient of the nth sheet,
d_n is the thickness of the nth section.
K is the wavenumber of the spaces between the sheets.

By specifying the frequency-dependent reflection coefficient $R(k)$, the physical parameters of the composite structure can be determined. Note that if each sheet were instead a slab of wavenumber k_n, then the k's in the exponentials would multiply each d_n.

3.4.3 *Phase recovery of wave-equation observables*

First, observe that there are several variants of spectral-domain parameter estimation, depending on the specific problem circumstances [7].

1. If the poles comprising $F_p(X)$ are unconstrained, then a P-pole spectrum would require $2P$ complex samples of $F(X)$, at a minimum, for the $2P$ complex parameters represented by P complex residues and P complex poles to be obtained.
2. When the residues and poles occur in conjugate pairs, as is the case for EM responses, then the number of required data samples can be halved, if knowledge that the parameters are complex conjugates can be exploited. It turns out for this particular situation that the coefficients of the numerator and denominator polynomials obtained by expanding the pole series in (2) in common-denominator form become pure real, thus halving the number of real unknowns and requiring P complex samples of $F(X)$.
3. A further variation arises when using magnitude-only samples of $F(X)$ are used in the parameter-estimation step which results in

$$|F_p(X)|^2 = \sum_{\alpha=1}^{P} \frac{R_\alpha}{(X - s_\alpha)} \sum_{\alpha=1}^{P} \frac{R_\alpha^*}{(X - s_\alpha)^*} = a_{P-1}^2 \frac{\prod_{\alpha=1}^{P-1} (X - z_\alpha)(X - z_\alpha)^*}{\prod_{\alpha=1}^{P-1} (X - s_\alpha)(X - s_\alpha)^*}$$

$$= \frac{N_{2P-2}(X)}{D_{2P}(X)}$$

where the z_α are zeros of the numerator polynomial, which depend on the poles and residues together, and a_{P-1} is a normalizing parameter. The number of unknowns is doubled when squaring the FM, with the poles and zeros reflecting about the $j\omega$ axis. Valid poles and zeros are those having negative real parts, however, so that upon factoring the numerator and denominator polynomials, only those roots having negative real parts are retained. Thus, the real and imaginary parts, or the magnitude and phase, of $F_p(X)$, can be estimated from $\sim 4P$ real samples of $|F_p(X)|^2$.

As also can happen in implementing approaches (1) and (2), a potential difficulty arises here in that it is only feasible to use a relatively small number of poles and zeros in the FM since the matrix being solved to obtain its parameters becomes increasingly ill-conditioned with increasing P. This means that it is necessary to window the data being modeled, which results in some ambiguity in the FM-parameter values. This is because the data in a given frequency window are affected by poles and zeros lying outside it and the data rank (i.e., P) is itself ambiguous (Miller, 1995). Finally, modeling in the SD has its primary goal that of providing an acceptable match to a transfer function between its data samples, which doesn't require that the poles and residues (or zeros) even be computed, nor imply how accurate they should be if they were computed. For the phase-recovery application, however, the poles and zeros are needed, since otherwise those arising from squaring the complex transfer function cannot be identified and removed. Note that it is often the case that measurements provide magnitude-only data rather than complex results.

3.4.4 Examples of Prony modeling of other EM observables

A book chapter by Miller [8] includes in substantial detail a variety of EM observables that are suitable for Prony processing. Among the topics discussed are physical problems having exponential form such as space/angle waveforms and angle spectra and sampling on straight lines and circular arcs and sampling over rectangular, circular, and spherical surfaces and sampling in three dimensions. Angle waveforms and space/angle spectra include sampling the far fields of linear, circular, planar, and 3D source distributions. Also included are frequency waveforms, spatial spectra, and the far field of a 3D source in frequency. Some results from [8] are summarized in Appendix A that emphasize some additional aspects of computational EM from a parameter-estimation perspective to demonstrate the variety of ways in which exponential-series and pole-series models can arise.

References

[1] E. K. Miller, "Model-Based Parameter Estimation in Electromagnetics, Part I: Background and Theoretical Development," *IEEE Antennas and Propagation Society Magazine*, Vol. 40, No. 1, pp. 42–52, 1998.

[2] E. K. Miller, "Model-Based Parameter Estimation in Electromagnetics, Part II: Applications to EM Observables," *IEEE Antennas and Propagation Society Magazine*, Vol. 40, No. 2, pp. 51–64, 1998.

[3] E. K. Miller, "Model-Based Parameter Estimation in Electromagnetics, Part III: Applications to EM Integral Equations," *IEEE Antennas and Propagation Society Magazine*, Vol. 40, No. 3, pp. 49–66, 1998.

[4] A.V. Oppenheim, and R. W. Schafer, *Digital Signal Processing*, Prentice-Hall, Inc., 1975.

[5] R. Prony, "Essai Expdrimental et Analytique sur les Lois de la Dilatabilite des Fluides glastiques et sur Celles de la Force Expansive de la Vapeur de 1'Alkool ii Differentes Tempgratures," *J. l'&cole Polytech. (Paris)*, Vol. 1, pp. 24–76, 1795.

[6] Claude E. Shannon, "Communication in the Presence of Noise," *Proceedings of the IRE*, Vol. 37, No. 1, pp. 10–21, 1949.

[7] Brittingham, J. N., E. K. Miller, and J. L. Willows, "Pole Extraction from Real-Frequency Information", *Proceedings of the IEEE*, Vol. 68, pp. 263–273, 1980.

[8] Miller, E. K., "Model-Based Parameter Estimation in Computational Electromagnetics," presented at the NATO Advanced Study Institute Electromagnetic Modelling and Measurements for Analysis and Synthesis Problems, Il Ciocco, Castelvecchie Pascoli, Tuscany, Italy, August 10–21, 1987 and Published in 1991 in *Electromagnetic Modeling and Measurements for Analysis and Synthesis Problems*, edited by B. de Neumann, Kluwer Academic Publishers, NATO ASI Series E: Applied Sciences Vol. 199, pp. 205–256l.

Chapter 4

Sampling strategies for effective implementation of Prony's method[1]

Abstract

Prony's method [1] belongs to a class of techniques in systems theory called "parameter estimators," which estimate constants appearing in mathematical models of physical phenomena. This method is interesting because with only modest computation, the exponents, or poles, of data described by an exponential series can be obtained. This chapter explores a moving-window approach for reducing the deleterious effects of noise to which Prony processing can be sensitive. A series of systematic computer experiments are conducted on analytically specified transient waveforms. In particular, the transmission of errors through the process is investigated by determining the accuracy of the input data, the characteristic-equation coefficients and roots, and the poles. The statistical distributions of these quantities are evaluated for various combinations of input data parameters. An especially important finding is that the pole accuracy resulting from a given number of window iterations and averaging the characteristic-equation coefficients generally provides pole accuracy comparable with averaging the poles themselves.

4.1 Introduction

Growing interest in processing transient electromagnetic data has developed over the years with the increasing activity in electromagnetic pulse coupling, short-pulse measurements, time-domain computations, and other techniques. These applications, plus the singularity expansion method concept that the electromagnetic behavior of an object can be described by a set of complex-frequency resonances, have spurred the study of Prony's method and related data-processing techniques. Prony's method has received particular attention because its derivation is so easily related to physical aspects of electromagnetic transients and because it was the first practical tool that could be applied to such data. Thus, whereas it is but one of a large class of parameter estimators, Prony's method finds continued use in processing transient electromagnetic data.

[1]Excerpted and edited from "*Prony's Method Revisited*," E. K. Miller, October, 1978, Lawrence Livermore Laboratory, UCRL 52590.

This chapter summarizes the results of computer experiments performed to find possibly more reliable ways to implement Prony's method and to understand its limitations, especially when it is applied to noisy data. Admittedly, other parameter estimation techniques may indeed prove more accurate, reliable, and convenient for estimating the poles and residues of transient data, but until the capabilities of Prony's method are better understood, the choice of the optimum technique for this purpose could remain uncertain. An objective comparison also requires that all competing methods be judged in the context of the same application.

Note that there is a significant advantage when dealing with data processing of electromagnetic phenomena because the model is already known, that is, the signal is a sum of exponentials. This fact greatly simplifies the task because the model itself need not be identified, as is generally the case in system identification. Instead, attention can be exclusively devoted to data processing or parameter estimation in which the poles (complex-frequency resonances) are of primary interest and the residues (mode amplitudes) are of secondary importance. Four essential points are established in this chapter:

1. The characteristic-equation (polynomial) coefficients are the least accurate parameters obtained, although the accuracy can be correlated closely to that of the real pole components.
2. Use of the moving-window procedure and overdetermined data improve the accuracy of extracted parameters.
3. **Averaging the coefficients can produce pole accuracies comparable to averaging the poles directly.**
4. **Pole accuracy is approximately equal to the peak-signal-to-peak-noise ratio of input data.**

Section 4.2 of this chapter briefly reviews the background of Prony's method and discusses in more detail the specific approach used here in its evaluation. Section 4.3 contains numerical results and a discussion of their significance, and concluding remarks and a summary follow in Section 4.4.

4.2 Background and approach

Interest in Prony's method stems from the fact that it can be used to estimate the poles (s_α) and residues (R_α) of the exponential signal [2]

$$f(t) = \sum_{\alpha=1}^{N_P} R_\alpha e^{s_\alpha t} \tag{4.1}$$

It should be noted that (4.1) represents a rigorous model for many physical processes, including those in acoustics and electromagnetics, so that the poles exhibit a physical significance beyond their use in curve fitting. Because the model is known, the simpler problem of parameter estimation is confronted rather than the more general problem of system identification in which the model must be developed. Prony's method estimates the pole and residue values in a minimum-squared error sense through the following basic steps (see [2] for details):

1. Obtain D samples of $f_i = f(it_p)$, where $i = 0, \ldots, D-1$, at D increments of t_p spaced uniformly over a window of width $T = t_p(D-1)$, where $D = 2N_p$; N_p is the number of poles being calculated.
2. From the data samples of (l), develop the linear system

$$\begin{bmatrix} f_0 & f_1 & f_2 & \cdots & f_{N_P-1} \\ f_1 & f_2 & f_3 & \cdots & f_{N_P} \\ \vdots & & & & \\ f_{N_P-1} & f_{N_P} & f_{N_P+1} & \cdots & f_{2N_P-2} \end{bmatrix} \begin{bmatrix} A_0 \\ A_1 \\ \vdots \\ A_{N_P-1} \end{bmatrix} = - \begin{bmatrix} f_{N_P} \\ f_{N_P+1} \\ \vdots \\ f_{2N_P-1} \end{bmatrix} \tag{4.2}$$

   and the vector A] represents the coefficients of the characteristic equation

$$\sum_{a=0}^{N_P} A_a X^a = 0; \& \;\; set \;\; A_{N_P} = 1. \tag{4.3a}$$

   Upon rewriting (4.3a) in terms of the roots X_α

$$(X - X_1)(X - X_2) \cdots (X - X_{N_P}) = 0; \;\; X = X_1, \cdots, X_{N_P} \tag{4.3b}$$

   it is obvious that A_α are complicated functions of the roots. Because all anti-diagonal elements are equal, M is called a Hankel matrix or a Vandermonde form.
3. Solve (4.2a) for the values of A. The roots X_α of (4.3) provide the values of the poles from the relationship

$$S_a = -\frac{1}{t_P} log(X_a), \tag{4.4}$$

Prony's method linearizes the problem of obtaining R_α and s_α in (4.1) by separating their solutions. After calculating the poles from (4.3), the residues are computed by returning to (4.1). The above steps comprise Prony's original method of 1795 and were described in 1956 by Hildebrand [2]. It is worth noting that (4.2) is a difference equation whose form in the Z-domain is that of an all-pole model [3].

In its simplest form, Prony's method requires that the user specify only two parameters, say t_P and N_P, from which D and T follow automatically. The result is a square system of linear equations with N_P unknowns, the coefficients of which are the sampled data, and with the general coefficient a_{ij} given by $fi{+}j{-}2$, where $i, j = 0, \ldots, N_P - 1$.

A unique solution can be found only when $N_P >= N_w$, where N_W is the number of poles in the actual data (4.l), and when $t_P < 1/2f_m$, where f_m, is the highest frequency component of $f(t)$. In general, because N_W may not be known *a priori*, experimentation using different values for N_P may be required to establish an acceptable choice for it. Similarly, to establish the best value of t_P, its variation to the extent permitted by the data may be necessary. Whereas t_P must be small enough to avoid aliasing (i.e. Le., t_P must satisfy the Nyquist criterion), it cannot be made too small without the risk of ill-conditioning the data matrix by decreasing the linear independency of neighboring rows and columns. Clearly, application of even the basic version of Prony's method requires care if acceptable results are to be obtained.

If more data points (D') were available than the minimum $2N_W$ values required to compute the N_W poles of $f(t)$, this would clearly result in an overdetermined system because the number of equations exceeds the number of unknowns. Two approaches to handling this situation are apparent. In the first, it is possible to continue using $2N'_P = D'$, that is, compute as many poles as permitted by the number of data samples available. This number of poles would, of course, be greater than that actually present in $f(t)$, and the data matrix in (4.2b) should then be singular unless as is invariably the case, the data are noisy or of otherwise limited accuracy. This approach has been followed in much of the early work in electromagnetics that employed Prony's method.

Alternatively, the original value of N_P could be used while reducing the number of equations from N_B to N_P by multiplying (4.2) by the transpose of the data matrix. This results in

$$[M']A] = -F']. \tag{4.5a}$$

where

$$[M'] = \begin{bmatrix} C_{00} & C_{01} & C_{02} & \cdots & C_{0,N_{P-1}} \\ C_{10} & C_{11} & C_{12} & \cdots & \cdot \\ C_{20} & C_{21} & C_{22} & \cdots & \cdot \\ \vdots & & & & \\ C_{N_{P-1,0}} & \cdots & & & C_{N_P-1,N_P-1} \end{bmatrix}, \tag{4.5b}$$

$$F'] = \begin{matrix} B_0 \\ B_1 \\ \vdots \\ B_{N_P-1} \end{matrix} \Bigg] \tag{4.5c}$$

$$C_{ij} = \sum_{k=0}^{N'_{P-1}} f_{i+k} f_{j+k}, \tag{4.5d}$$

and

$$B_i = \sum_{k=0}^{N'_{P-1}} f_{i+k} f_{N_{P+k}} \tag{4.5e}$$

The two approaches could be combined by choosing any value of N'_P between N_P and $(D')/2$.

Unlike M, M' is not a Hankel matrix but has the form of an auto-covariance matrix, the usual starting point for digital signal processing of transient waveforms. This is achieved by varying the model parameters. Prony's method does not consider this error explicitly, however, and, as originally formulated, it is essentially a point-matching procedure. The process of developing the new data matrix in (4.5b) through multiplying the original matrix by its transpose also minimizes the squared error; this is sometimes called regularization but is referred to more often as generating the pseudo inverse.

An important point to be noted is that Prony's method contains features identical to a technique employed in digital signal processing called all-pole modeling. Furthermore, a spectral estimation procedure called the maximum-entropy method is also based on an all-pole model and is, therefore, equivalent to Prony's method. It is thus a little puzzling to hear frequent claims that Prony's method is more ill-conditioned, more sensitive to noise, etc., than more "modern" methods. The critics evidently overlook or deny the links of these other techniques to Prony's method because, as originally formulated, Prony's method did not include the pseudo-inverse approach. Clearly, however, Prony's method is the forerunner of such techniques and merits recognition as such.

There are additional ways to handle an overdetermined system. One that has been found to be most useful, called the moving-window technique (Figure 4.1), processes the data using several sequential time windows that "move" along the waveform by $t_w = nt_P$. In this way, the data are not used all at one time, and several sets of estimates are obtained for the poles. Authentic poles should cluster around an average value while spurious poles due to noise and curve fitting (if $N_P > N_W$) will appear randomly. Superimposing the plots of all the pole sets on one graph helps identify the valid poles. An interesting alternate procedure used by Van Blaricum {5} is to process the data in both forward and reverse directions with respect to time. The result is that genuine poles are reflected about the $j\omega$ axis, and artifacts remain predominantly on the negative half-plane.

Although the number of decisions required to implement Prony's method seems small, the number of resulting alternatives can be large. The results that follow might provide some guidance to produce more effective and reliable application of the technique. This is done by examining the effects of various tradeoffs on the accuracy obtained for the derived pole values. All results reported are produced using specified sets of pole and residue values to generate a signal waveform. Noise effects are simulated by adding zero-mean, uniformly distributed random numbers to the signal; the maximum value of these numbers is a specified fraction of the peak signal level. The accuracies of the various quantities of interest are compared with their true values and expressed in bits of agreement as

$$B(Y) \equiv -\log_2 \left| \frac{Y_{true} - Y_{computed}}{Y_{true}} \right| \tag{4.6}$$

where Y is a parameter. Generally, the B-values for all equivalent parameters in a set (e.g., all real or imaginary poles) were averaged to obtain

$$\bar{B}(Y) \equiv \frac{1}{N} \sum_{i=1}^{N} B(Y_i) \tag{4.7}$$

where N is the total number of parameters of type Y in the set.

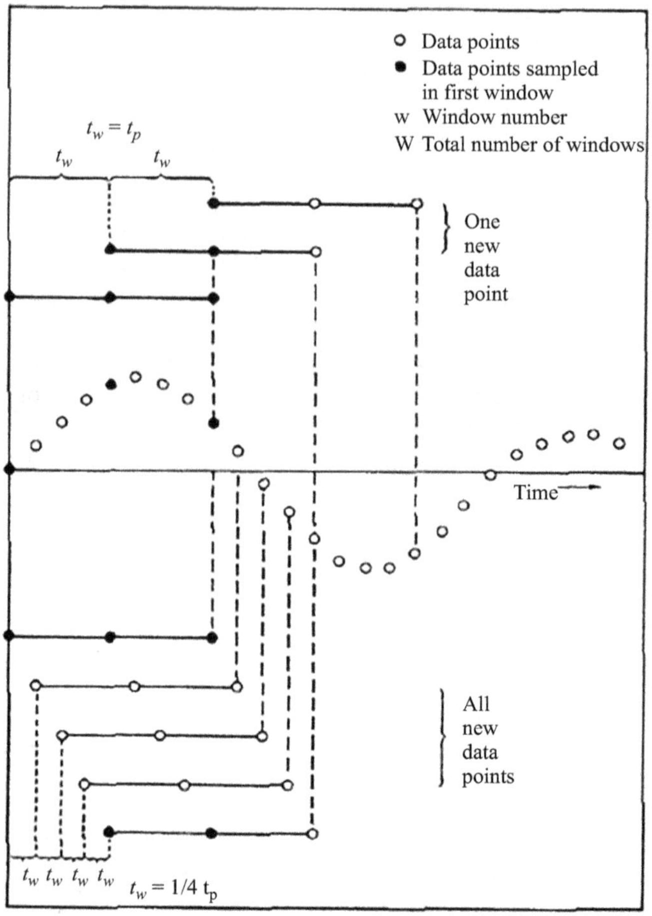

Figure 4.1 *Moving-window procedure. The data encompassed by each window are processed separately to obtain a number, W, of estimates of the associated parameters. (a) If the window moves an interval of $t_W = t_P$, only one new data point is added to those of the preceding window. (b) If the window is by any other interval (e.g., $t_W = \frac{1}{4} t_P$), all new data points result.*

When dealing with noisy data, a single run produces too few results to be useful. Each run is repeated (nominally 29 times), and the expected values of the quantity of interest are computed. The corresponding value of the accuracy is then

$$\langle \tilde{B}(Y_i) \rangle \equiv \frac{1}{R} \sum_{r=1}^{R} B_r(Y_i) \tag{4.8}$$

where $B_r(Y_i)$ is the bit value for parameter Y_i from run r and R is the total number of runs. The average that results from using the moving-window technique is defined as

$$\tilde{B}(Y_i) \equiv \frac{1}{W}\sum_{w=1}^{W} B_w(Y_i) \tag{4.9}$$

where W is the total number of windows used and $B_w(Y_i)$ is the bit value for parameter Y_i from window w. All three averages may occur either separately or in combination.

Finally, the cumulative bit accuracy is defined by

$$\tilde{b}(w, Y_i) \equiv \frac{1}{w}\{[w-1]\tilde{b}(w-1, Y_i) + B(Y_i)|_w\} \tag{4.10}$$

The quantities $\tilde{b}(w, Y_i)$ and $B(Y_i)$ provide similar kinds of information in that both show the influence of window averaging on the accuracy of the extracted parameter Y_i. These quantities differ in that $B(Y_i)$ is computed for only a specified number of windows [i.e., if W in (4.9) is increased by one, then the accuracies of a new set containing $W + 1$ windows is computed] and that $\tilde{b}(w, Y_i)$ adds one new window to increase w, the window number, by one. The advantage of using $\tilde{b}(w, Y_i)$ is that it shows the result of adding one new window to previously computed results for each individual run. In the limit of an ensemble average of a large number of runs, these accuracy measurements should approach the same limit.

4.3 Numerical results

A large number of Monte Carlo computer experiments were conducted with the results presented in approximately the same order as the experiments were actually performed, in hopes of providing added insight into the cases studied. Although it would be preferable that general conclusions for each of the questions explored could be provided, the significance of many results depends on the context in which they were generated.

4.3.1 Initial moving-window tests

Early work using Prony's method encountered serious difficulties in dealing with noisy data. A significant step in circumventing this shortcoming was to process two or more sets of data successively by moving the data window along the waveform by $t_w = nt_p$. All pole values thus obtained were plotted on the same s-plane diagram. In general, true poles present in the original data formed clusters due to noise, curve fitting, and the fact that $N_P > N_W$ were randomly scattered. The actual data pole values were then estimated by averaging the clusters.

The use of this moving-window technique prompts several questions. Are the successive pole estimates improved in accuracy by averaging, and, if so, how much? How many separate windows are justified? Rather than averaging poles, what would be the result of averaging the roots or characteristic equation coefficients and computing the poles from those averages? How does the relative size of the windows and the interval between successive windows affect the accuracy of the extracted parameters? To explore such questions, a set of ten runs was performed using the ten-pole set given in Table 4.1 and the parameters $t_p = 0.0467$, $t_w = t_p/10$, and $S_P = 10^3$, where S_P is the peak-signal-to-peak-noise ratio. The bits of agreement of the input data and computed coefficients, roots, and poles were calculated for each window; in each of the ten runs, a cumulative average of the windows used was compiled up to the maximum of ten ($W_{max} = 10$).

Representative results obtained from this set of ten runs are presented in Figures 4.2–4.5. The cumulative accuracy of the running pole average for each run

Table 4.1 The 10-Pole Set Used for Moving-Window Tests

α^a	S_α
1	$-0.5 \pm j1.0$
2	$-0.5 \pm j2.5$
3	$-0.5 \pm j5.0$
4	$-0.5 \pm j7.5$
5	$-0.5 \pm j10$

[a]Residue $R_\alpha = 1 \pm j0$

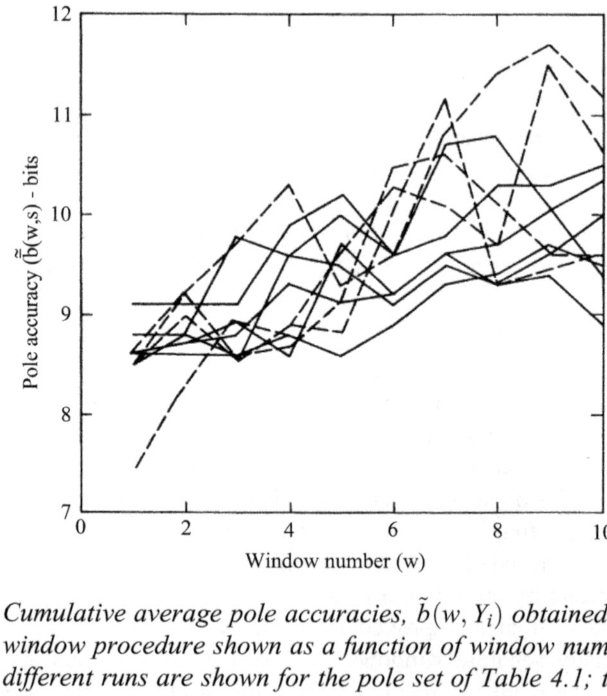

Figure 4.2 Cumulative average pole accuracies, $\tilde{b}(w, Y_i)$ obtained from moving-window procedure shown as a function of window number. Ten different runs are shown for the pole set of Table 4.1; the parameters used are $N_W = N_P = 10$, $D = 20$, $t_P = 10t_W = 0.0486$ s, and $S_P = 10^3$. Accuracy improvement is not monotonic from window to window but exhibits an average value of 1.45 bits over ten windows.

is shown as a function of the number of windows, w, in Figure 4.2. Run-to-run variation of pole accuracy improvement is clearly evident, ranging from 0.3 to 3.3 bits with an average improvement of 1.5 bits. This value, it should be noted, approximately equals the square root of ten, consistent with the noise model employed.

Figure 4.3 shows results that describe the single window accuracies of the input data, coefficients, roots, and poles for one of the ten runs. The average bit accuracies of all of these quantities from each of the ten windows are plotted on the same axes, and the data points from each separate window are connected by straight lines to highlight the trends. Somewhat unexpectedly, the coefficients and data exhibit less average accuracy than do the roots and poles. This result implies that the errors are not amplified through the successive steps of the computation. Also unexpected is the curious result that the input data seem to be less accurate on average than some of the computed results. This behavior is observed to continue in Figure 4.4, which compares the ten-window averages for each parameter for all ten runs.

For each of the ten-window runs, Figure 4.5 displays both the accuracies of the average pole values as computed directly and those obtained by using the average

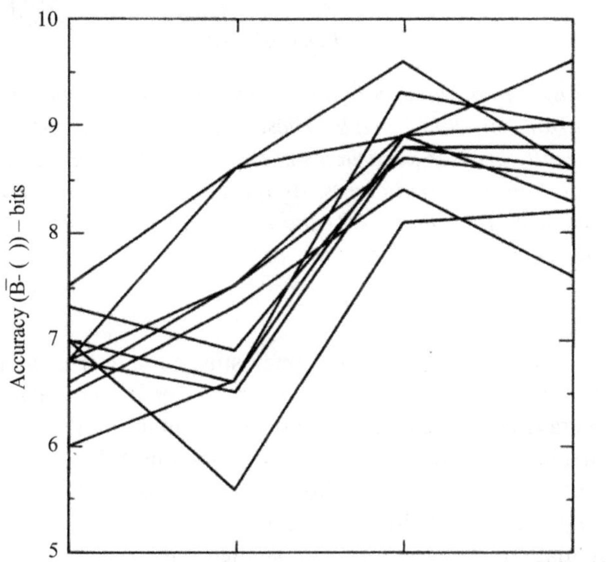

Figure 4.3 Average input data, coefficient, root, and pole accuracies for ten one-window runs. Parameter values are given in Figure 4.2. Results show that the roots and poles are more accurate than the input data, according to the accuracy definitions of (4.6–4.8). Lines are drawn connecting the results of individual runs to help visualize the accuracy trends.

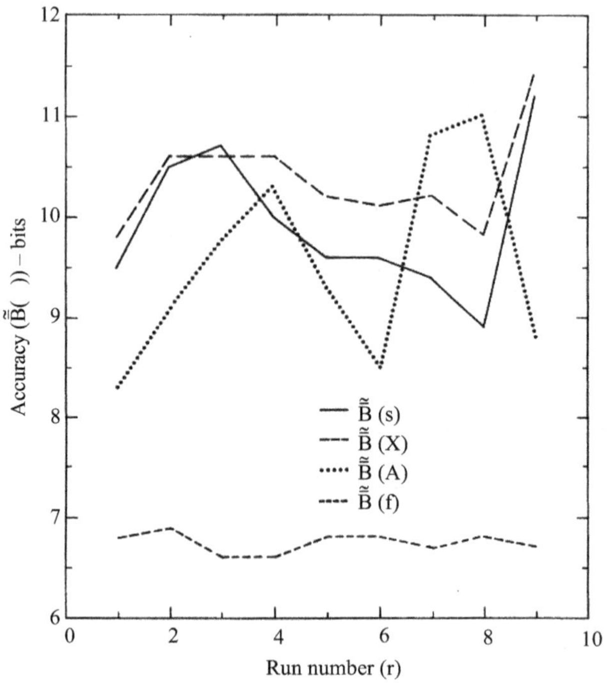

Figure 4.4 Average input data, coefficient, root and pole accuracies for ten separate ten-window runs. These results also demonstrate the generally greater accuracies of the roots and poles compared with the input data and coefficients. Again the other parameter values are those used for Figure 4.2, and f, A, X, and s denote input data, coefficients, roots, and poles, respectively.

root and average coefficient values. It is interesting to note that the pole accuracy for all three sets of values is comparable, an observation of particular importance in the use of the average coefficients; solving the polynomial for its roots a single time is sufficient, whereas pole averaging requires a solution of the polynomial for each window. Without multiple solutions, however, the possibility of identifying the actual data poles by means of clusters is lost.

Similar calculations were performed a single time for $W_{max} = 121$ on five pole sets as described in Table 4.2, and Figure 4.6 displays the running pole averages for $w = 1, 11, 21, ..., 121$. The improvement in pole accuracy was again approximately equal to the square root of W_{max}. It was also observed that the average coefficients generally provide pole values of equal or even better accuracy than the averaged poles themselves. On the basis of these results, it can be concluded that the moving-window method can improve the accuracy of pole values obtained from Prony's method, and that averaging coefficients can be a worthwhile procedure.

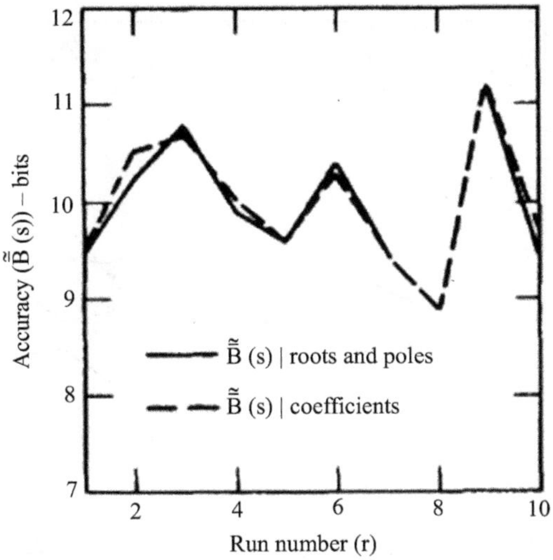

Figure 4.5 Accuracies obtained from pole averages compared with pole accuracies provided- by average values of coefficients and average roots. Using the moving-window technique and ten windows for the set of parameters in Figure 4.2, all three methods of obtaining the pole values yield similar accuracies.

Table 4.2 Pole Sets Used for Results of Section 4.3.2

Pole set α[a]	S_α
1	$-0.5 \pm j$ 1, 2.5, 5, 7.5, 10
2	$-0.5 \pm j$ 2.5, 5, 7.5, 10
3	$-0.5 \pm j$ 3.3, 6.7, 10
4	$-0.5 \pm j$ 5, 10
5	$-0.5 \pm j$ 10

[a]Residue $R_\alpha = 1 \pm j0$

Consequently, the results that follow are based on coefficient averaging whenever the moving-window procedure is employed unless indicated otherwise.

4.3.2 Pole calculations using $N_P > N_W$

The following series of calculations was based on the set of poles described in Table 4.2, except that set 1 was replaced by pole set 1': $-0.5 \pm j$ 2, 4, 6, 8, 10. Twenty-nine 10-window runs were conducted for noisy waveforms generated from these pole sets, where $N_P = N_W, N_W + 2, \ldots, 10$ and where $D = 2N_P$. The average pole accuracies in bits as obtained from the average coefficients for these runs are shown in Figure 4.7 as a function of N_P. In these calculations, $t_w = t_p$, whereas for

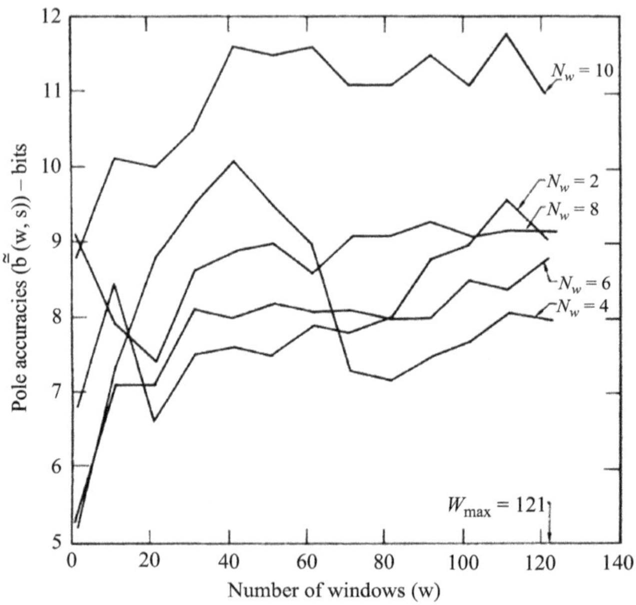

Figure 4.6 Cumulative average pole accuracies, b(w,s), obtained by moving-window technique for one run as a function of the number of windows for pole sets of Table 4.2. A maximum of 121 windows is used and N, = N,. Other parameters are t = 12lt, = 0.0467 s, D = 2N$_P$, and S$_P$ = 10^3'. The increase in accuracy is again on the order of the square root of W.

the case considered previously, $t_w = t_p/10$; this represents a tenfold decrease in window width.

Increasing N_P beyond N_W affords a significant improvement in accuracy, I_A, which is given approximately by

$$I_A = 0.75(N_P - N_W). \tag{4.11}$$

Except for the two-pole case, the pole accuracy is ~3.3 bits when $N_P = N_W$ and increases to over 6 bits within the range of N_P values examined here. Because the peak-signal-to-peak-noise ratio is 100 (~6.7 bits), these results are considered to be quite good.

What Figure 4.7 demonstrates is the improvement that results from using $N_P >$ N_W. In order to compare this improvement with that obtained from the moving-window technique, the pole accuracies for N_W = 6 and N_P = 6, 8, and 10 as a function of W are presented here where pole set 3 from Table 4.2 was used. The curves in Figure 4.8(a) show that using $N_P > N_W$ to overdetermine the system appears more fruitful to improve accuracy than employing the moving-window procedure. The latter technique provides improved accuracy but only for $N_P = N_W$ if $W <= 5$.

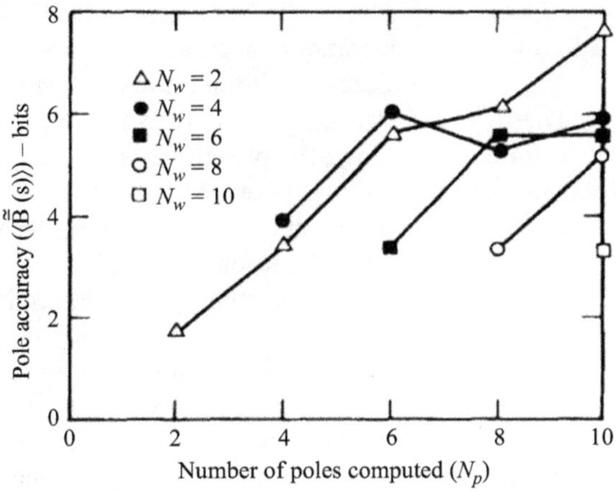

Figure 4.7 *Effect of overdetermining data by using $N_P > N_W$, demonstrated for pole sets of Table 4.2. Other parameters arc $t_p = t_w = 0.0467$, $S_P = 10^2$, $W = 10$, $R \sim 29$, and $D = 2N_P$. A significant increase in pole accuracy is obtained by computing more poles than actually contained in the data.*

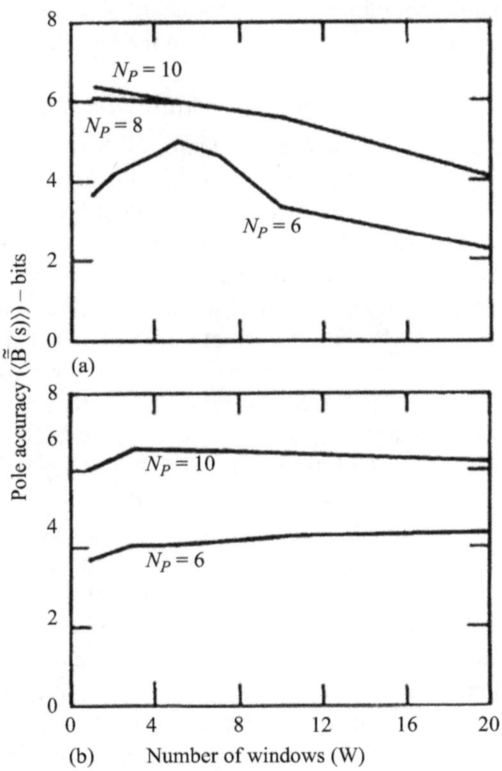

Figure 4.8 *(a) Pole accuracies obtained for pole set 3 of Table 4.2 ($N_w = 6$) shown as a function of the number of windows using three values of N_P*

The behavior of the accuracy that results from increasing W for $N_P = N_W = 6$, – a maximum value followed by a decrease in accuracy – differs from that observed previously for $N_P = N_W = 10$ in Figure 4.6 although, in the former case, $t_w = t_p/10$ rather than $t_w = t_p$. The former approach uses, in effect, completely new data (at least up to $W <= 10$) for each window; the present approach introduces only one new datum for each successive window. If two of the cases shown in Figure 4.8 (a) are recalculated, using $N_P = 6$, 10 and $t_w = t_p/10$, then the results shown in Figure 4.8(b) are obtained. In contrast to the results in Figure 4.8(a), increasing W for both values of N_P gives improved accuracies, although a slight decline does occur for $N_P = 10$ for $W >= 3$.

Comparison of Figures 4.8(a) and 4.8(b) reveals the surprising fact that, for values of $W <= 7$ where $N_p = N_w$, greater accuracy occurs when $t_w = t_p$, even though fewer data samples are available. This outcome cannot be readily explained except to postulate that it results from the use of coefficient averaging.

Although using the moving-window technique improved the accuracy for $N_P = N_W$, overdetermining the data by employing larger values of N_P is more effective. Parameter values are $S_P = 100$, $t_p = t_w$, $D = 2N_P$, and $R \sim 29$. Result of reducing time interval between successive windows (t_w) relative to data sample interval (t_p) is shown in 4.8(b). The parameters of Figure 4.8(a) are used except that $t_w = t_p/10$, and only the cases in which $N_P = 6$, 10 are included. Some improvement in pole accuracy is obtained both for $N_P = 10$, a result not observed in Figure 4.8(a) for $t_w = t_p$, and for $N_P = 6$, although the improvement in the latter case is not as great.

4.3.3 Pole calculations using the pseudo inverse

Because improved accuracy was obtained using $N_P > N_W$, it was suspected that solving the overdetermined system using the pseudo inverse, or use of an auto-covariance matrix, might also lead to increased pole accuracy. To test this possibility, the six-pole set of Table 4.2 was again employed, and the effect of varying either the number of windows or the number of data points was studied while using $N_P = 6$. These results are shown in Figures 4.9(a) and 4.9(b), respectively.

In Figure 4.9(a), where $D = 20$, the effect of increasing W differs substantially, depending on whether $t_w = t_p$ or $t_w = t_p/10$, which are the two cases considered. Both curves produce greater pole accuracy initially as W increases from 1. When $W > 5$, the accuracy decreases if $t_w = t_p$, but if $t_w = t_p/10$, the pole accuracy increases to a maximum of ~7.2 bits at about $W = 10$ but beyond $W = 10$ it declines slightly. Interestingly enough, at $W = 1$, these results compare well with those of Figure 4.8(a) for $N_P = 8$ and 10, showing that, in this case at least, comparable accuracies are realized whether $N_P > N_W$ or the pseudo inverse is employed to exploit supplemental data. The results on which this observation is based are too limited, however, to permit any general conclusions to be inferred.

Figure 4.9(b) displays the effect of increasing D beyond $2N_P$ for which two cases are shown where $t_w = t_p/10$ for $W = 1$ and $W = 10$. A significant improvement in the pole accuracy is obtained as D increases until about $D = 20$, at which value the accuracy begins to decline. The accuracy trend as a function of the number of

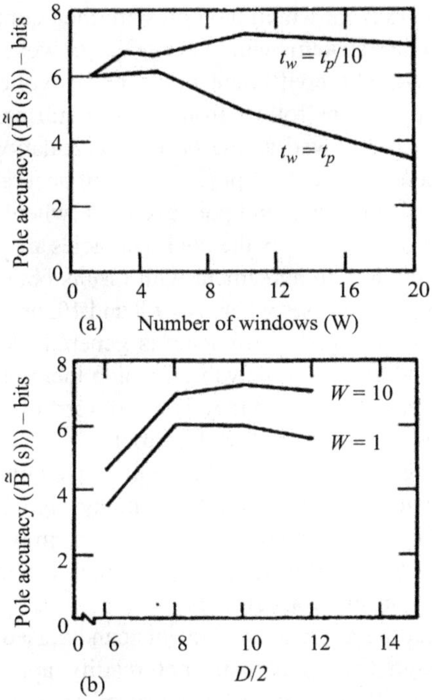

Figure 4.9 *(a) Result of using pseudo-inverse technique with overdetermined data for two values of t_w. Parameter values are $D = 20$, $N_P = N_W = 6$, $S_P = 100$, $t_p = 0.0467$ s, and $R \sim 29$. The pseudo-inverse approach gives slightly better accuracy when using the same number of data points, compared to the curve in Figure 4.8(b) where $N_P = 10$. (b) Effect of varying number of data points when using pseudo-inverse technique demonstrated for $W = 1$ and $W = 10$, $t_p = 10t_w = 0.0467$ s, $N_p = N_w = 6$, $S_P = 100$, and $R \sim 29$. Increasing D beyond 20 appears to be nonproductive for which a possible explanation may be that, as the waveform amplitude decreases with time, additional data samples are increasingly inaccurate and their inclusion produces less accurate parameter estimates.*

data points is similar to that in Figure 4.7 and demonstrates the usefulness of the additional information made available when $D > 2N_P$. Figures 4.9(a) and 4.9(b) exhibit a maximum accuracy of about 7 bits, a value comparable to the peak-signal-to-peak-noise ratio, which being 100 is −6.7 bits.

4.3.4 Additional examination of the coefficients

It was mentioned above that when using the moving-window technique, pole values of comparable accuracies were obtained either by averaging the poles directly or by averaging the coefficients from which the poles are then computed. This result was established only for square systems where $N_P = N_W$, however, and now explored is the case where $N_P > N_W$. The coefficient behavior is explicitly examined as well because interesting implications follow from these inquiries.

The six-pole system employed as the basis of calculations in the previous two sections is continued here. Figure 4.10 presents the bit accuracies obtained for poles computed from the coefficient, root, and pole averages (where $W = 10$ and $t_w = t_p/10$) for several runs. For the case of $N_P = 6$, the pole accuracies are nearly equivalent for all three methods of computation, in agreement with results obtained earlier for the ten-pole system (Figure 4.5). In the cases where $N_P = 8$ and 10, however, the pole accuracy resulting from using the averaged coefficients is generally less than that calculated from the average root and average pole values, which themselves agree closely. This result suggests that increased pole accuracy is the reward of the extra computational effort needed to find the characteristic-equation roots.

Because most of the results presented here are based on coefficient averaging, the estimated pole accuracies extracted from noisy signals can be regarded as somewhat conservative. For example, the accuracy implied by the results of Figure 4.10 where $N_P = 8$ and 10 is about 8 bits, which is about one bit higher than the highest value shown in previous figures.

It is remarkable that the average coefficients can be used to calculate any poles when $N_P > N_W$, although this is perhaps not readily appreciated upon viewing Figure 4.10. If (4.3) is written explicitly in terms of its roots, X_α, it becomes

$$\prod_{\alpha=1}^{N_P}(-X_\alpha) + \sum_{\beta=1}^{N_P}\prod_{\alpha=1}^{N_P}{}^{\beta}(-X_\alpha)X + \frac{1}{2}\sum_{\gamma=1}^{N_P}\sum_{\beta=1}^{N_P}{}^{\gamma}\prod_{\alpha=1}^{N_P}{}^{\beta,\gamma}(-X_\alpha)X^2$$
$$+\frac{1}{3}\sum_{\delta=1}^{N_P}\sum_{\gamma=1}^{N_P}{}^{\delta}\sum_{\beta=1}^{N_P}{}^{\gamma,\delta}\prod_{\alpha=1}^{N_P}{}^{\beta,\gamma,\delta}(-X_\alpha)X^3 + \ldots + \sum_{\alpha=1}^{N_P}(-X_\alpha)X^{N_P-1}$$
$$+X^{N_P} = \sum_{\alpha=0}^{N_P}A_\alpha X^\alpha \tag{4.12}$$

Equation (4.11) demonstrates that the coefficients in the characteristic equation are complicated functions of its roots. The only relatively simple terms are A_0, a product of all the roots, and A_{N_p-1}, a sum of all the roots. It is thus difficult to understand how errors in the coefficients transform to errors in the roots, even when $N_P = N_W$. When $N_P > N_W$, the situation becomes even more complex because the curve-fitting poles that are inevitably generated influence all coefficients (and vice versa) in non-trivial ways.

Plotted in Figure 4.11(a) are six individual nonaveraged A_i coefficient values as functions of run number for a single value of $N_P = 6$. In Figure 4.11(b), the A_0 coefficient for three values of $N_P = 6$, 8, and 10 is shown. Large variations from run to run are observed in both cases, but those for $N_P > N_W$ are more extreme and even exhibit sign changes. Despite this deviation, however, the bit accuracies for the

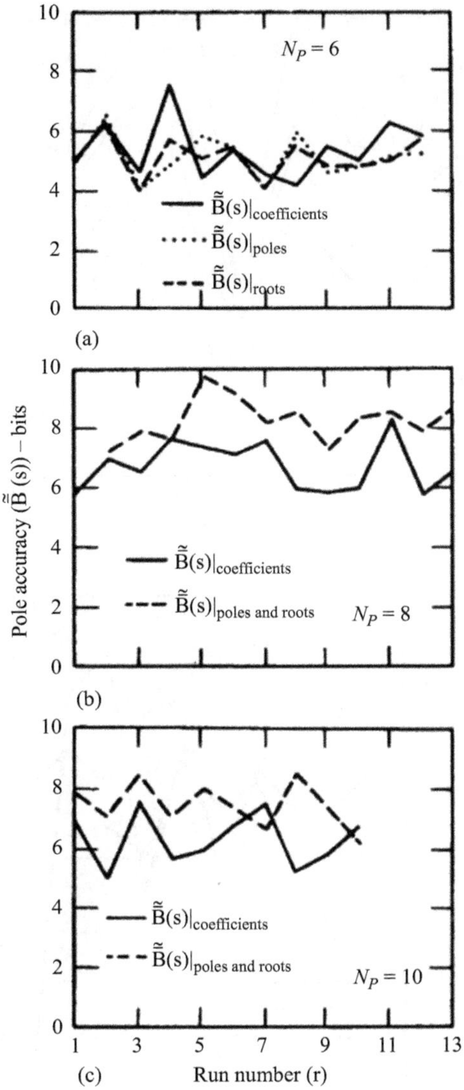

Figure 4.10 *Pole accuracies obtained for individual runs from the average coefficient, root, and pole values. (a) $N_P = 6$, (b) $N_P = 8$, and (c) $N_P = 10$ with $N_W = 6$ in all cases. Parameter values are $W = 10$, $t_p = 10t_w = 0.0467$ s, $D = 2N_P$, and $S_P = 100$. When $N_P = N_W$, the poles obtained from the averaged coefficients are generally somewhat less accurate. For (b) and (c), the pole accuracies produced by the average roots and average pole values are essentially identical and are represented by the same line.*

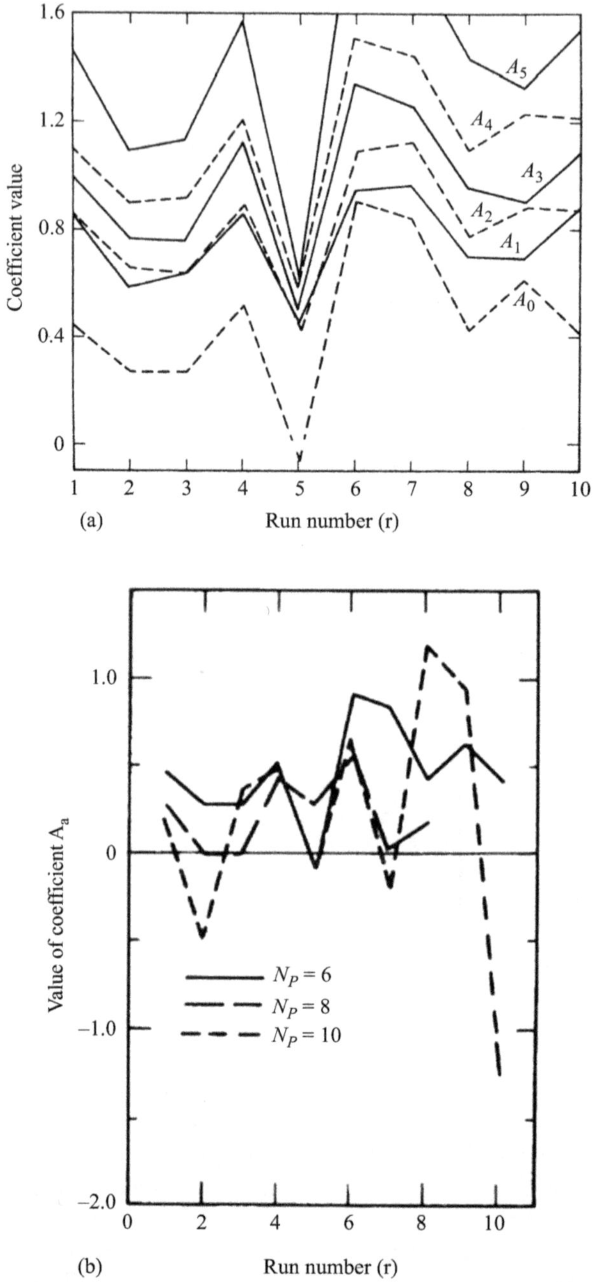

Figure 4.11 *(a) Variability of six coefficients for individual runs using the pole system in which $N_P = N_W = 6$. (b) The A_0 coefficient for cases in which $N_P = 6,8,10$ and $N_W = 6$. The confidences vary notably from run to run, especially for overdetermined cases in which the excursions are largest and sign changes most frequent. Other parameters are $S_P = 100$, $t_p = 0.0467$ s, $W = 1$, and $D = 2N_P$.*

poles calculated by averaging these coefficients for $N_P = 8$ and $N_P = 10$ are 7.0 and 7.5, respectively, for $S_P = 100$. That reasonable pole values should result from these data seems, at the least, unexpected.

Scatter plots offer an additional method for comparing the accuracies of the coefficients and other parameters. Using the same six-pole set as before, a comparison of the average bit accuracies of combinations of the input data, coefficients, roots, and poles in is obtained as shown in Figure 4.12. Of these parameters, the roots and poles can be most closely correlated with each other (Figure 4.12(c)).

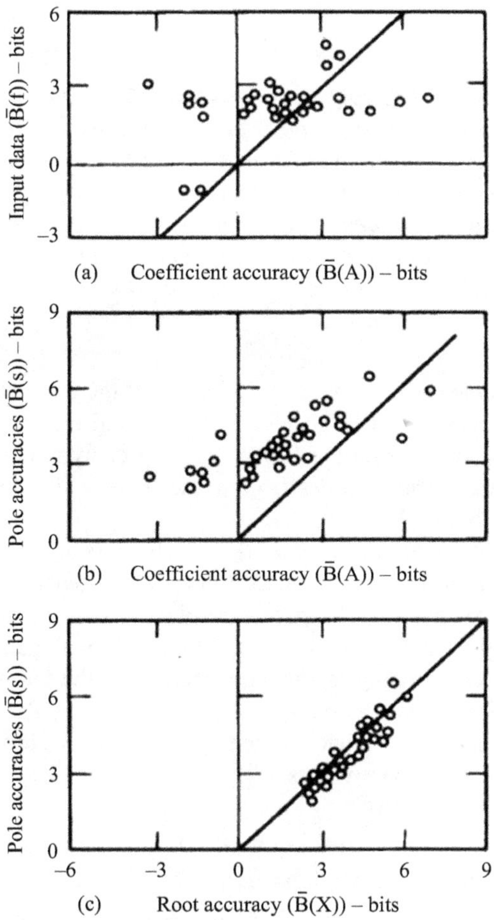

Figure 4.12 Scatter plots illustrating the relationships existing among input data, coefficients, roots, and poles. (a) Input data and coefficient accuracies appear essentially uncorrelated. (b) Pole accuracies prove significantly better than coefficient accuracies. (c) Pole and root accuracies are closely correlated. Parameter values are $N_P = N_W = 10$, $S_P = 100$, $t_p = 0.0467s$, $W = 1$, $D = 2N_P$, and $R = 1$.

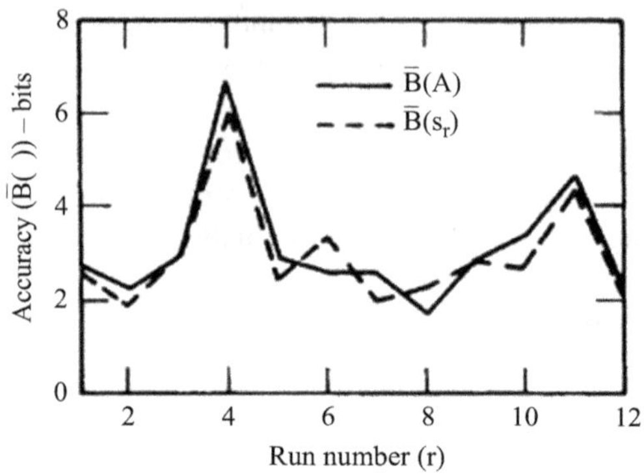

Figure 4.13 *Comparison of coefficient and real pole (s_r) accuracy for individual runs. The close agreement between values demonstrates that the coefficients are not independent of all subsequent parameters. The imaginary component is found to be consistently more accurate than the real component, which provides the basis for the greater overall pole accuracy compared to that of the coefficients. Parameter values are the same as those for Figure 4.10.*

The coefficient accuracy seems almost independent of the input data (Figure 4.12(a)), whereas in agreement with earlier results, the roots (and hence the poles) show significantly greater accuracy (Figure 4.12(b)). It is especially interesting to see that the coefficient bit accuracy can at times become negative while the poles continue to maintain accuracies of 2–3 bits.

It is interesting to examine whether the coefficients are correlated with any of the extracted parameters? In Figure 4.13, the average coefficient values and average values of the real parts of the poles are plotted as a function of run number. These curves clearly show that the coefficients and real parts of the poles are closely correlated. This finding is also consistent with the knowledge that the real parts of the poles are generally obtained with less accuracy than are the imaginary parts. It is reassuring to learn that the coefficients appear correlated with at least one of the other extracted parameters.

4.3.5 *Additional parameter variations*

The results presented in the sections above provide at best a glimpse of how Prony's method works and what can be expected of from its application. Primarily examined here were the effects of using the moving-window technique compared with overdetermination of the data using $N_P > N_W$ or $2D > N_P$. Other features of the data and the means by which it is processed influence the utility of Prony's method, several of which are discussed here. The six-pole system of Table 4.2 was again used for all calculations.

4.3.6 *Variation of t_p*

The need to sample transient data at least twice per period of the highest frequency component is well known. To minimize any chance of aliasing, there is sometimes a consequent tendency to sample much more rapidly than the Nyquist rate requires. Unless the number of data points is increased correspondingly, however, there is the risk that the data may be underdetermined even though $D = 2N_P$. This can occur because the linear independence of the data samples decreases as t decreases.

Previous results demonstrated this effect using noise-free data [4,5]. The influence of noise is demonstrated in Figure 4.14 where the pole bit accuracies are shown as a function of t_p for $N_P = 6$ and 10. Although about the same maximum accuracy is realized in both cases, the range of best results for t_p is much narrower for $N_P = 6$ than for $N_P = 10$. Because the data window width, T, is wider for the ten-pole case, the linear independence of the data presumably holds up at smaller values to t_o which leads to a greater range of acceptable results. It should be noted that up to this point, all results were calculated using the largest value of t_p given in Figure 4.14. The accuracies determined are thus likely to be less than the maximum which could have been achieved if t_p had been optimized.

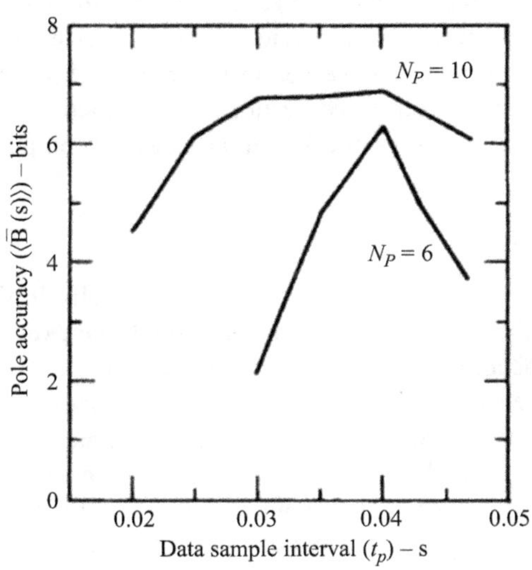

Figure 4.14 *Pole accuracy as function of data sample interval, t. Two curves are included for $N_P = 6$ and 10 with $N_W = 6$ in each case. The former curve shows much greater sensitivity to t_p than the latter, demonstrating the benefit of overdetermining the data. With decreasing t_p, the values of both curves decline at about the same data-window width. Parameter values are $S_P = 100$, $W = 1$, $D = 2N_P$, and $R \sim 29$.*

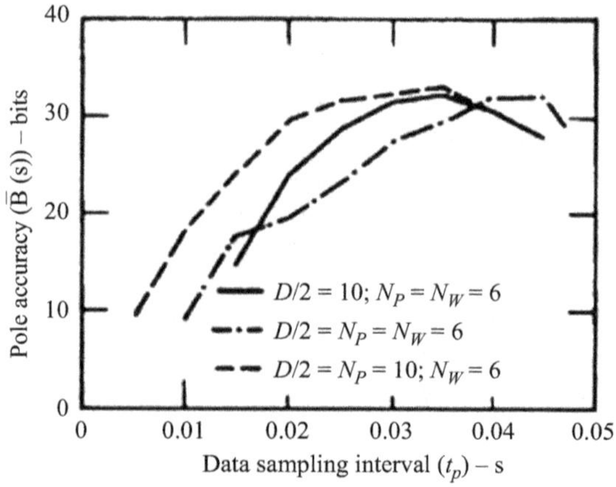

Figure 4.15 Effect of variations of t_p on noise-free data. The pole accuracy is most responsive to changes in t_p when $N_P = N_W = 6$. Setting $N_P = 10$ provides the greatest latitude in the sampling interval. These results resemble those in Figure 4.14. Parameter values are $W = 1$ and $R = 1$.

The influence of varying t_p for the noise-free case is illustrated in Figure 4.15. The three curves show the results for $N_P = 6$ and 10 with $D = 2N_P$, and $N_P = 6$ with $D = 20$. The overdetermined data provides good results over a wider range of t_p, and a behavior resembling that observed in Figure 4.14. It is interesting to discover that, although both overdetermined results are obtained using exactly the same data, the method that employs the larger number of computed poles yields the better performance.

4.3.7 Variation of s_r

Because the waveform amplitude decreases exponentially in time in proportion to the magnitudes of the real-pole components, it might be expected that their values will influence the accuracy of the extracted parameters quite strongly. Figure 4.16 demonstrates this sensitivity, and several interesting features can be observed: (1) the accuracy of the imaginary component always exceeds that of the real component; (2) the accuracy of the imaginary component increases monotonically with decreasing value of the real component and exceeds nine bits at the maximum shown; and (3) the accuracy of the real component peaks at about five bits when S_R has a value of −0.25 to −0.5.

It is reasonable that increased damping of the transient signal reduces the accuracies of the extracted parameters because the consequently more rapid attenuation of the signal decreases the total quantity of available information. The falloff in accuracy of the extracted real component as s decreases in value does not have so obvious an origin; however, it arises more likely from the definition used for accuracy (Eq. (4.6)), which relates the extracted parameter value to the known value. If the error magnitude remains constant and

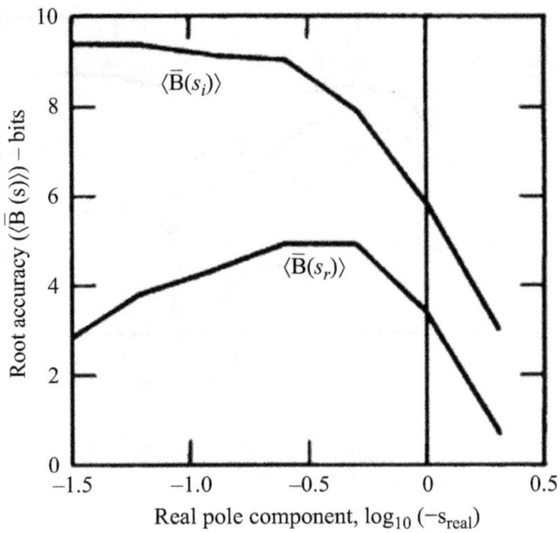

Figure 4.16 Sensitivity of pole accuracy to variations of real pole component. Parameters are: $N_P = 8$, $N_w = 6$, $S_p = 10$, $t = 0.0467$ s, $D = 2N_p$, $R \sim 29$, and $W = 1$. The real (damping) component is always obtained with fewer bits of accuracy than the imaginary (oscillatory) component, and the combined accuracy is maximal at s, -0.4/s.

independent of the actual value, then the number of bits of accuracy as defined by (4.6) will decrease as the input parameter is made smaller, even though the absolute error does not increase. Thus, the accuracy definitions that have been used in here, though apparently reasonable, can lead to results which require care in interpretation.

Another result illustrating the effect of changing the real pole component is depicted in Figure 4.17. These calculations showed that the moving-window technique is quite sensitive to the magnitude of the real pole component. This observation generally agrees with that found in Figure 4.16 and further demonstrates the decreasing information content of a signal as the data window is moved past the signal peak.

4.3.8 Variation of noise level

The majority of results presented to this point in which noise was present in the data have employed a peak-signal-to-peak-noise ratio of 100. In Figure 4.18, the effects of varying the noise level upon the pole accuracy are demonstrated for the six-pole system processed using the moving-window technique. Similar behavior is observed for all three noise levels included in this figure, but the maximum broadens and the improvement in accuracy increases as the noise level decreases.

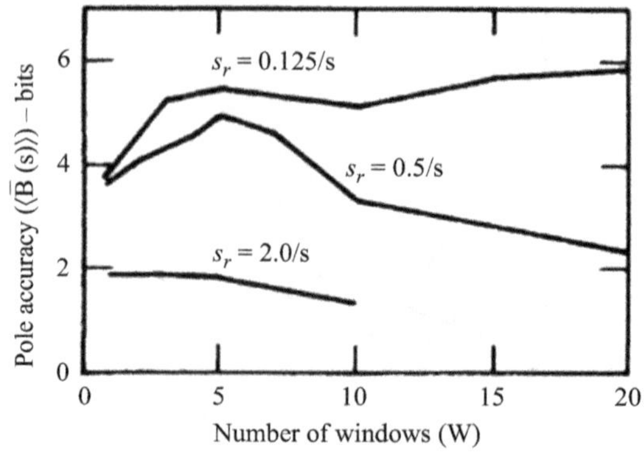

Figure 4.17 Value of pole accuracy as a function of the number of windows with several loss parameters denoted by s_r. As the damping rate decreases, more windows can be used before the pole accuracy attains its maximum value. The parameters are: $S_P = 100$, $N_P = N_W = D/2 = 6$, $R \sim 29$, and $t_p = t_w = 0.0467$ s.

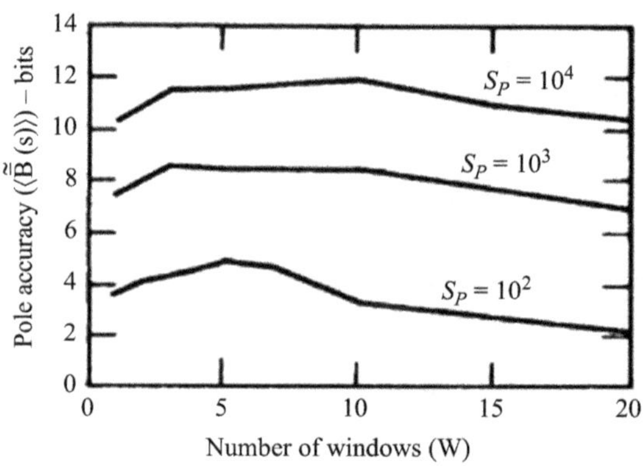

Figure 4.18 Pole accuracy as a function of the number of windows for three peak-signal-to-peak-noise ratios. The increase in accuracy realized by the moving-window technique is relatively unaffected by changes in S_P. As noise is reduced, accuracy increases and a broader maximum in peak accuracy is observed. Parameter values are $N_P = N_W = D/2 = 6$, $R \sim 29$, and $t_p = t_w = 0.0467$ s.

The pole bit accuracy of the six-pole system as a function of the peak noise level for $N_P = 8$ and without using the moving-window technique is presented in Figure 4.19. Significantly, the accuracy and noise level are very closely correlated and are nearly equal. This shows the ability of Prony's method to extract parameters from noisy data with an accuracy comparable to that implied by the noise level.

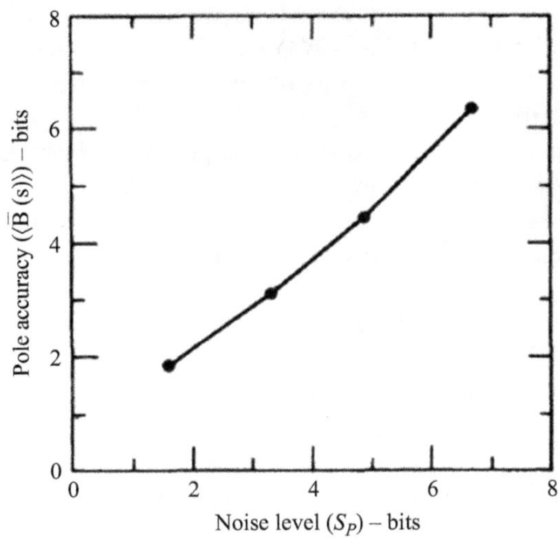

Figure 4.19 *Extracted pole accuracy as a function of peak-signal-to-peak-noise ratio. This correlation serves as a useful measure of the overall performance of Prony's method in estimating the parameters of noisy exponential data. The agreement between input information, defined by the input data accuracy, and output information, represented by the bit accuracies of the poles, suggests that Prony's method preserves essentially all of the information during its transformation of data samples to parameter estimates. Processing parameters used are* $N_P = D/2 = 8$, $N_W = 6$, $W = 1$, $R \sim 29$, *and* $t_p = 0.0467$ *s.*

4.4 Conclusions

Prony's method has been further studied [4] here from the perspective that it is mathematically an information estimator in that it enables the information content of a signal to be extracted in terms of the data's underlying parameters. The technique was applied to noisy exponential data, and the input data was transformed to characteristic equation coefficients, to roots, to poles, and to residues, although residues were not explicitly considered here. Also examined were methods for reducing the effects of noise: (1) by moving a window through the waveform and averaging the clusters of the pole estimates thus obtained and (2) by over-determining the data through the use of a pseudo inverse or by increasing the number of poles being computed from the given data.

The results presented here are informative in demonstrating that Prony's method can be a valuable signal-processing tool when used carefully and appropriately. The important findings include:

1. The composite (real and imaginary) root and pole accuracies are consistently greater than the coefficient accuracies, although a close correlation exists between the accuracy of the real pole component and that of the coefficients.

2. The moving-window technique can improve the pole accuracy, but, to be effec tive, it generally requires $N_P \sim N_W$, i.e., that the estimation rank (number of poles being computed, N_P) approximates the waveform rank (number of poles in the waveform, N_W).
3. If the moving-window technique is used, pole values obtained from averaged values for the coefficients are generally nearly as accurate as averaging the pole values directly. This is true even if the estimation rank exceeds the waveform rank, which causes coefficients to vary widely from window to window.
4. Pole accuracy can be improved by overdetermining the data, both by increasing the estimation rank and, for the same number of samples, by reducing the estimation rank through the use of the pseudo inverse. The former method was found to be more effective.
5. The pole accuracy and peak-signal-to-peak-noise ratios are closely correlated, which demonstrates the ability of Prony's method to preserve information.

References

[1] R. Prony, "Essai Expdrimental et Analytique sur les Lois de la Dilatabilite des Fluides glastiques et sur Celles de la Force Expansive de la Vapeur de 1'Alkool ii Differentes Tempgratures," *J. l'&cole Polytech. (Paris)*, Vol. 1, pp. 24–76, 1795.

[2] F. B. Hildebrand, *Introduction to Numerical Analysis*, New York, McGraw-Hill, 1956, pp. 378–382.

[3] J. Makhoul, "Linear Prediction: A Tutorial Review," *Proceedings. of the IEEE*, Vol. 63, No. 4, pp. 561–580, 1975.

[4] E. K. Miller, and D. L. Lager, "Information Extraction Using Prony's Method," Lawrence Livermore Laboratory Report No. UCRL-52329, August 23.

[5] M. L. Van Blaricum, "Techniques for extracting the complex resonances of a system directly from its transient response," PhD Dissertation, Electrical Engineering Department, U. of Illinois, December 1975.

Chapter 5

Conserving waveform information content in the spectral domain using Prony's method[1]

Abstract

Prony's method (PM) is a technique for determining the coefficients (poles and residues in the frequency domain) of an exponential series. Initial applications of this method in electromagnetic began in the 1970s, because it offers the possibility of obtaining the singularity expansion method poles of conducting objects from their transient response. In addition, PM offers insight into the general problem of extracting useful information from one-dimensional time-series or related data.

In particular, Prony's method (PM) can be viewed as a transformer of data from one form to another. The amount of information extractable from transformed data can be equivalent to or less than that in the original data. Clearly, retention of the greatest possible amount of information in the transformed data is desirable. Thus, in order to evaluate the information-transformation process, one must determine the amount of information in the original data and in the transformed data.

This chapter considers these questions of information content and data transformation from the perspective of conservation of information. To define information content, measures of data rank (number of poles in the exponential data) and data precision are introduced. Then, to study data transformation, a waveform is defined in terms of specified poles and residues and sampled to obtain the associated data, and PM is used to transform the data and compare it with the original information content. Other numerical processes, such as matrix inversion, can be viewed in the same way. It is shown that problems of ill conditioning that may be encountered are related more to the data than to the process. Finally, briefly considered are applications of PM to other physical problems with exponential solutions.

5.1 Introduction

System identification is defined in general as "the determination from experimental data of a set of unknown parameters in a mathematical model of a physical process" [1]. One specific tool that was found useful in systems identification for

[1]Excerpted and adapted from "*Information Extraction Using Prony's Method*," Lawrence Livermore Laboratory Report UCRL-52329, by E. K. Miller and D. L. Lager, August, 1977.

electromagnetic data is PM [2], a technique for obtaining the parameters (exponents and coefficients) of a complex exponential series. Its use in electromagnetic was instigated by the development of the singularity expansion method (SEM) [3], which depends on knowledge of the exponents or poles associated with a given object. PM can be used to find the SEM poles from transient data [4], and a frequency-domain counterpart can similarly be used for spectral data [5]. Actually, these techniques are useful not only for experimental data, as implied above, but also for analytical or numerical results. Thus, they provide alternatives to procedures first used in SEM for finding the poles. Because so much of physics can be described via complex exponentials, potential applications of PM are very widespread. Although PM has been found beneficial in electromagnetics and related areas in the above references, it has not been as well received in general [6]. Evidently, this is because results obtained from its use can be quite variable, depending on the assumed rank of the data, the distribution of poles in the complex plane, the presence of noise in the data, etc. Mathematically, PM is regarded as ill-conditioned because, in some circumstances, small changes in the input data produce large changes in the output. It is this aspect of Prony processing that is particularly to be discussed here. Our experience in its use leads us to view it more favorably than some others have had processing complex exponential data.

The purpose of this discussion is to present some observations and results based on various computer experiments performed in this investigation. In the next section, the approach used is described and the kinds of computations that have been performed. Typical results from these computations are summarized. A concluding section summarizes our findings to date and includes recommendations for future study.

5.2 Approach

5.2.1 Information-content analysis of the data

Evaluation of PM that follows is from an information-theoretic viewpoint. That is, an attempt is made to characterize its performance as an information transformer that accepts information of one kind (a transient waveform, for example) as an input and provides other information (the extracted poles and residues) as an output. The specific transient of concern is

$$f(t) = \sum_{\alpha=1}^{N} R_\alpha \exp(s_\alpha t)$$

where R_α and s_α are the residues and poles, respectively. Because $f(t)$ is real, R_α and s_α occur in conjugate pairs, so only $2N$ real numbers are required to characterize the data, N of these associated with the poles and N with the residues.

Observe that the term information is used here in the general sense of knowledge acquired from data rather than as defined by Shannon [7] because of the difficulty in quantitatively assessing the information content of data on a probabilistic basis. Specifically, the goal is to determine how much information might be preserved or lost when PM is used to process the data, and to judge its performance on this basis.

One approach would be to define the total information content of the input data as

$$I_{in} \equiv \frac{N}{M}\sum_{i=1}^{M} D(f_i) \tag{5.1}$$

where $D(f_i)$ is the number of correct digits in input sample f_i and M is the number of data samples. A more convenient procedure is, however, to use instead the average number of correct digits per data sample as given by[2]

$$\bar{D}(f) = \frac{1}{M}\sum_{i=1}^{M} D(f_i) \tag{5.2}$$

to obtain a rank-independent measure of input information.

Either of these measures of input information might be suitable when the residues are all of comparable amplitude. When the residues are significantly different, however, so that they do not contribute equally to f_i, then (5.2) may not be appropriate. In that case, it might instead be more appropriate to define I_{in} or $\bar{D}(f)$ from the parameters that determine f_i rather than from f_i itself.

Similarly, the total output information content of the extracted parameters would be similarly defined by

$$I_{out} \equiv \sum_{i=1}^{N} D(\tilde{P}_i)$$

where $D(\tilde{P}_i)$ is the number of correct digits in the extracted parameter $\tilde{P}_i$, the set of which includes both poles and residues. But this also involves the rank of the system, as does (la), so a more suitable alternate is used here instead.

$$\bar{D}(\tilde{P}) = \frac{1}{N}\sum_{i=1}^{N} D(\tilde{P}_i)$$

which is the average number of correct digits per parameter extracted. Since it is sometimes more useful to evaluate separately the pole and residue results given by

$$\bar{D}(\tilde{s}) = \frac{2}{N}\sum_{i=1}^{N/2} D(\tilde{s}_i)$$

$$\bar{D}(\tilde{R}) = \frac{2}{N}\sum_{i=1}^{N/2} D(\tilde{R}_i)$$

[2]In Chapter 4, parameter accuracy was measured in bits, B. Here a digit measure is used instead since D digits accuracy then simply converts to a numerical value of 10^{-X}. The conversion between the two is $B = 2,3xD$.

which are the average number of correct digits per pole and residue values extracted.

Finally, the performance of PM as an information processor is defined as

$$P \equiv \frac{I_{in} - I_{out}}{N} = \bar{D}(f) - \bar{D}(\tilde{P})$$

which represents the average number of digits of information lost per parameter.

5.3 Eigenvalue analysis of the data

It is also instructive to examine the eigenvalue characteristics of the data as they appear in the process [8]. A matrix of data values f_i is first obtained that, for $M = 2N$, has the form,

$$\bar{F} = \begin{bmatrix} f_1 & f_2 & f_3 & \cdots & f_N \\ f_2 & f_3 & f_4 & \cdots & f_{N+1} \\ \ddots & & & & \\ & \ddots & & & \\ f_N & f_{N+1} & f_{N+2} & \cdots & f_{2N-1} \end{bmatrix}$$

the solution of which yields the coefficients of a polynomial whose roots give the poles.

One might expect, then, that the vector of data values $\bar{f}$ strongly influences the success of the overall process. In essence, the samples f_i constitute a basis set that is to represent all the waveform $f(t)$, including the part not sampled, i.e., the sample space must span the waveform space. Therefore, it should be possible somehow to test the data matrix $\bar{F}$ for its suitability in this regard. This is where the eigenvalue procedure is applicable.

The eigenvalues λ and eigenvectors $\bar{m}$ of a matrix $\overline{\overline{M}}$ are defined by this relation [9]

$$\overline{\overline{M}} \cdot \tilde{m} = \lambda_i \tilde{m}, i = 1, \ldots, N$$

where λ_i are solutions of the determinantal equation

$$\overline{\overline{M}} - \lambda_i \overline{\overline{I}} = 0$$

For the given matrix $\overline{\overline{M}}$, the eigenvectors $\bar{m}$ represent a basis set for expanding the rows of $\overline{\overline{M}}$, and the eigenvalues represent their corresponding contribution to that expansion. For example, the dominance of the rows of $\overline{\overline{M}}$ by one function, an exponential in our case, would result in only one eigenvalue of significant magnitude. Thus, an eigenvalue analysis of the matrix can provide the test mentioned above.

5.4 Outline of computer experiments

The above quantities have been evaluated for data obtained from a specified pole set as a function of the various parameters involved. A ten-pole set having the nominal parameters$(s = \pm i\omega + \sigma)$ was employed.

α	ω_α	σ_α	R_α
1	1	−0.1	1.0
2	2.5	−0.25	1.0
3	5	−0.5	1.0
4	7.5	−0.75	1.0
5	10	−1.0	1.0

Among the parameters that can be varied are two distinct sets. The one associated with the Prony process itself comprises:

1) Δt, the intersample spacing;
2) T, the total data window, which equals $(M - l)\Delta t$;
3) N_p, the number of poles used in the processor, which for a square system results in $N_p = M/2$.

The other, much larger set of parameters is associated with the data being processed and includes:

1) N, the number of poles in the data
2) R_α, the set of residues
3) s_α, the set of poles
4) $D(f_i)$, the specified accuracy of the input data.

In the study being reported here, a series of computer experiments involving changes in both sets of parameters have been conducted. These are:

1) For the nominal pole set, use $\Delta t_N/3 \leq \Delta t \leq 2\Delta t_N$ for a fixed value of $D(f_i)$, where Δt_N is the Nyquist rate sample spacing.
2) For the Δt that produces the best $\bar{D}(\tilde{P})$ from (1), systematically vary $D(f_i)$ over the range of word sizes permitted by the computer (~12–14 digits) by truncating the exact values of f_i to obtain a fixed $D(f)$.
3) Repeat (2) for the set of residues varying as $R_\alpha = 10^{1-\alpha}$ and $R = 10^{\alpha-5}$.
4) Using the same Δt as (2), systematically vary $D(f_i)$ by truncating the exact values of f_i at a fixed level relative to the maximum to obtain an effective signal-to-noise ratio of $D(f)_{max}$.
5) Repeat (4) as in (3).

6) Using the same Δt as (2), systematically shift ω_α using $\omega_\alpha = \omega_5 - \Delta\omega_\alpha (5 - \alpha)$, and increase the data window T by increasing M while keeping $N_P = M/2$.
7) Repeat (4) and (5), using the moving-window procedure.

These experiments were performed to obtain a systematic set of results that might provide some insight into the Prony process as an information transformer. Experiment 1 simply establishes a baseline behavior for the process, while experiment 2 evaluates its input-output characteristics over a wider range of the input data. Experiment 3 examines the effects of widely varying residues on the output, because previous experience has indicated this to be an important aspect of the effect of noise on Prony processing. Experiments 4 and 5 approximate the effects of noise on Prony's performance. Experiment 6 determines how the pole locations, relative to the origin and to each other, might affect the data sampling that produces the best results.

Finally, in experiment 7 some aspects of iterative data processing were quantitatively examined using a succession of overlapping data windows. Clearly, many other experiments could be considered as well. In particular, the more general effect of noise on Prony's performance is of great significance. Another area of concern is the effective data rank, and how the data might best be sampled. These issues would benefit from further detailed study. The motivation and justification for doing so may be provided by the initial results presented below.

5.5 Numerical results

Some results obtained from experiment 1 are given in Figures 5.1–5.3. The quantities $\bar{D}(\tilde{s})$ and $\bar{D}(\tilde{R})$ are shown as functions of Δt in Figure 5.1. They both increase monotonically to $D(f_i)$ as Δt approaches Δt_N. (Actually, sampling with $\Delta t \geq \Delta t_N$ can continue to produce good results if the poles whose frequencies exceed one-half the sampling frequency can be unfolded.) An explanation for this result may be deduced from Figure 5.2, where the eigenvalues of the data matrix are plotted for three values of Δt. The ratio of largest to smallest eigenvalue is a minimum for $\Delta t = 0.29$, but very much larger for the smaller Δt values, showing that the effective rank of the data is very dependent on the intersample spacing.

This dependence is demonstrated in a different way in Figure 5.3, where $\bar{D}(\tilde{s})$ and $\bar{D}(\tilde{R})$ are plotted as functions of the eigenvalue ratio, E_R. In this case, E_R, which is sometimes used as a condition number, provides a clear indication of the data's information content. The increase of $\bar{D}(\tilde{s})$ and $\bar{D}(\tilde{R})$ and the decrease of E_R with increasing Δt might be surmised to come from the increasing linear independence of f_i that results. The reason for the downturn of both curves that occurs with decreasing E_R past the peak is unclear. Since the numerical procedure employed here is the same for all Δt used, and only f_i are changed, it is reasonable to conclude that in this case the data and not the Prony process is the limiting factor.

Upon selecting from experiment 1 $\Delta t = 0.95\Delta t_N$ as the optimum value for subsequent calculations, the results shown in Figure 5.4 from experiment 2 are found, where $D(f_i)$ is systematically varied. The relationship between input and

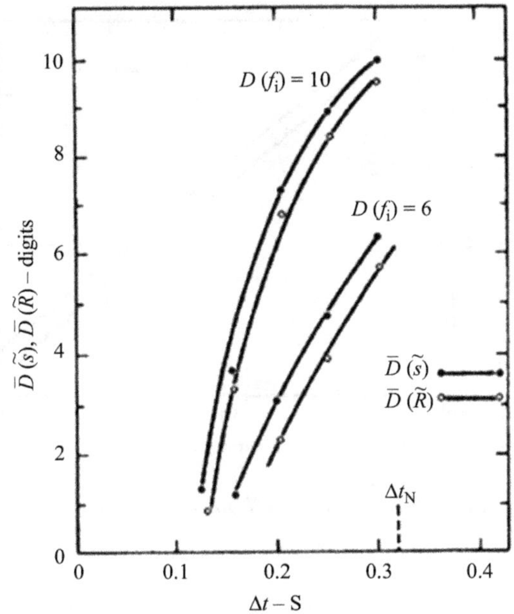

Figure 5.1 *The accuracy of Prony's method is very sensitive to the time interval Δt between data sampies, as illustrated here for noise-free data. Shown here are the pole and residue accuracies as a function of Δt for two values of $D(f_i)$. In this and all following cases, the solution is for a square system, where $M = 2N$.*

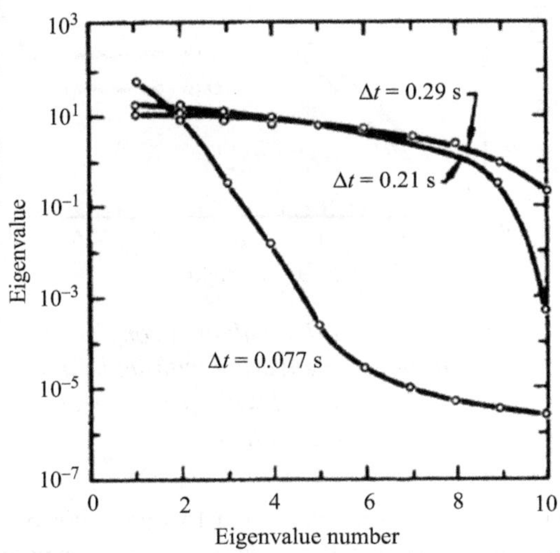

Figure 5.2 *Eigenvalues of the data matrix for three values of Δt demonstrate the ill-conditioning of the data that can occur when the samples are too closely spaced*

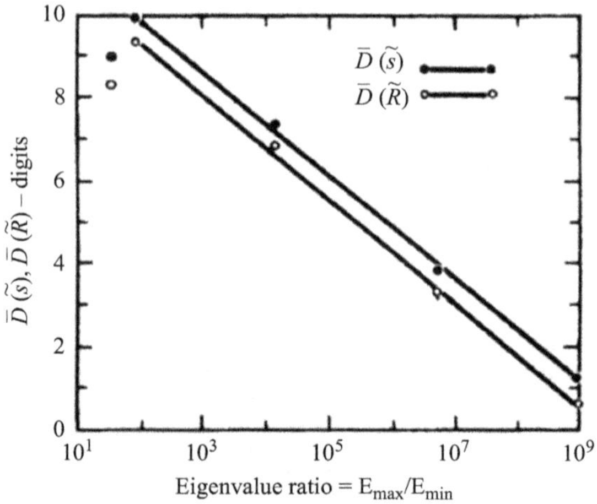

Figure 5.3 The accuracy of Prony's method depends directly on the eigenvalues ratio

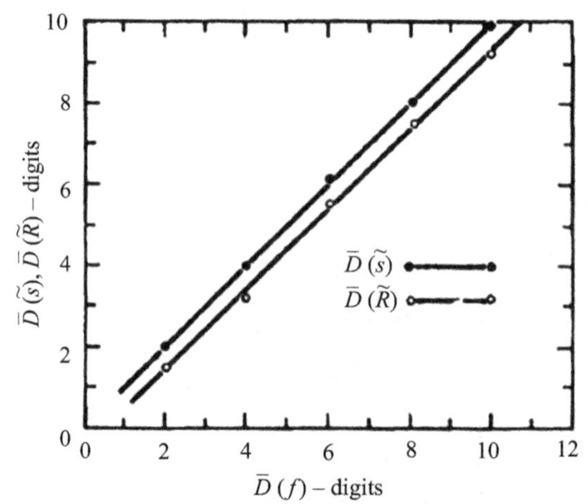

Figure 5.4 In terms of pole and residue values, Prony's method can provide output information commensurate with input information. The pole values are typically obtained with greater accuracy as these than are the residue values, results indicate.

output information content is essentially one to one. This outcome, which was somewhat surprising, indicates that Prony preserves the original information content of the sampled data in the transformed pole-residue output. Conversely, this result also implies that no further information could be extracted by, for example,

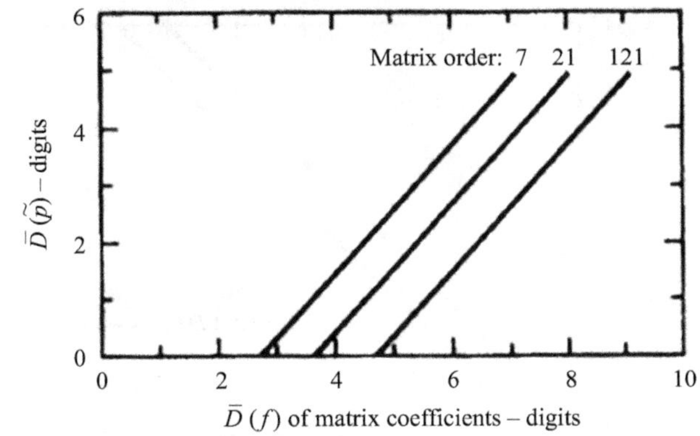

Figure 5.5 By contrast to Prony's method, matrix inversion demonstrates a loss of information when evaluated in terms of pole and residue values

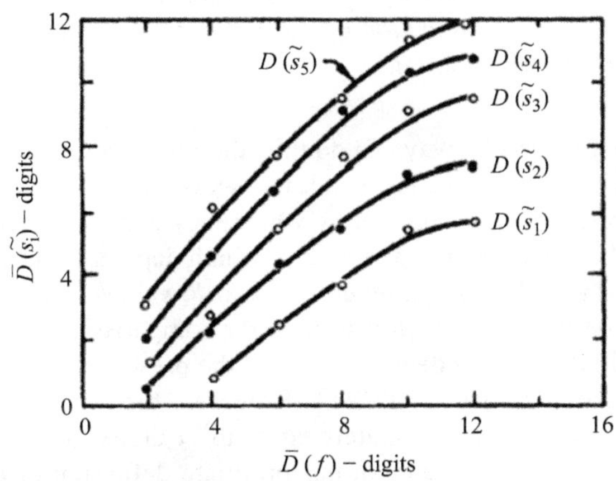

Figure 5.6 The accuracy with which the poles are obtained is directly proportional to the amplitudes of their associated residues, as these results demonstrate for noise-free data

using more data samples. By contrast, matrix inversion does not imply that, as shown in Figure 5.5 [10].

Experiment 3, in which R_i differ by factors of ten, yields results like those in Figure 5.6, where the individual $\bar{D}(\tilde{s}_i)$ are shown as a function of $D(f_i)$. As expected from previous computations with noisy data, the poles of terms having the largest residues are computed most accurately. As a matter of fact, the decrease in relative accuracy is almost linearly dependent on the decrease in relative residue value.

The above results were obtained from, in effect, noise-free data. Now examined is a more interesting and more realistic situation in which the data is

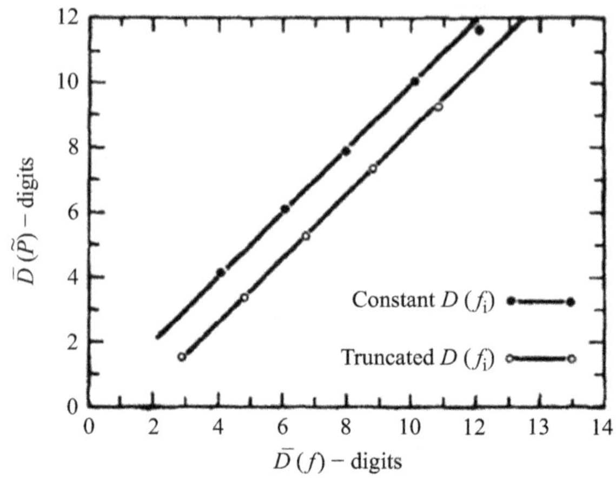

Figure 5.7 The output information remains proportional to the input information when the data is noisy, as shown here. In this case, the noise was simulated by truncation of the input at a given level relative to the peak signal amplitude.

contaminated in a noise-like way. To do this, the input $D(f_i)$ is truncated at a fixed point relative to the maximum. The result is a set of input samples whose accuracy is directly proportional to the amount by which they exceed the truncation point or "noise" level. The outcome of experiment 4, which depicts the influence of varying the truncation, or signal-to-noise, level for equal R_i values, is shown in Figure 5.7. The result of experiment 2 is included there for comparison.

$\bar{D}(\tilde{P})$ is as linearly dependent on $\bar{\mathrm{D}}(\mathrm{f})$ for the present case of a truncated $D(f_i)$ as it was for a constant value of $D(f_i)$. However, the present results are offset downward by an amount approximately equal to ¾ digits in $\bar{D}(\tilde{P}_i)$. This apparent loss of information may be due to an inappropriate definition of $\bar{\mathrm{D}}(\mathrm{f})$ used in this case. Recall that $D(f_i)$ is the number of correct digits in sample f_i. For this experiment, all the digits are counted that exceed the truncation level in computing $\bar{D}(f)$. But the information conveyed by a one-digit sample may be less than, for example, $1/N$th that of an N-digit sample. In other words, the utility of the data may decrease as the $D(f_i)$ are made unequal. Certainly, the information content of f_i cannot exceed that as defined here, but it could be significantly less.

Repetition of experiment 3 using truncated or noisy data as discussed above leads to the results shown in Figure 5.8. These resemble the noise-free results of experiment 3 (Figure 5.6), but are shifted downward in accuracy by about one digit. Again, it can be observed that the poles are lost in the inverse order of their associated residue size. The minimum loss is approximately one digit for the largest residue pole and increases monotonically with decreasing residue value, which agrees approximately with the finding of experiment 4. Noise effects like these might be alleviated in several ways. A moving-window technique (experiment 7) has been found useful, but requires further study to quantitatively assess it.

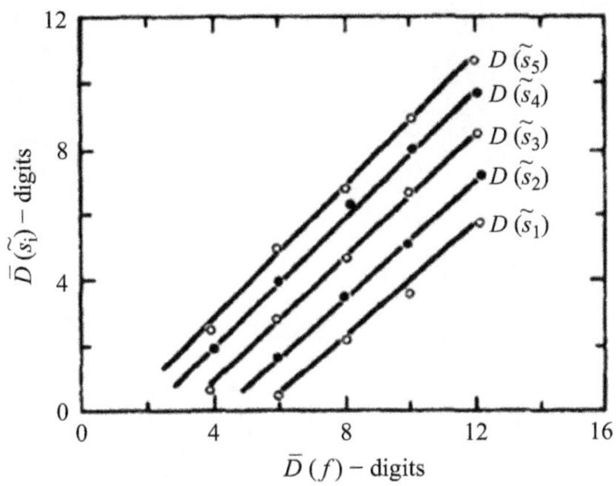

Figure 5.8 The proportionality of the pole accuracy to residue amplitude is retained when noisy data like that of Figure 5.7 is used

Now examined are the effects on the Prony process of varying the imaginary components of the pole locations. This was done to assess the possible influence of the frequency separation of adjacent poles upon the data sampling required for achieving maximum accuracy in the extracted poles. One might anticipate, for example, that in addition to the Nyquist sampling criterion, which is dictated by the maximum frequency component in the waveform, there is a sampling requirement related to some aspect of its minimum frequency behavior.

This is indeed the case, as demonstrated in Figure 5.9. These results were obtained by fixing the frequency, ω_5, of the highest-frequency pole, while varying the uniform interval $\Delta\omega$ among all the poles. The Δt from experiment 2 was used throughout, and the data window, T, was systematically widened by increasing $M = 2N_p$. As the plots of Figures 5.9–5.11 show, in order to achieve maximum accuracy, it is required that $T \geq 2\pi/\Delta\omega$, i.e., accuracy requires an observation time on the order of the minimum beat frequency. This is not surprising because the closer together two poles are, the longer the waveform must be observed to resolve their frequency difference, a well-known result in other applications. It was found, however, that the minimum frequency, even when purely imaginary, has no such corresponding influence on T. The eigenvalue results are also useful in showing when the data window is wide enough, as demonstrated in Figures 5.10 and 5.11. Because the moving-window procedure has proven quite useful in reducing noise effects, [11] this section is concluded with application of the moving window to data with white noise and a signal-to-noise ratio of 200. An example from this exercise is included in Figure 5.12. A significant improvement in the pole values obtained from the noisy data occurs as a result of the moving-window technique. By plotting the average pole values as a function of the number of iterations, a convergence toward the actual noise-free value occurs, as demonstrated in Figure 5.12.

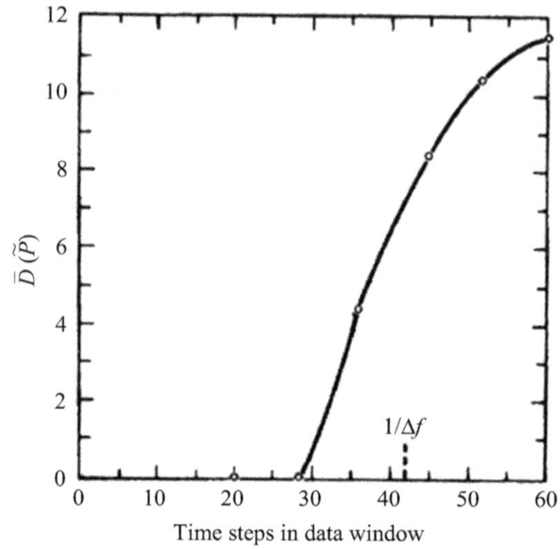

Figure 5.9 When applied to exponential data, the accuracy of Prony's method is very sensitive not only to the interval between data samples but also to the total data window available. As shown here, the data window must be comparable to the inverse beat frequency to obtain maximum accuracy.

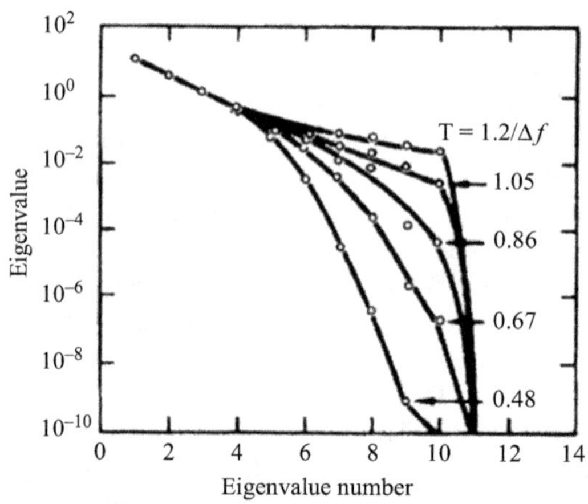

Figure 5.10 The data window influences the range of eigenvalues of the data matrix

The effect of the moving window is essentially to average the data within a single waveform. Because the Prony process provides a model for the data, each separate application of PM yields a smoothed representation of the waveform. A number, I, of such applications provides the equivalent of I individual

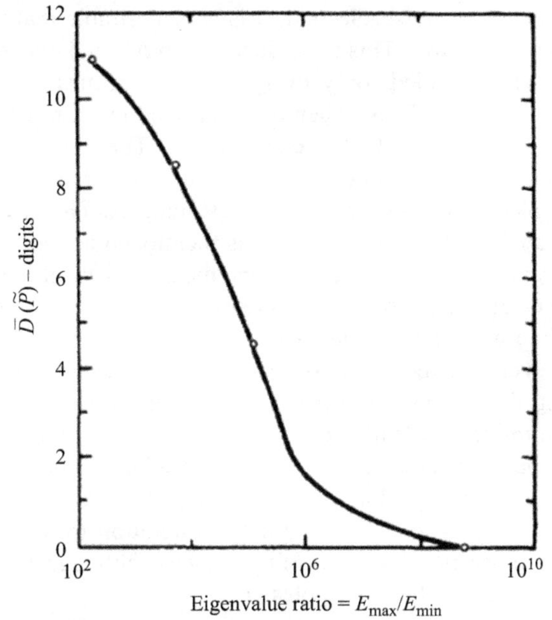

Figure 5.11 The eigenvalue ratio indicates when the data window is adequate

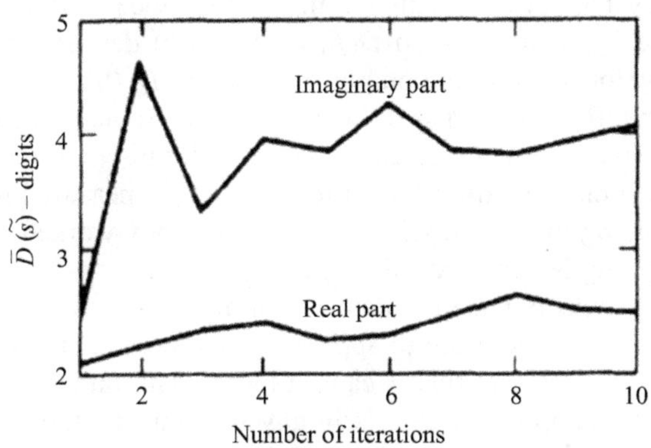

Figure 5.12 Multiple processing can increase the accuracy of results obtained from noisy data. In-the case shown here, the data window was moved one data-sample interval for each calculation

representations for the same waveform, whose average leads to an improved model for the data. Averaging is done here with respect to the poles, although other averages, such as the individual reconstructed waveforms, could also be used.

It is important to mention that success of this technique depends significantly on the data and how it is handled. First, it is clear that the method is best suited to data having uncorrelated noise. Second, it is very desirable that the data be oversampled, the more the better. This is because, when $\Delta t_N > \delta$, where δ is the sample interval of the data provided, only every second, or third, or ... data value is needed to, implement PM. Thus, when the data window is moved forward by δ, a completely new set of samples is encountered. This can have the effect of decomposing the single measured waveform into two or more individual waveforms, which is like having two or more measurements. The benefit of this could not easily be attained without PM, which, as mentioned above, permits averaging these separate waveforms extracted from one measurement. It is thus obviously to our benefit to oversample the minimum possible in using PM but to oversample the maximum possible when taking the data.

Of course, even if every successive data point must be used because when $\Delta t_N \sim \delta$, moving the window is still worthwhile because of the new data sample each iteration introduces. Clearly, one new data sample cannot convey as much information as could 2N new samples, but a sufficient number of iterations may ultimately be as effective. In this case, however, even when the noise is uncorrelated, the small change in the data vector from iteration to iteration can result in curve fitting and noise poles also forming clusters, although their variances will generally exceed those of the valid poles [12].

5.6 Observations and conclusions

This study has systematically examined the performance of PM as an information transformer. The input information $D(f_i)$ is the set of data samples of a complex exponential waveform, while the output information $D(\tilde{P}_i)$ is the set of poles and residues extracted. Because PM specifically, and the problem of fitting exponential data generally, have been considered mathematically ill-conditioned, attention has been concentrated on evaluating PM in terms of $D(f_i)$ measured relative to $D(\tilde{P}_i)$. The goal was finding the circumstances for which Prony worked best in this sense, and in understanding better what might cause it to fail.

Based upon results obtained to date as summarized above, it can be concluded first that PM can indeed be made to appear ill-conditioned. But this outcome leads to the further conclusion that this is caused by the data rather than the processing technique. If, for example, the data is in any way underdetermined, Prony's performance will deteriorate. Underdetermined data can result from oversampling (too small a sampling interval), undersampling (too small a data window), or noise. In the first two cases, the problem is due to data of insufficient linear independency and inadequate rank. In the third case, reducing the information content per data sample can cause the problem. In all three cases, small changes in the data (or coefficients of the polynomial that gives the poles) can produce large changes in the poles.

A second conclusion is that Prony essentially can preserve the information content of data that is appropriately sampled and processed. Appropriate data sampling methods have already been mentioned. By appropriate processing is meant for the moving-window technique and others that attempt to increase the total amount of information extractable from the data. Unambiguously defining $D(f_i)$ is difficult, so the above conclusions could benefit from further investigation to confirm them.

Two main areas needing attention remain sampling strategies and noise effects. These are areas of general concern in systems identification, and their general resolution may not be realizable. But our sole interest is the handling of exponential phenomena, a specialization that can greatly reduce the breadth of the problem. It is helpful to have the advantage of working in the context of a physical application, which provides a significant amount of insight and guidance into what could otherwise be a less productive mathematical exercise.

Finally, it should be emphasized that PM is only one of many ways of treating exponential data, and that other techniques may be found to improve its performance. For now, however, it remains a leading candidate for treating the kinds of data encountered in electromagnetic applications.

Acknowledgments

This work was motivated by other efforts involving Prony's method that continue to be performed by the Wave Theory and Applications Group of LLL's Engineering Research Division. Helpful discussions were held in this regard with A. J. Poggio, T. H. Lehman, J. V. Candy, J. N. Brittingham, and D. J. Dudley of the University of Arizona. A particular debt of gratitude is owed to G. J. Hudson, now of Air Force Weapons Laboratory, who, with D. L. Lager, developed the moving-window technique described here. We also especially appreciate the valuable comments from S. Aggarwal.

References

[1] M. P. Polis, and R. E. Goodson, "Parameter Identification in Distributed Systems: A Synthesizing Overview," *Proceedings of the IEEE*, Vol. 64, pp. 45–61, 1976.

[2] F. Hildebrand, *Introduction to Numerical Analysis*, McGraw-Hill, New York, 1956.

[3] C. E. Baum, "On the Singularity Expansion Method for the Solution of Electromagnetic Interaction Problems," AFWL Interaction Note, No. 88, 1971.

[4] M. L. Van Blaricum, and R. Mittra, "A Technique for Extracting the Poles and Residues of a System Directly from Its Transient Response," *IEEE Transactions on Antennas and Propagation, AP*, Vol. 23, pp. 777–781, 1975.

[5] J. N. Brittingham, E. K. Miller, and J. L. Willows, *The Derivation of Simple Poles in a Transfer Function from Real Frequency Information, Lawrence Livermore Laboratory*, Livermore, California, UCRL-52050, 1976.

[6] N. K. Gupta, and J. F. Bohn, *A Technique for Measuring Rotorcraft Dynamic Stability in the 40 x 80 Foot Wind Tunnel*, NASA Contractor Rept. NASACR-151955, 1977.

[7] C. E. Shannon, *A Mathematical Theory of Communication*, University of Illinois Press, Urbana, IL, 1949.

[8] M. L. Van Blaricum, "Techniques for Extracting the Comp2ex Resonances of a System Directly from Its Transient Response," Ph.D. dissertation, University of Illinois, Urbana, Ill., 1975.

[9] A. Ralston, *A First Course in Numerical Analysis*, McGraw-Hill, New York, 1965.

[10] E. K. Miller, and F. J. Deadrick, "Some Computational Aspects of Thin-Wire Modeling," in *Ch. 4 in Numerical and Asymptotic Techniques in Electromagnetics*, R. Mittra (ed.), Springer-Verlag, New York, 1975.

[11] G. J. Hudson, and D. L. Lager, *Observations on the Operation of the SEM-PEX Code*, Lawrence Livermore Laboratory, Livermore, California, UCID-17440, 1976.

[12] E. K. Miller, J. N. Brittingham, and J. L. Willows, *The Derivation of Simple Poles in a Transfer Function from Real Frequency Information. Part 111: Object Classification and Identification, Lawrence Livermore Laboratory*, Livermore, California, UCRL-52211, 1977.

Chapter 6

Minimizing the number of frequency samples needed to represent a transfer function using adaptive sampling[1]

Abstract

One of the most commonly encountered problems in wave-equation applications arising in acoustics and electromagnetics is that of estimating transfer functions from discrete frequency samples of a first-principles, or generating model (GM) such as NEC, FERM, and E-PATCH. This problem has been approached in a mostly ad hoc way wherein the GM samples are spaced uniformly and closely enough such that linear interpolation yields a reasonably smooth graphical representation of the continuous response. Often, additional sampling between the original ones is done subsequently to improve the result, a process that may be repeated several times until an apparently satisfactory outcome is obtained. This approach will usually result in substantial oversampling, with a concomitant increase in the associated cost, while offering no assurance that important details of the true GM transfer function have not been missed. At the same time, there is no quantitative measure of how much the linearly interpolated estimate might differ from the actual response between the sampled values. The procedure described here uses model-based parameter estimation with rational-function, overlapping fitting models (FMs) to automatically determine where the GM needs to be sampled to reduce the mismatch between the FM and GM below a specified value.

6.1 A summary of the basic idea

As demonstrated elsewhere [1,2], hereafter referred to as RI and RII respectively, a rational function, as a generalization of a pole series, is a good choice for approximating an electromagnetic frequency response. The possibility of exploiting that

[1]E. K. Miller, R. S. Roberts and S. Chakrabarti (1992), "Using Adaptive Frequency Sampling for More Efficient Determination of Broad Band Transfer Functions," in *Boundary Element Technology VII*, Ed. C. A. Brebbia and M. S. Ingber, Computational Mechanics Publications, Southampton & Boston, pp. 745–756; S. Chakrabarti, K. R. Demarest and E. K. Miller (1993), "An Extended Frequency-Domain Prony's Method for Transfer Function Parameter Estimation," *International Journal of Numerical Modelling*, 6(4), pp. 269–282.

idea to automate sampling of a first-principles, or generating model (GM) subject to a specified estimation error was also considered; a detailed description of one approach is given here. The basic idea is to use a series of rational-function fitting models (FMs), $M_i(n_i,d_i)$, of index number i and numerator and denominator polynomial orders n_i and d_i respectively, $i = 1, \ldots, N$, where some of the GM samples for each M_i are shared by FMs M_{i-1} and M_{i+1}, etc.

A small number of GM samples are computed across the bandwidth of interest to begin the process with a different subset of these samples assigned to each of the initial, low-order FMs employed. A sequence of more closely spaced FM estimates is then generated from each FM. In the frequency range where two (or more) FMs overlap, the differences between their estimates are computed. The minimum match in digits, E_i, is then determined for each FM. If any of these is less than a specified estimation error, EE, a new GM sample is computed at the frequency where $\mathrm{ME}_1 \equiv \min[E_1, E_2, \ldots, E_N]$ occurs. The FMs that contain the new GM sample are then increased by one in order (alternating between increasing n and d as new GM samples are added). In order to avoid ill conditioning, when the order of one or more of the affected FMs reaches some specified maximum value, it may be split into two lower-order, overlapping, new FMs. New E_is are computed over the bandwidth spanned by the FMs thus affected and ME_2 is determined. This process continues with ME_3, ME_4, etc. until all E_i's exceed EE. A final GM sample density of about three per resonance can be achieved without initial knowledge of where these resonances are located for an EE of 10^{-2} or a minimum FM match of 2 digits.

6.2 Rationale for using a rational-function fitting model

The above-described procedure is known as model-based parameter estimation (MBPE). In contemplating using MBPE for electromagnetic (EM) or any other phenomena, the first issue to consider is what FMs might be appropriate. Unless an FM is physically based, MBPE is basically a curve-fitting procedure. While curve fitting can be a useful tool, MBPE is potentially much more powerful when the problem physics are explicitly involved in the process and might therefore be described as "smart" curve fitting. As discussed in RI and RII, exponential and pole series occur in numerous ways in EM, and thus provide obvious choices for FMs. Furthermore, the two most frequently used domains for formulating and solving EM problems, and for many other physical phenomena as well, are the time domain (TD) and frequency domain (FD), for which generic descriptions are given by exponential and pole series, respectively, which together form a Laplace-transform pair. As mentioned in RI, the same transform relationship exists between other observable pairs that are also described by exponential and pole series (see Table I of RI). Thus, the terms "waveform domain" (WD) and "spectral domain" (SD) respectively, are used for phenomena that are described by exponential series and pole series, as a generalization of their more specific and familiar TD and FD forms. While a pole series does not fully describe an electromagnetic transfer function, it can be a good approximation to that response and provide a concise way to represent it. For purposes of the present discussion, which is to reduce the number of samples of EM observables that are needed in the first place, the FM does not have to be exact so long as it leads to an acceptably accurate and parsimonious representation.

The generic WD and SD FMs can be expressed as

$$f(x) = f_p(x) + f_{np}(x) = \sum_{\alpha=1}^{P} R_\alpha e^{s_\alpha x} + f_{np}(x) \tag{6.1}$$

and

$$F(x) = F_p(X) + F_{np}(X) = \sum_{\alpha=1}^{P} \frac{R_\alpha}{(X - s_\alpha)} + F_{np}(X) \tag{6.2}$$

where "x" represents the WD independent variable and "X" is its SD, or transformed, counterpart. For the time–frequency transform pair, x would be the time variable t and X would be the complex frequency s, in general, but for most purposes would be limited to the radian frequency $j\omega$. The exponential or pole series contributions are designated, respectively, by $f_p(x)$ and $F_p(X)$, and represent what might be called the "resonant" response, with the non-pole part denoted by $f_{np}(x)$ and $F_{np}(X)$. The FM parameters, the complex resonances (or poles), "s_α," and the modal amplitudes (or residues), "R_α," (or their polynomial counterparts) are quantified by fitting samples of the relevant observable to the desired model, $f_p(x)$ or $F_p(X)$. The various FMs and their estimated parameters provide a mathematically concise and physically insightful way to characterize electromagnetic and other wave-equation phenomena. Subsequent attention is here limited to the spectral domain.

6.3 Sampling in the spectral domain

Spectral-domain function sampling begins with the FM given by (6.2) and assumes the availability of samples denoted by

$$F_i = F(X_i) = \sum_{\alpha=1}^{P} \frac{R_\alpha}{(X_i - s_\alpha)} + F_{np}(X_i) \tag{6.3}$$

where, in contrast to waveform sampling, there is no requirement that the samples X_i be uniformly spaced (RI). However, also in contrast to waveform sampling where the FM can be a purely exponential series at late times, spectral sampling cannot avoid the presence of the generally unknown non-pole term. The possibility of using variable numerator and denominator polynomial orders provides a way to approximate the effect of the non-pole term by simply increasing the order of the numerator polynomial. For example, an increase of one in the numerator order has the effect of representing F_{np} by a constant, which, when absorbed into the rational function, results in equal numerator and denominator orders. If F_{np} is represented by a constant and a term linear in X, this has the effect of making the numerator order one greater than the denominator. Thus, by varying the relative orders of the polynomials that comprise the FM, various approximations of the non-pole, SD contribution are included. Therefore, in general, the SD FM is given by

$$F(X) = \frac{N(X)}{D(X)} \tag{6.4a}$$

where the numerator and denominator polynomials $N(X)$ and $D(X)$ are given by

$$N(X) = N_0 + N_1X + N_2X^2 + \cdots + N_nX^n \tag{6.4b}$$

and

$$D(X) = D_0 + D_1X + D_2X^2 + \cdots + D_dX^d. \tag{6.4c}$$

The coefficients of the SD FM can be obtained from sampled values of the response, i.e., from function sampling. How this is done is easy to see by rewriting (6.4) as

$$F_iD_i = N_i, i = 0, \cdots, D-1 \tag{6.5a}$$

$$\text{with } F_i = F(X_i) \text{ the function samples and} \tag{6.5b}$$

$$D_i = D_0 + D_1X_i + D_2(X_i)^2 + \ldots + D_d(X_i)d, \tag{6.5c}$$

$$N_i = N_0 + N_1X_i + N_2(X_i)^2 + \cdots + N_n(X_i)^n. \tag{6.5d}$$

There are $d + n + 2$ unknown coefficients in the set of homogeneous equations given in (5a), requiring a constraint or additional condition to make the sampled equations solvable. As discussed in RI, there is no unique choice for this constraint, but for $D_d = 1$, then the following equations result:

$$F_0D_0 + X_0F_0D_1 + \cdots + F_0(X_0)^{d-1}D_{d-1} - N_0 - X_0N_1 - \cdots - (X_0)^nN_n$$
$$= -(X_0)^dF_0$$

$$F_1D_0 + X_1F_1D_1 + \cdots + F_1(X_1)^{d-1}D_{d-1} - N_0 - X_1N_1 - \cdots - (X_1)^nN_n$$
$$= -(X_1)^dF_1$$

$$F_{d-1}D_0 + X_{d-1}F_{d-1}D_1 + \cdots + F_{d-1}(X_{d-1})^{d-1}D_{d-1} - N_0 - X_{d-1}N_1 - \cdots$$
$$-(X_{d-1})^nN_n = -(X_{d-1})^dF_{d-1} \tag{6.6}$$

where $D \geq n + d + 1$ is required for a solution. Note that the matrix coefficients comprise a product of a data sample and the frequency at which the sample is taken raised to a power. The exponentiation of the sampling frequencies suggests that large dynamic numerical ranges in the matrix coefficients may result if d and n are very large. One way to avoid this is to scale the frequency, so that, for example, if the sampling range is centered at 1 GHz, a scaling of 10^9 in the frequencies leads to nominal scaled values near unity. It is also possible to center the SD model about a frequency in the interval of interest so that terms like $(X_s - X_{ref})^n$ result. Combining scaling and translation similarly produces terms like $[(X_s - X_{ref})/X_{ref}]^n$.

An over-sampled system, i.e., one where $D > n + d + 1$, can be handled in various ways, one of which is to employ a pseudo inverse for the solution, but which has the disadvantage of approximately squaring the condition number. Another approach would be to employ overlapping windows of different data sets to compare performance of their respective, lower-order FMs.

Besides quantifying the FM using function sampling, derivatives of the response can be used, as can various combinations of function and derivative sampling. One result of using derivatives is that the samples can then be spaced more widely in frequency than using function sampling alone. A more important consideration is that in some circumstances, a derivative sample can be obtained for a computation operation count that is of order $1/X_s$ of the first function sample alone. If a derivative provides information concerning the response from it is obtained that is equivalent to that provided by a function sample, an obvious computation advantage is achieved. Using derivative sampling in the context of a frequency-domain integral equation is discussed more fully in RII. Thus, a given FM, whether an exponential series or a pole series, can be quantified by a variety of data and also by using various approaches. Although determining the time or frequency derivatives of some response may not be an obvious observable in computational electromagnetics (CEM) and such derivatives are unlikely to be measurable, they can be obtained from computation.

6.4 Adapting and optimizing sampling of the First Principles Model

A major advantage of MBPE, and a prime motivating factor for its use, is its potential for minimizing the operation count needed for computing various EM (and other wave-equation phenomena) responses This is especially the case concerning the time–frequency domain transform pair, where there is more flexibility with respect to sample placement and model order in frequency than in time. As noted in RI in connection with waveform sampling, it is necessary that a polynomial representation of the sampled data be developed, which leads to a need for uniform sample spacing. Actually, this requirement is not inconsistent with how most time-domain computations are performed, where equal time steps are most often used throughout the time interval for which the model is run. However, it also means that if subsequent evaluation of the time response shows that it was undersampled; it is not very practical to add new samples to the original result; instead, the model computation must be entirely repeated with a new time step. This is not the case for all WD models, however, as is demonstrated by sampling the far field of a source distribution (model 4 in Table I of RI), where new observation angles can be added to an existing set without repeating any previous computation. This

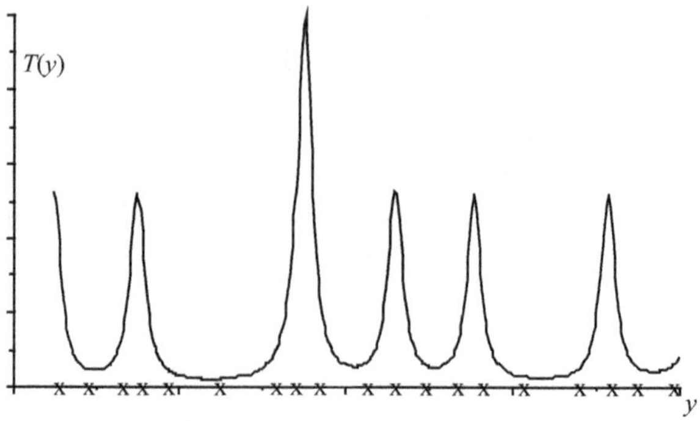

Figure 6.1 Conceptual diagram to illustrate optimal placement of frequency samples in developing a fitting-model representation of a transfer function T(y). Since there is no computational penalty in doing so, the samples can generally be non-uniformly spaced therefore maximizing the information provided by each.

advantage is always true in the FD, where adding one, or several, new function or derivative samples to an existing set can be done while fully retaining the benefit of whatever samples have already been obtained.

Ideally, the goal of minimizing the operation count required to obtain an EM transfer function would involve choosing sample locations and derivative orders (if used) such that the new information provided by each new sample is the maximum that can be provided over the range of interest relative to the cost of obtaining that sample. If this goal were to be realized for each of the samples ultimately needed for computing the FM parameters while satisfying a specified error criterion, then the overall cost should be a minimum for the chosen error measure. This idea is illustrated conceptually in Figure 6.1.

There seems to be no obvious or unique approach for achieving this goal, since n and d, the number of samples D and the frequency range they span, are all free parameters in developing an FM. This means that some computer experimentation will usually be required to determine suitable numerical values for them. A few of the several sampling strategies that seem worth considering are discussed immediately below, followed by a discussion of possible error measures.

6.4.1 Adaptive sampling strategies

Adaptive sampling strategies form a basis for much numerical analysis. Three possibilities that might be applicable to transfer-function representation are listed below.

(1) A strategy similar to one found useful for adaptive numerical quadrature based on Romberg's method (RM) [3] might be adapted to the spectral-estimation problem. It involves choosing a starting subinterval over which five successive uniformly spaced samples of an integrand are developed upon which RM is

applied to the three trapezoidal-rule quadrature values that are obtained (samples 1 and 5, 1, 3 and 5, and all five samples). An error estimate provided by the RM shows whether new samples are required in either half of the original subinterval. If this is the case, new samples are added where indicated halfway between the original ones and the process is successively repeated. On the other hand, if the initial error is smaller than specified, the process is repeated while doubling the subinterval size. This kind of adaption handles sharply peaked integrands, yielding maximum-to-minimum sample spacings of 10^6, but it may not be as well suited to spectral estimation since for many EM applications, the resonances are relatively uniformly spaced.

(2) The first two samples could be taken at the endpoints of the variable range with additional samples subsequently developed by evenly subdividing this range into subintervals until the error criterion has been satisfied over all subintervals that are thus formed.

(3) If an error measure is available as a continuous function of the independent variable over the range of interest, then placing a new sample wherever that measure is a maximum seems to be an obvious choice as a way to achieve the greatest amount of additional information from a single additional sample. This approach is used for the procedure implemented here.

6.4.2 Estimating FM error or uncertainty

An adaptive process can be only as effective as the error measure used for estimating the degree to which the FM (or its equivalent) differs from the GM (or whatever process whose results the FM is to approximate). This observation is a general one that applies to all manner of numerical processes having the goal to minimize the number of accurate samples that are needed as a means of reducing the cost of developing a sampled representation of some process over a specified range of independent variable(s). For an SD application, it appears desirable to use lower-order FMs over subintervals of the spectral range to be covered to avoid possible ill conditioning. It then follows that two or more FMs will be needed to span the spectral range of interest, leading to the situation illustrated in Figure 6.2. By using overlapping FM's that share common data, the differences between them can then be used to estimate FM uncertainty, or minimum match in digits as a function of frequency, as given by

$$MM_{i,j}(f) \equiv -\log_{10}\left(\min\left[\frac{|M_i(f) - M_j(f)|}{|M_i(f)| + |M_j(f)|}\right]\right) \text{ digits.} \tag{6.7}$$

The minus sign is needed to express the results in positive digits and is computed for each pair of overlapping models as a function of frequency. Subsequent sample placement and type would then be chosen to maximize the information acquired from each sample by adding each new sample at the frequency where the minimum match for all FMs occurs. Sampling of the GM would be concluded when the specified error criterion is satisfied.

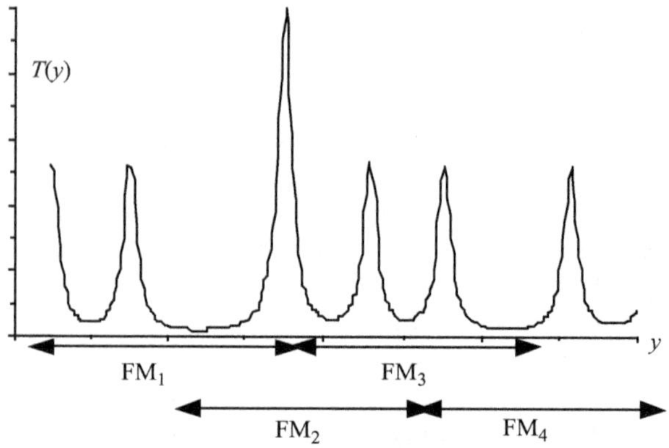

Figure 6.2 The possibility of developing a fitting model (FM) representation of a transfer function over a wide frequency interval by employing a number of subintervals, lower-order FM, M_i, is illustrated here

Also note that, alternatively, an exact error measure results from comparing an FM result with a GM sample $G(f_i)$ as given by

$$GM_{i,k}(f) \equiv -\log_{10}\left(\min\left[\frac{|G(f_i) - M_k(f_i)|}{|G(f_i)| + |M_k(f_i)|}\right]\right) \text{ digits.} \tag{6.8}$$

However, doing this potentially would require more GM samples with a consequent increased computation time, while providing, in addition, only a point-wise error measure in f. Thus, $MM_{i,j}(f)$ requires only FM computation and yields a global, but approximate, error measure while $GM_{i,k}$ requires GM computation and yields a point-wise, but exact, error measure.

There at least three kinds of errors that the FM might produce relative to GM results:

(1) Non-physical results, e.g., negative conductance or resistance;
(2) Amplitude shifts in transfer functions; and
(3) Resonance shifts between transfer functions;

with the latter two occurring either between two (or more) FMs or between FM and the GM result, for which different sampling decisions might be made. For non-physical errors, an additional sample at the peak in negative conductance would seem most appropriate. For baseline-shift errors, an additional sample could be placed at the location of the maximum difference if it exceeds the specified error criterion. For resonance-shift errors, an additional sample could be put between the response peaks, again if the shift exceeds the specified error criterion.

Observe that each FM represents the response over a frequency interval $F_i = F_{i,u} - F_{i,l}$ where $F_{i,u}$ and $F_{i,l}$ are the upper and lower frequencies of FM_i with which overlapping adjacent FM's share some frequency samples. The $MM_{i,j}(f)$ of Eq. (6.7) provides a normalized, mutual-error estimate between FM's i and j. By evaluating $MM_{i,j}(f)$ as a function of frequency, new samples, either function or derivative, can be placed where $MM_{i,j}(f)$ does not satisfy a specified value and where, therefore, the corresponding FM results are most uncertain and further GM are needed.

To summarize, for a FM having numerator and denominator polynomials of order n and d, respectively, $n + d + 1$ samples of the GM are needed, which can range, on the one hand, from using $n + d + 1$ different frequencies to, on the other hand, using one function sample and $n + d$ derivatives at a single frequency. The "best" approach would be that which minimizes the GM operation count required to achieve some specified accuracy or uncertainty in the transfer function being estimated, using an appropriate mix of function and derivative samples. By varying the n and d parameters of the rational-function FM, the effects of non-pole contributions and poles that lie outside the frequency interval of interest can be approximated, although it is worth considering other approaches to this problem.

6.5 Adaptive sampling of a simulated transfer

A pole series provides a good test for an adaptive sampling strategy because many actual transfer functions exhibit a marked resonance structure and its properties are controllable. A 20-pole series having poles at $s_i = -(i^{1/2})/20 + j{*}i$, $i = 1,2, \ldots, 20$ was used as a GM with $j = (-1)^{1/2}$. The initial GM sample was placed at 0.5 intervals from $s = j{*}3.5$ to $= j{*}11$, numbered from 1 to 16. Six FMs were employed spanning GM sample numbers 1–6, 1–8, 5–10, 7–12, 9–16 and 11–16, respectively. Thus, four of the FMs (1,3,4,6) were initially of order 6, using $n = 3$ and $d = 2$, while two (2 and 5) were of order 8, using $n = 3$ and $d = 4$. The latter two FMs are arranged to include the endpoints so that there are a minimum of two overlapping FMs across the entire frequency range of interest. The estimation error, EE, way set at 10^{-2}, or, when measured in digits, EE = 2. Some results for applying the adaptive procedure above are shown in Figures 6.3–6.5.

The real part of the final FM is shown in Figure 6.3, on which are also indicated the original and additional GM samples, whose values in order of sampling are 6.2, 7.2, 9.2, 5.3, 8.2, 9.35, 4.15, 6.25, 5.05, 1.05, 8.3, 7.05, 4.2, and 10.95. It is also of interest to observe the behavior of the minimum-match values for each of the FMs, E_i, as the process of model modification continues, a result that is shown in Figure 6.4. Also shown is ME_i, the minimum-match value for all six FMs, as a function of the model iteration number. An E_i can remain constant for several iterations if none of the added GM samples is contained in its frequency span and its E_i is unchanged as well. Although this behavior is not guaranteed, ME_i can be seen in this example to exhibit a monotonic increase. Finally, it is significant to

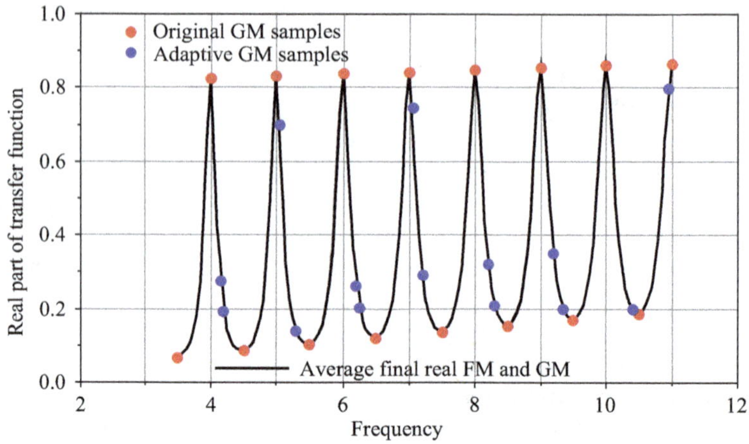

Figure 6.3 Result obtained using adaptive sampling of the pole-series GM described in the text. The original 16 samples of the real part of the GM are indicated by the solid red circles and the additional 14 samples needed to obtain a 0.01 normalized match between overlapping FMs across the 3.5 to 11 frequency range are shown as black circles. The final FM, obtained from averaging the individual FMs in their regions of overlap, is shown by the solid line. The final FM- GM match is also better than 0.01 (see Figure 6.6 for further discussion).

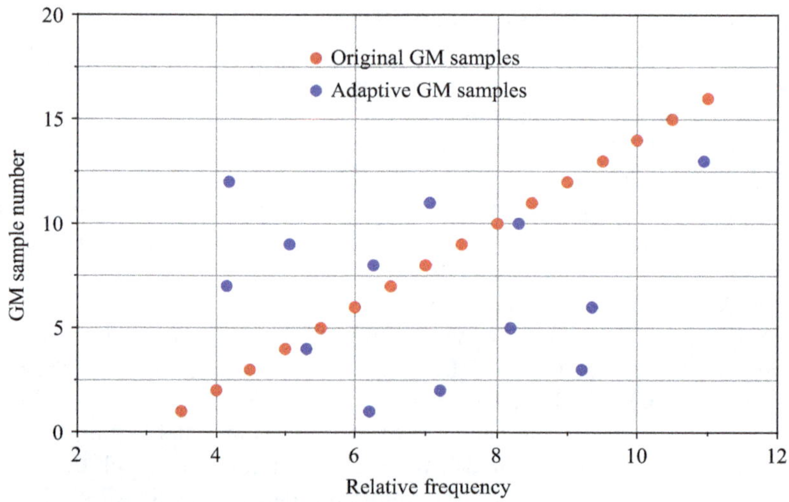

Figure 6.4 The numerical sequences of the original GM samples and the adaptively added GM samples needed to develop the converged frequency spectrum of Figure 6.3

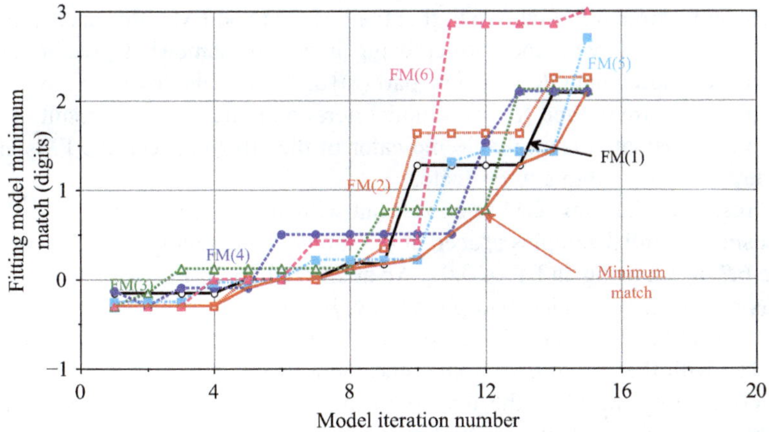

Figure 6.5 An illustration of how the FM parameter ME_k can vary with model iteration number k. The E_i curve for each FM is indicated by FM(i), where the initial set of FMs is number 1 on the horizontal axis. For this particular case, ME increases monotonically, from less than 0 for the first several models (indicating that the normalized errors are greater than unity) to the two-digit specified estimation error at k = 15.

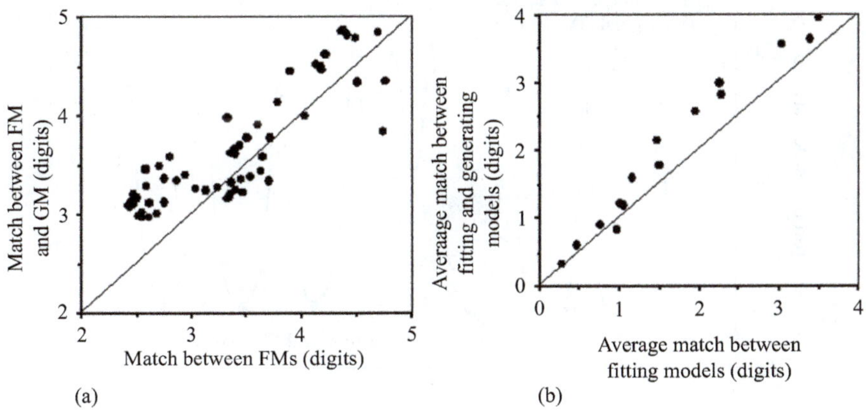

Figure 6.6 The average minimum-match values between the FMs and GM data samples for the 16th model iteration of the above example (a) and the frequency-averaged values of these quantities for each of the 16 FM sets (b). In both cases, the FM–FM minimum match is seen to be generally less than the FM–GM value, showing that the former is a conservative error estimate of the accuracy achieved by the FMs.

note, as demonstrated in Figure 6.6, that the FM–FM and FM–GM mismatch errors are well correlated, with the former being generally somewhat greater than the latter, for the individual GM samples [part (a)] and where the frequency averages of these quantities are plotted for each model iteration [part (b)]. This result indicates that the FM mismatch is a reliable indicator of the error between the FMs and the GM they are intended to approximate.

A result for the same GM as above, but with an added pole at $s = -(i^{1/2})/20 + j*7.5$ using 16 initial samples spaced at 0.5 intervals beginning at 3.75 is shown in Figure 6.7, again using an EE of 0.01. An additional 16 GM samples were required for this example, two more than previously, because of the added pole. Although the locations of the original GM samples are shifted relative to the poles in this case compared with the previous example, and there is an added pole in the frequency range covered by the FMs, the final performance is comparably good, indicating the robustness of the adaptive-sampling approach.

Results for another smaller set of specified poles randomly located are illustrated in Figures 6.8 and 6.9. While the original GM samples are uniformly spaced in frequency, something that is displayed more clearly in Figure 6.9, the adaptive GM samples are located rather randomly and for the most part are near regions of more rapid current variations. The "scatter' plot of their adaptively added frequencies in Figure 6.9 exhibits this behavior in a different way, something that is also used for other problems in this chapter.

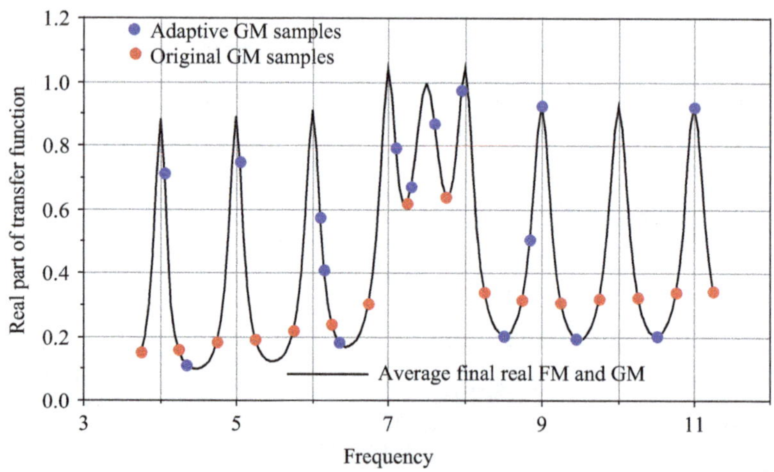

Figure 6.7 Results for the real component of the 20-pole GM used above but with an added pole at $s = -(i^{1/2})/20 + j*7.5$ and using original GM samples shifted by 0.25 in frequency. The average FM is for a 0.01 specified estimation error as obtained at the 16th additional GM sample

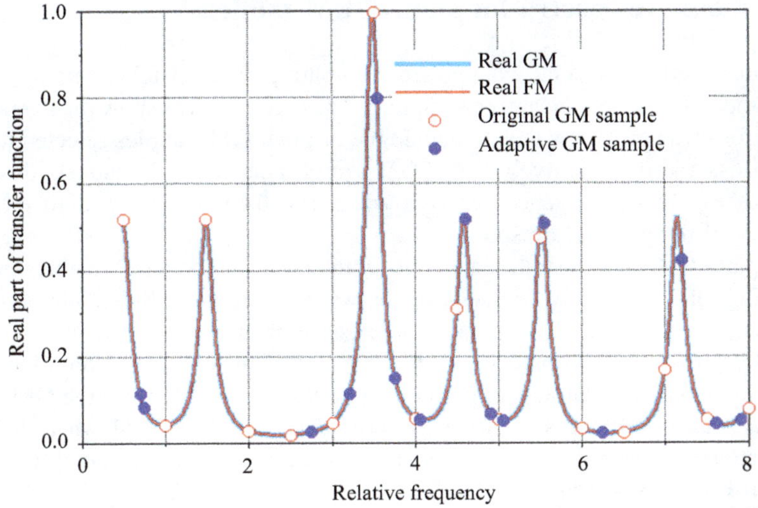

Figure 6.8 Comparison of the GM and FM results for the real part of the transfer function due to several randomly located poles

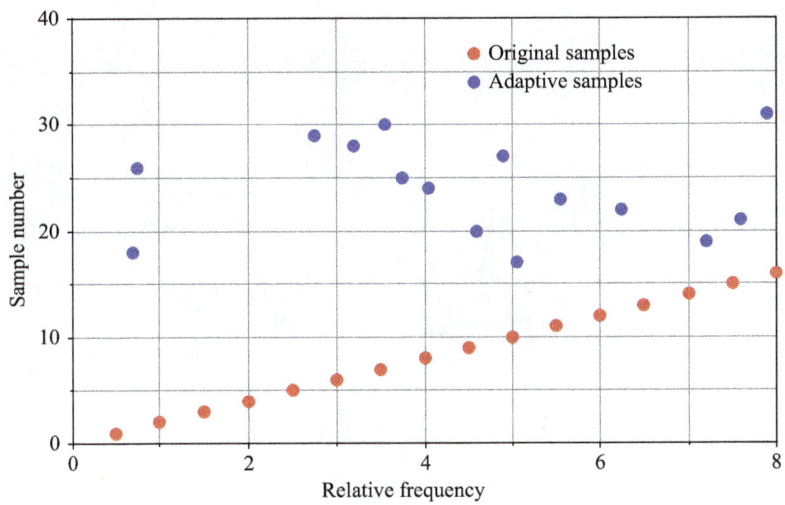

Figure 6.9 A scatter plot showing the frequencies of the one evenly distributed original GM samples and the adaptively added samples needed to achieve a two-digit minimum match between overlapping FMs

6.6 Adaptive spectra for some CEM models

Next presented are several examples of estimating spectral responses for various EM computations using adaptive sampling. These results will show the average FM results for these problems which include the original GM samples selected to begin the process and the adaptively added GM samples chosen as previously described. In all cases, the convergence criterion was set at 0.01, for a two-digit minimum accuracy of the peak GM response.

Figure 6.10 demonstrates a result from a classic boundary-value problem, scattering of a normally incident plane wave from an infinite, circular, PEC cylinder. The GM finely sampled complex current on the front of the cylinder is plotted over several resonances as a function of the cylinder circumference in wavelength. The FM results are graphically indistinguishable from the GM curves. This particular problem is discussed in more detail in Chapter 11 where MBPE is demonstrated to provide a way of estimating the underlying accuracy of various kinds of EM phenomena.

There are 12 equally spaced original GM samples of the conductance and susceptance and ten adaptive samples obtained from NEC needed to achieve the maximum specified estimation error of 0.01 shown in Figure 6.11. A scatter plot of the GM sampling sequence for each set of GM samples beginning with sample 1 is presented in Figure 6.12. The uniform spacing of the original samples is clear, while that of the adaptive samples is quite random since it is determined by the successive locations of the maximum FM–GM differences.

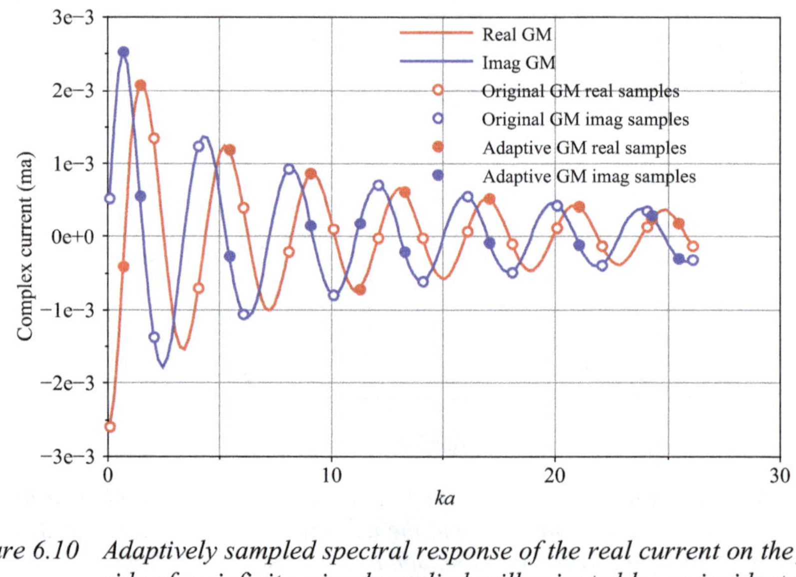

Figure 6.10 Adaptively sampled spectral response of the real current on the front side of an infinite, circular cylinder illuminated by an incident plane wave

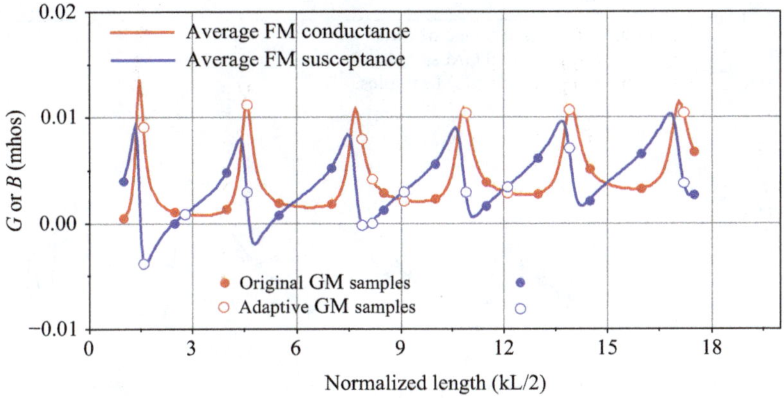

Figure 6.11 *The input admittance for a center-excited dipole antenna as a function of its normalized length in. units of $\pi L/\lambda$ with the original GM samples plotted as solid circles and the added GM samples as open circles*

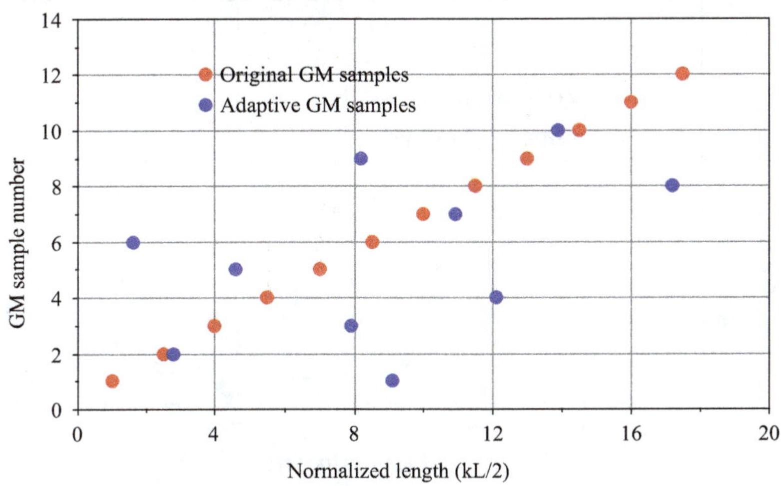

Figure 6.12 *A scatter plot of the locations in dipole length where the successive original and adaptive GM samples are located*

Similar results for a goal-post antenna modeled using NEC are presented in Figures 6.13 and 6.14. This antenna has the shape of a standard football goal post but with one of the vertical arms about 3%, or one segment shorter than the other. This results in smaller resonance peaks between the major peaks of Figure 6.13.

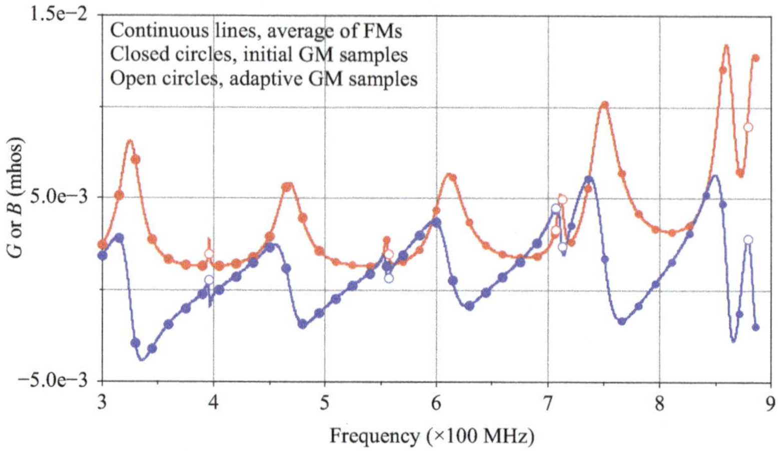

Figure 6.13 The frequency-dependent admittance of a goal-post antenna determined from NEC and using adaptive sampling

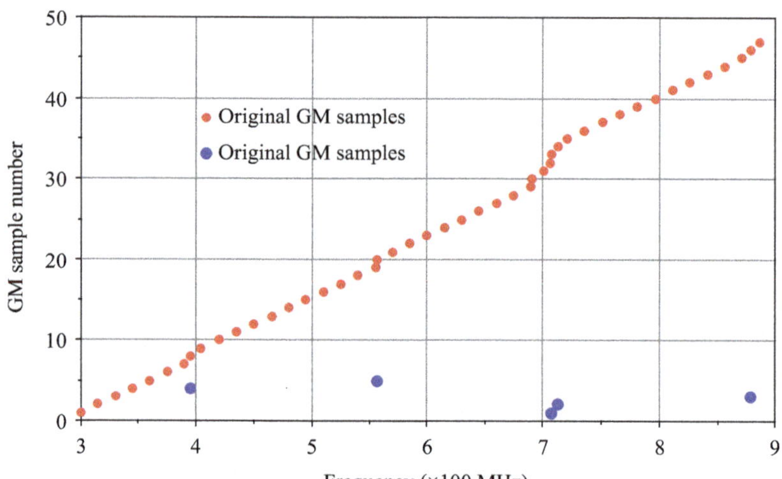

Figure 6.14 The GM samples used to determine the estimated spectrum of Figure 6.13

A total of seven overlapping FMs were used to cover the frequency range shown in the adaptive spectrum estimate of Figure 6.13. This accounts for the deviation from a straight plot of the original GM samples of Figure 6.14. Only five adaptive GM samples were needed to finalize the estimated spectrum, all of which are located at rapid changes or "glitches" in the frequency response.

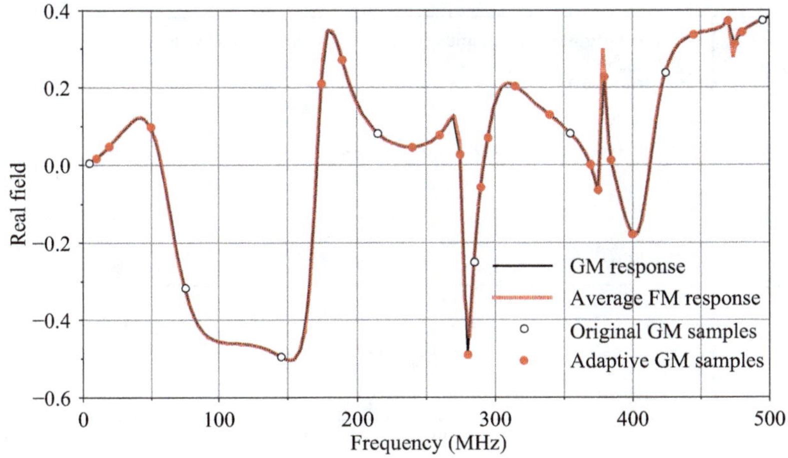

Figure 6.15 The real backscattered field for the dielectric cylinder

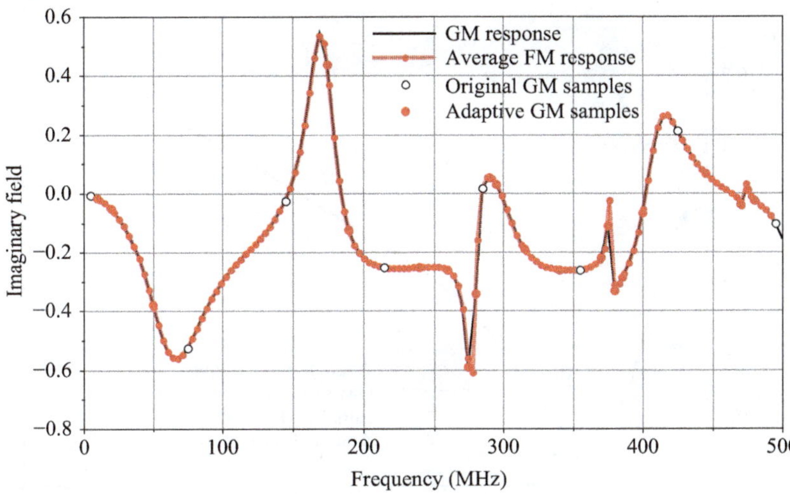

Figure 6.16 The imaginary backscattered field for the dielectric cylinder

The results shown in Figures 6.15–6.20 are derived from the scattered field of a penetrable infinite circular cylinder [4]. The relative permittivity of the cylinder of Figures 6.15 and 6.16 is 16 and for Figures 6.17 and 6.18 it is lossy with a permittivity of $16 - j16$. The estimated spectra use relatively few of the data from [4] and were used as a test of how effective MBPE might be for analyzing presampled results. These same results are further analyzed in Chapter 13 where MBPE is used to determine the accuracy, or uncertainty, of data, whether computed or measured.

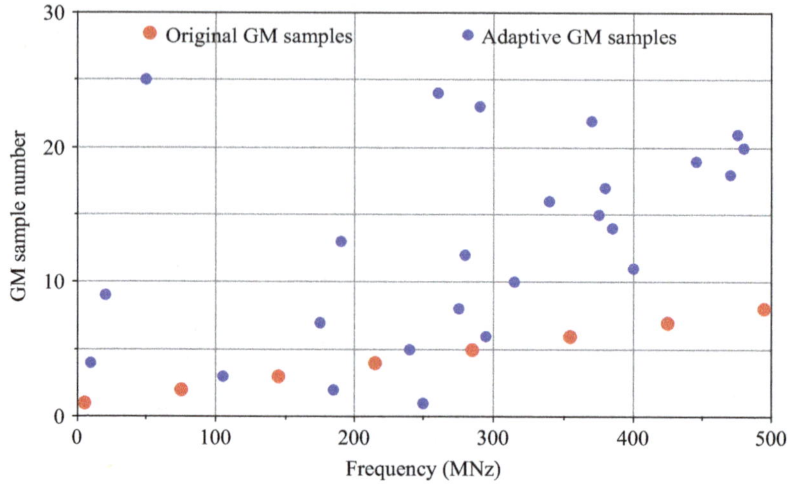

Figure 6.17 The GM samples used to determine the spectra of the lossless or dielectric cylinder for Figures 6.12 and 6.13

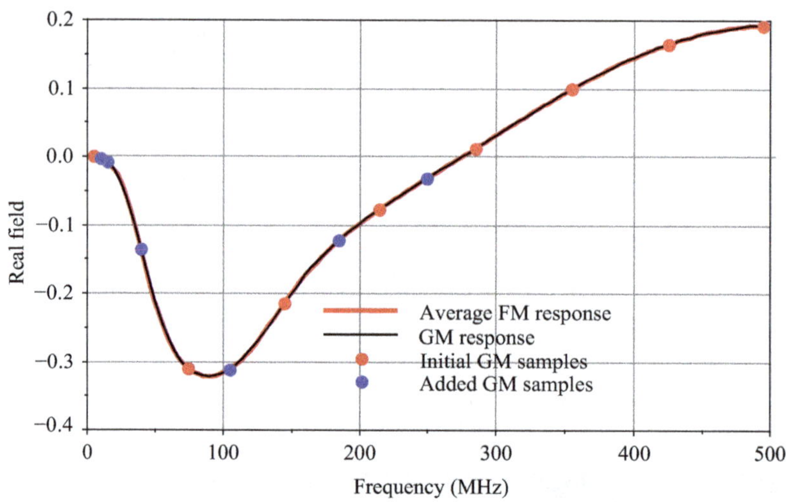

Figure 6.18 The real backscattered field for the lossy cylinder having a relative permittivity of 16 – j16

The complex spectra for this data are plotted separately for clarity. Observe that the resonances of the dielectric cylinder are substantially reduced when the dielectric is lossy. This, in turn, greatly reduces the number of GM samples required to obtain a converged result.

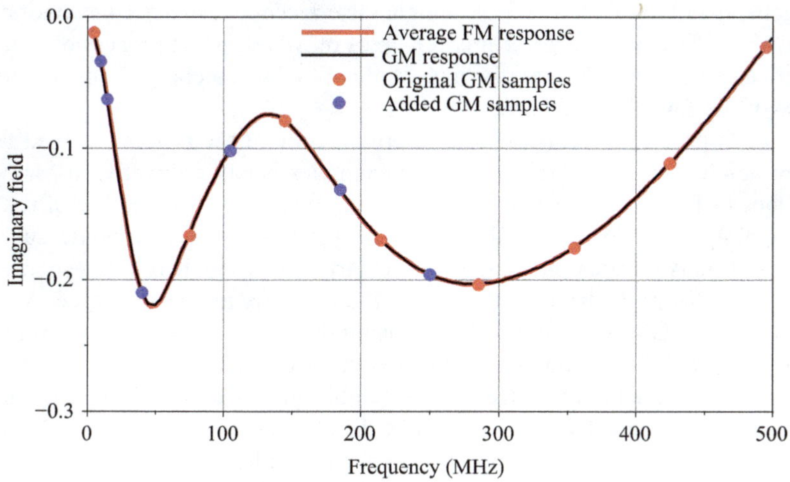

Figure 6.19 The imaginary backscattered field for the lossy cylinder of relative permittivity 16 – j16

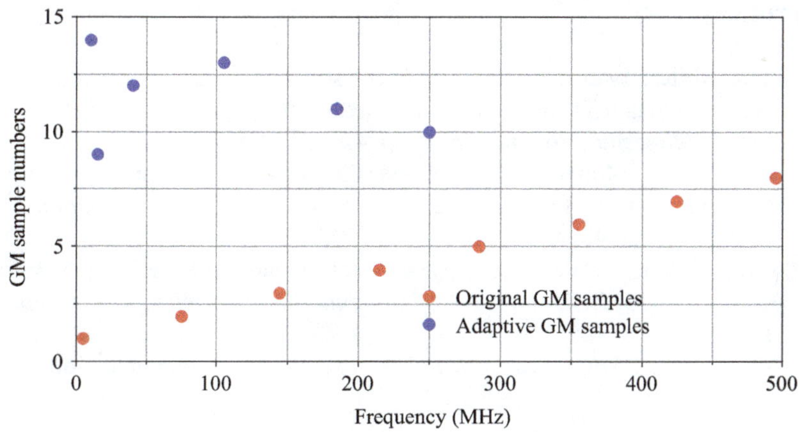

Figure 6.20 The GM samples used to determine the spectra of the lossy dielectric cylinder for Figures 6.18 and and 6.19

6.7 Concluding discussion

This discussion has illustrated the adaptive sampling of a pole-based transfer function like those that arise in EM frequency responses using MBPE. A useful FM for this purpose, the "model-based" part of MBPE, is a rational function that serves as a generalization of a pole series that characterizes the resonance behavior of EM observables. The "parameter-estimation" part of MBPE is the process of obtaining numerical values for the coefficients of the FM by matching them to sampled values of the EM observable of interest.

An FM provides a reduced-order representation of the results obtained from a first-principles, or GM which for electromagnetics is based directly on Maxwell's equations or from experimental measurements. It offers the possibility of reducing the sampling required of the GM when using it to develop a response such as a broadband transfer function. This is particularly the case where the response contains details that might be missed if the GM samples are too widely space. Adaptive sampling of the GM is made possible by using several, low-order, overlapping FMs that share initial GM samples chosen to cover a given frequency range. The FM results are compared at other frequencies within their shared bandwidths and new GM samples are added wherever the FM differences exceed a specified level. A final set of FMs is thus developed for the GM samples that provide a continuous representation over a given frequency range while meeting desired accuracy, or uncertainty, specification. Such adaptive sampling can not only greatly reduce the number of GM samples that would otherwise normally be used but also ensure that fine details of the response are not missed.

References

[1] E. K. Miller, "Model-Based Parameter Estimation in Electromagnetics: Part I. Background and Theoretical Development," *IEEE Antennas and Propagation Society Magazine*, Vol. 40, No. 1, pp. 42–52, 1998.

[2] E. K. Miller, "Model-Based Parameter Estimation in Electromagnetics: Part II. Applications to EM Observables," *IEEE Antennas and Propagation Society Magazine*, Vol. 40, No. 2, pp. 51–64, 1998.

[3] E. K. Miller, "A Variable Interval Width Quadrature Technique Based on Romberg's Method," *Journal of Computational Physics*, Vol. 5, pp. 265–279, 1070.

[4] M. Kluskens, *Naval Research Laboratory*, Private Communication, 1995.

Chapter 7

Using Prony's method to design arrays that produce the patterns of continuous source distributions and prescribed radiation patterns[1]

Abstract

Antenna pattern synthesis remains an important topic in electromagnetics. One goal of continuing interest has been that of developing an array of discrete sources that replicate the radiation pattern of continuous sources such as an aperture radiator or some current distribution. Another is to determine a discrete array that produces a prescribed radiation pattern. This chapter outlines a procedure based on Prony's method for finding discrete-source arrays to accomplish these goals. A variety of problems are described to which Prony synthesis has been applied that demonstrate its performance.

7.1 Introduction

A wide variety of methods have been explored to design linear arrays of current elements that produce a given radiation pattern [1–8], a problem that is one aspect of antenna pattern synthesis. Two issues related to this problem are considered here. This chapter reviews one way of designing a discrete array that generates a radiation pattern of a specified continuous source distribution. The other, discussed in the next chapter, is to explore whether a given discrete array can be modified to generate a pattern that improves upon the one that it was designed to produce. An approach based on spatial poles is used for both of these applications. For purposes of the present discussion, a candidate pattern is considered "unrealizable" if it cannot be produced to within some prescribed accuracy by an array of isotropic current sources.

This chapter summarizes the pole-based approach and demonstrates its application. Various sources and patterns are analyzed to determine the feasibility of approximating them with a discrete array derived from a pole-based representation. For illustrative purposes, the discrete-source model is also applied to various other

[1]Edited and excerpted from Edmund K. Miller, "Using Prony's Method to Synthesize Discrete Arrays for Prescribed Source Distributions and Exponentiated Patterns," *IEEE Antennas and Propagation Society Magazine*, 57 (1), February, pp. 147–163, 2015.

patterns produced by prescribed current sources such as sinusoidal and uniform currents.

The next chapter introduces a new synthesis concept in which a pattern of interest is developed by exponentiating or raising a given pattern to an integer power M. This yields a new pattern having a reduced main beamwidth and decreased sidelobe levels. The second step in that approach is to determine whether pole-based synthesis will yield the array geometry and currents that will radiate the exponentiated pattern.

7.2 Synthesis of a discrete source mode

A discrete source approximation of a given pattern $P_{DSA}(\theta)$ can be developed using Prony's method (or its equivalent) [9,10]

$$P_{DSA}(\theta) = \sum_{\alpha=1}^{N} S_\alpha e^{kz_\alpha \cos(\theta)} \tag{7.1}$$

to determine the strengths $S_\alpha, \alpha = 1, \ldots, N$ of N point sources located at position $z_\alpha = z_{\alpha,real} + iz_{\alpha,imag}$ along the axis of a linear array lying in the complex z-plane. In complex-variability notation, these are usually referred to as residues and poles, respectively. The far-field observation angle θ is measured from the array axis. Note that $z_{\alpha,imag}$ in (7.1) corresponds to the real space location of source α while a non-zero real part implies that the source has a directivity D_α given by

$$D_\alpha = e^{kz_{\alpha,real} \cos(\theta)} \tag{7.2}$$

a situation that would not normally be of interest since most practical arrays would either employ isotropic radiators or ones having at most a sinusoidal pattern.

Since there are $2N$ unknown, possibly complex, parameters in (7.1), the Prony discrete-source array requires $2N$ far-field samples of the pattern of interest for their computation. Note that these $2N$ samples need to be spaced uniformly in cosθ to form the polynomial needed by Prony's method [9,10]. Some numerical issues that arise in using Prony's method in connection with (7.1) and (7.2) are illustrated in Section 7.2 that follows, with representative applications demonstrated in the following sections. While Prony's method seems to be viewed skeptically among some numerical analysts, as will be demonstrated by the results presented here, it can produce valid results when used prudently.

7.2.1 Numerical implementation of Prony's method

Numerical implementation of Prony's method involves three interrelated parameters: (1) the number of poles, N; (2) the angle sampling interval, Δcosθ; and (3) the total angle observation window, W, measured in units of cosθ. It has been found from computations that N must be at least one more than twice the source size in wavelengths for a valid fitting model (FM) to be obtained. Unless otherwise indicated, the FM is considered to be valid when its pattern agrees with the generating model (GM) to at least 0.1 dB. But N must also at least equal the number of degrees of freedom or rank, R, of the data to be modeled while the sampling interval must be small enough to avoid aliasing. Finally, W must be wide enough to produce a reasonably well-conditioned data matrix. These requirements can be summarized as

$$N - 1 \geq WL = W/\Delta\cos\theta \tag{7.3a}$$

and

$$N \geq R \tag{7.3b}$$

where L is the current or aperture size in wavelengths. Some experimentation is usually needed to select appropriate values for these parameters. In order to minimize the possible effects of ill conditioning, following computations were done in 24-digit precision unless stated otherwise.

An effective way to establish the rank of a pattern is to perform a singular-value decomposition of the data matrix associated with Prony's method. As a simple example, consider the pattern factor of a modified sinusoidal current filament (*SCF*) multiplied by sinθ and centered on the z-axis as given by [11]

$$P_{MSCF} = \cos\left(\frac{kL}{2}\cos\theta\right) - \cos\left(\frac{kL}{2}\right) \tag{7.4}$$

where θ is the observation angle measured from the axis of the *SCF* and k is the wave number in free space.

Although it may not be obvious, (7.4) is the far field of at most 3 point sources. As shown below, sources of strength 0.5 are located at each end of the modified *SCF*. A third source is located at the middle whose value oscillates between zero and one as L varies in length between an odd and even integer of half wavelengths. A pattern obtained from an analytical expression for the pattern as in (7.4) is denoted here as a GM. Similarly, the pattern obtained using the Prony model of (7.1) is denoted as an FM.

Results for the GM of (7.4) and its Prony FM are shown in Figure 7.1 for $L = 5$. Two pattern sets are included there for cosθ sampling windows of -0.999 to $+0.999$ and -0.05 to $+0.05$ (ranging in degrees from 177.4 to 2.56 and 92.87 to 87.1, respectively, with the latter indicated by the arrows). The GM and FM results for both windows are graphically indistinguishable. It is interesting to note that a data from a window less than 6° wide is able to reconstruct the entire pattern, something made possible by the pattern's low-order rank.

Singular-value spectra obtained from the data matrices that result from applying Prony FMs to (7.4) for observation windows of various widths as plotted in Figure 7.2 demonstrate that the pattern rank is 3. The cosθ window widths used in Figure 7.2 vary from -0.05 to 0.05 to -0.999 to 0.999. For each window width N was increased until a converged FM was obtained. Increasing N beyond that value produced no further change in the FM. All spectra are seen to fall to a noise floor between 10^{-23} and 10^{-24} beyond the third singular value, which is consistent

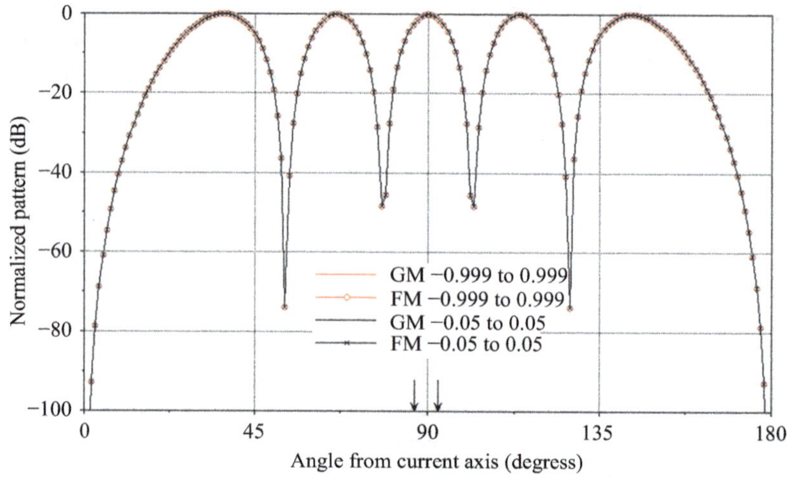

Figure 7.1 *Radiation patterns for a 5-wavelength modified sinusoidal current filament obtained from (7.4) as designated by GM and from the Prony model as designated by FM. Results are shown for two different angle windows, one for a* cos θ *observation window extending from −0.999 to +0.999 and the other from −0.1 to +0.1 whose limits are indicated by the two arrows*

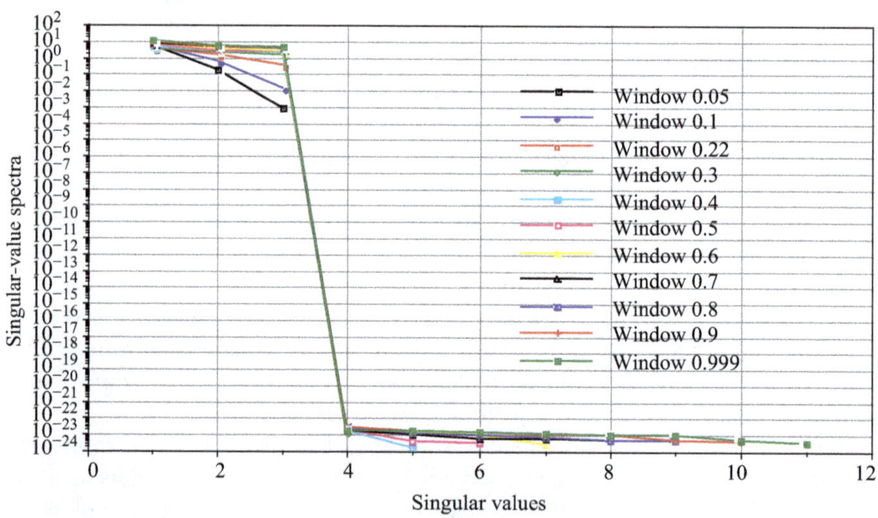

Figure 7.2 *Singular-value spectra for the pattern of the modified sinusoidal current filament in (7.4) with the observation window a parameter increasing from −0.05 and +0.05 to −0.999 and +0.999*

with a compute precision of 24 digits and a source that consists of three resolvable point sources.

The FMs that yield the patterns of Figure 7.1 are exhibited in Figure 7.3 on the complex exponent plane of (7.1). The vertical-axis thus represents a source coordinate in real space and the horizontal-axis represents a source directivity as indicated by (7.2). The solid squares and arrows show the locations of the FM poles located at $z_\alpha = z_{\alpha,real} + iz_{\alpha,imag}$ and the large X's on the vertical axis depict

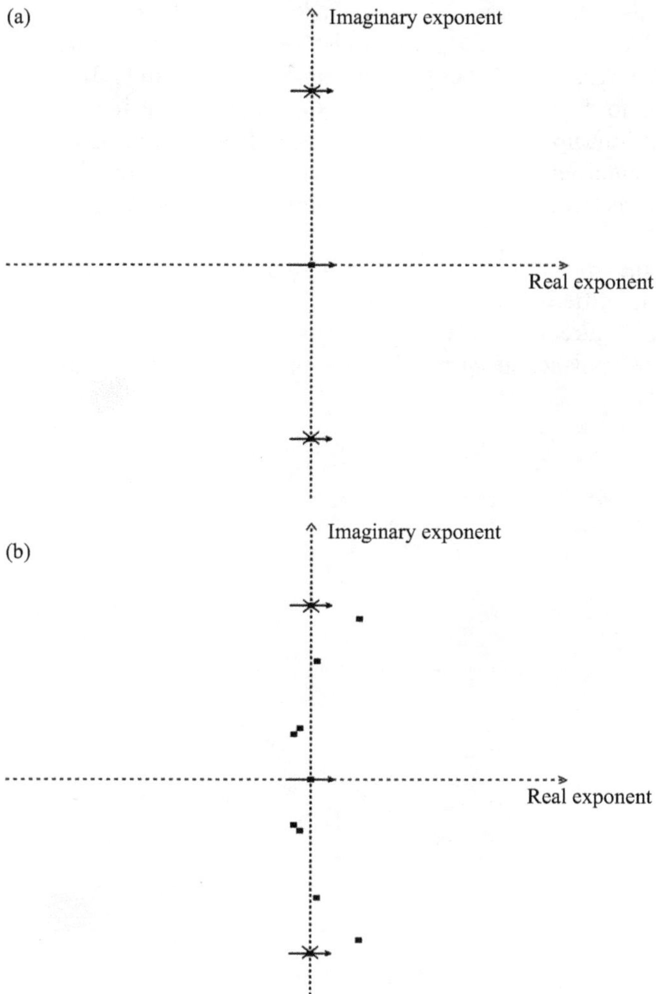

Figure 7.3 Source locations z_α of the fitting model of (7.1) for a 5-wavelength modified sinusoidal current filament sampled using cos θ *windows extending from −0.1 to +0.1 with three fitting models in (a) and from −0.999 to +0.999 with 11 fitting models in (b)*

the physical ends of the source being modeled. The arrows, centered at the discrete source locations, show the source strength on a three-decade log scale with the source phase indicated by the angle that the source arrow makes with the positive real axis of the figures. The phases of the three sources are all zero and their locations are on the imaginary axis, that is, $z_{\alpha,imag} = 0.$, showing that they are isotropic radiators.

Three sources are observed for each case in Figure 7.3, one at each end of the modified sinusoidal current filament and one at its center. The plot of Figure 7.3(a) demonstrates that using the -0.05 to $+0.05$ window with $N = 3$ yields a valid solution for the three sources. Using a window width of -0.999 to $+0.999$ requires a total of 11 terms in the FM, with eight having negligible residues, to yield the same three sources. This is because, as remarked above in (7.3a), N must increase in proportion to the observation window size to avoid aliasing.

The relationship between the number of FMs and the window width for the 5-wavelength modified sinusoidal current is plotted in Figure 7.4. The minimum number of FMs needed to achieve a converged result is in agreement with (7.3a) and (7.3b).

A complex-plane plot of the fitting-model sources determined for a 5.5 wavelength modified sinusoidal current using a window width of -0.05 to $+0.05$ is presented in Figure 7.5. Only two sources are needed for this case because the center source vanishes at current lengths that are an odd-integer number of half wavelengths.

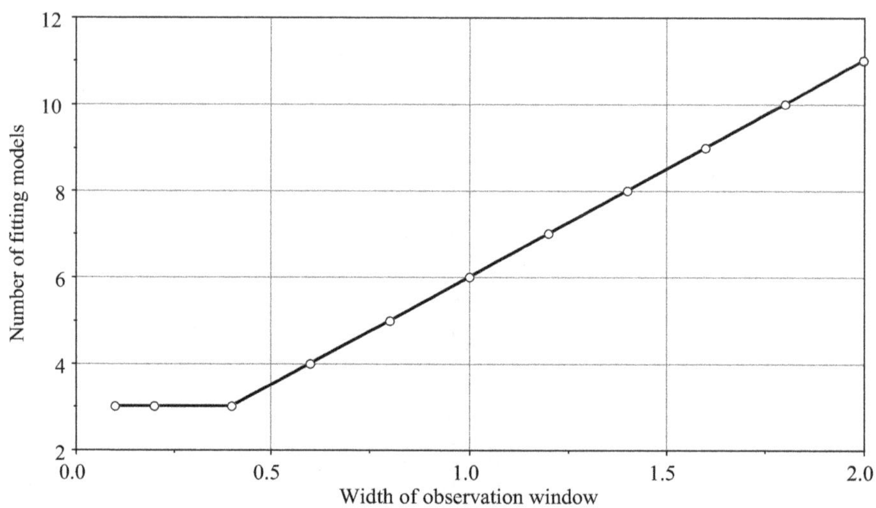

Figure 7.4 The number of FMs, N, needed to obtain a converged result for a modified sinusoidal current 5 wavelengths long as a function of the window width, W, in units of cos θ *is shown here to confirm (7.3a) and (7.3b)*

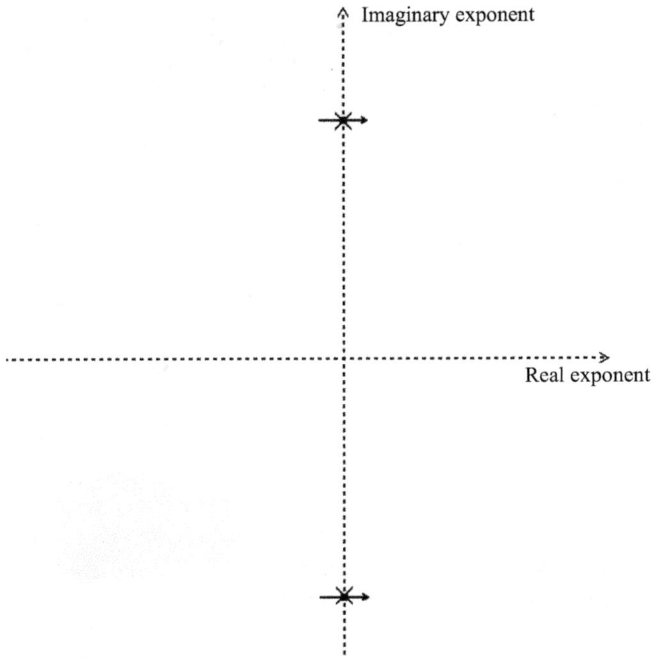

Figure 7.5 *Source locations z_α of the FM of (7.1) for a 5.5-wavelength modified sinusoidal current filament sampled using a* cos θ *window extending from −0.05 to +0.05*

7.3 Other patterns and source distributions

Several other sources and patterns are explored here to further demonstrate application of the Prony FM of (7.1). In all cases, the width of the data window used for the FM computation varied from −0.999 to +0.999 in cos θ.

7.3.1 A sinusoidal current filament

An unmodified, sinusoidal current has a pattern factor is given by [11]

$$P_{SCF} = \frac{\cos\left(\frac{kL}{2}\cos\theta\right) - \cos\left(\frac{kL}{2}\right)}{\sin\theta} \tag{7.5}$$

which differs from (7.4) only by the $\sin\theta$ term in the denominator. Equation (7.5) yields the results of Figures 7.6 and 7.7 for the far-field pattern and source geometry where the FM parameters are $N = 11$ with the sampling window in $\cos\theta$ extending from −0.999 to +0.999. Except within a few degrees of the axial directions, the FM and GM patterns are essentially indistinguishable. Only nine poles are shown in Figure 7.7 as two numerically insignificant poles lie outside the area covered by the graph. The fact that six of the poles have real components shows

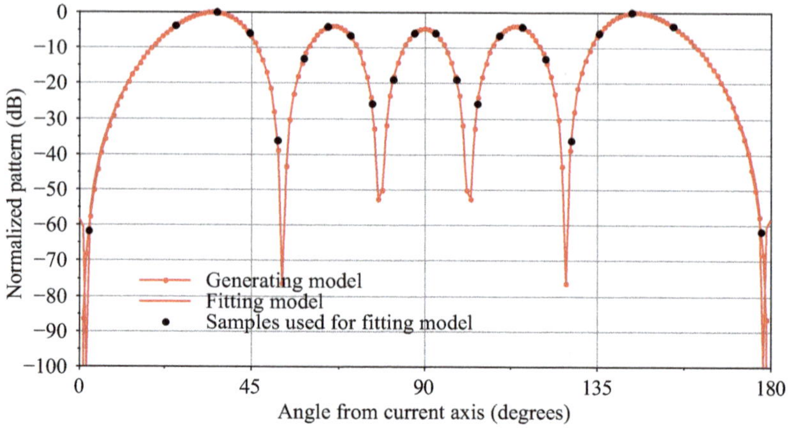

Figure 7.6 A sinusoidal-current filament produces a far-field pattern (in black) that is graphically indistinguishable from an 11-term fitting-model pattern (in red) down to −60 dB beyond which they differ in the near-axial directions. The generating-model samples used to compute the fitting-model parameters are indicated by the black dots

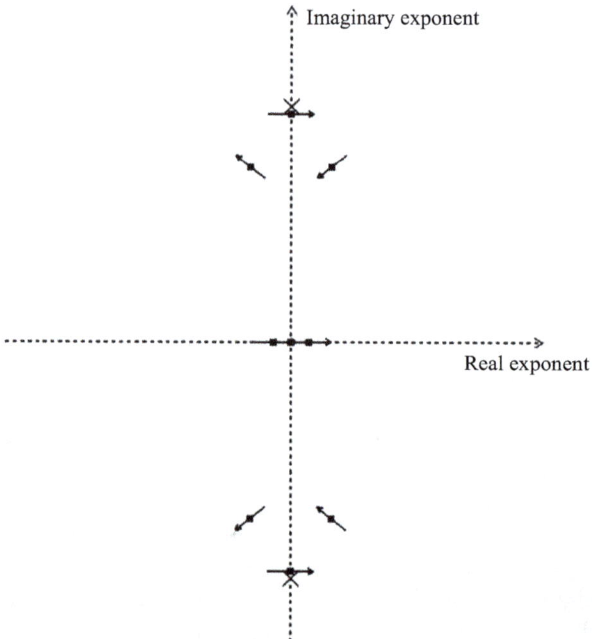

Figure 7.7 Source locations z_α of the fitting model of (7.1) for a sinusoidal current filament 5 wavelengths long using a sampling window extending from −0.999 to +0.999 in $\cos\theta$

that these sources are not isotropic radiators. This implies that the Prony FM would not be realizable from a practical perspective.

7.3.2 The principal-plane pattern of a rectangular aperture

Results are presented in Figures 7.8–7.10 for a 5-wavelength long aperture having a uniform negative electric field on its left half and a positive field on its right half whose pattern factor is given by [12]

$$P_{\pm} = L\left(\frac{1 - \cos\left(\frac{\pi L}{\lambda}\cos\theta\right)}{\frac{\pi L}{\lambda}\cos\theta}\right) \tag{7.6}$$

with θ measured from the plane of the aperture. The radiation pattern in Figure 7.8 is produced by the discrete array whose pole geometry is shown in Figure 7.9. That the separation between adjacent current elements is non-uniform may not be apparent in Figure 7.9, so this derived parameter is plotted in Figure 7.10. The solid circles are plotted halfway between elements whose separation they represent. The element separations are found to vary from about 0.3 to about 0.6 wavelengths between the ends and center of the array. This result emphasizes the fact that the Prony FM is not restricted to uniform arrays. It is interesting to observe in Figure 7.9 that the outermost current elements are not quite as far apart as the 5-wavelength size of the slot.

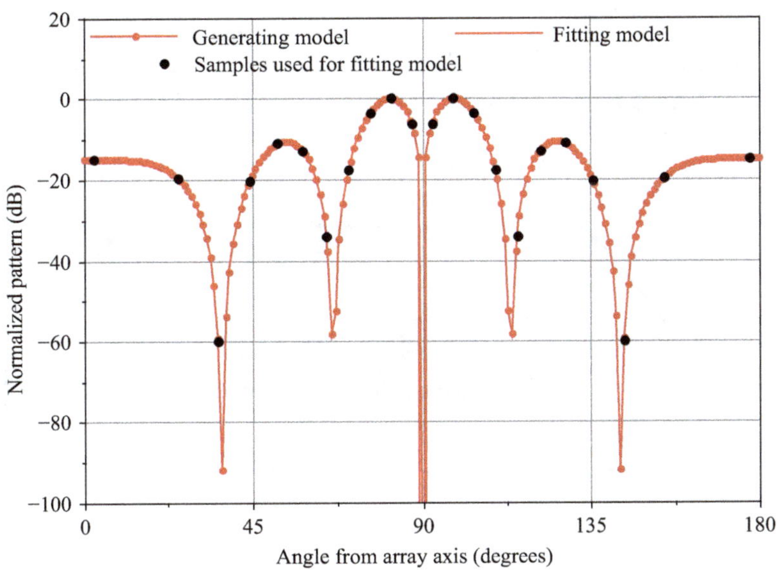

Figure 7.8 A ± square-wave aperture produces a far-field pattern (in black) that is graphically indistinguishable from an 11-term fitting-model pattern (in red). The generating-model samples used to compute the fitting-model parameters are indicated by the black dots

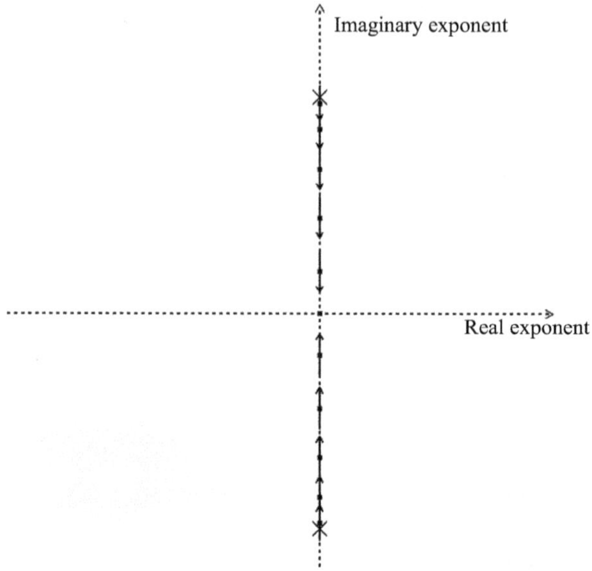

Figure 7.9 *Source locations z_α of the fitting model of (7.1) for a ± square-wave aperture 5 wavelengths long using a sampling window extending from −0.999 to +0.999 in* $\cos\theta$

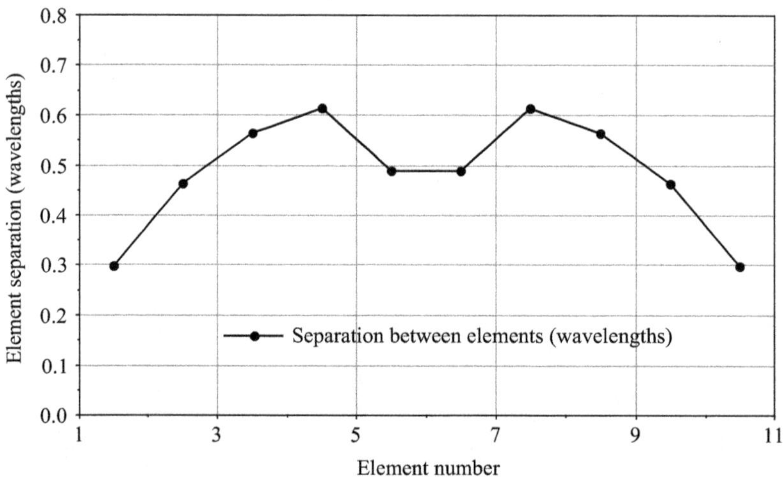

Figure 7.10 *The separations between the 11 elements in the fitting model of Figure 7.9 for a ±square-wave aperture 5 wavelengths long*

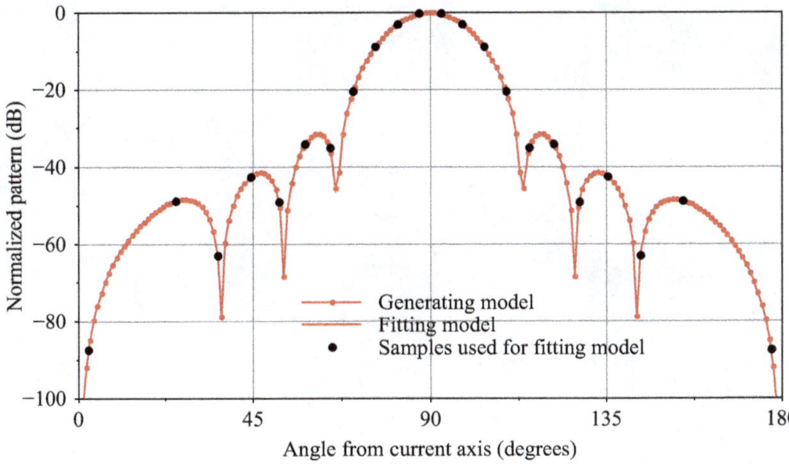

Figure 7.11 The generating model used here is the pattern (in black) of a $\cos^2(\pi z/L)$ current 5 wavelengths long whose 11-term fitting model of (7.1) (in red) is graphically indistinguishable and where the black dots depict the generating samples used for the computation

7.3.3 A cosine-squared current

A cosine-squared current of length L of the form

$$I = I_0 \cos^2\left(\frac{\pi z}{L}\right) \tag{7.7a}$$

has the pattern factor [12]

$$P_{\cos^2} = \frac{\sin(u)}{u}\left[\frac{\pi^2}{\pi^2 - u^2}\right] \tag{7.7b}$$

where $u = (\pi L/\lambda)\sin\,\theta$ results for which are presented in Figure 7.11–7.13. The current length is 5 wavelengths and an FM with $N = 11$ was again used with the sampling window extending from −0.999 to +0.999. The GM and FM are seen to be graphically indistinguishable in Figure 7.11. The corresponding element locations and their separations are illustrated in Figures 7.12 and 7.13.

7.3.4 Modified uniform current filament

A uniform current filament of length L multiplied by $(\sin\theta)^P$ has a pattern factor given by [12]

$$P_{UCF} = (\sin\theta)^P \frac{\sin(kL\cos\theta)}{kL\cos\theta}. \tag{7.8}$$

Results for a 5-wavelength modified uniform current filament are presented in Figures 7.14–7.16 for the pattern for values of P varying from 0 to 3, and the FM

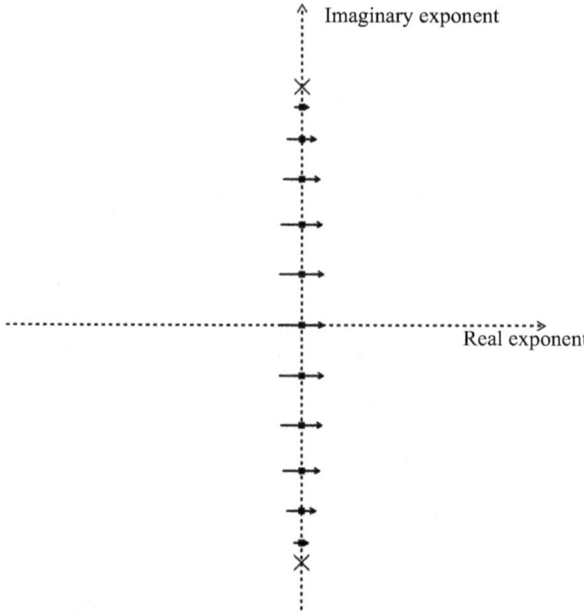

Figure 7.12 Source locations z_α for the 5-wavelength $\cos^2(\pi z/L)$ *current*

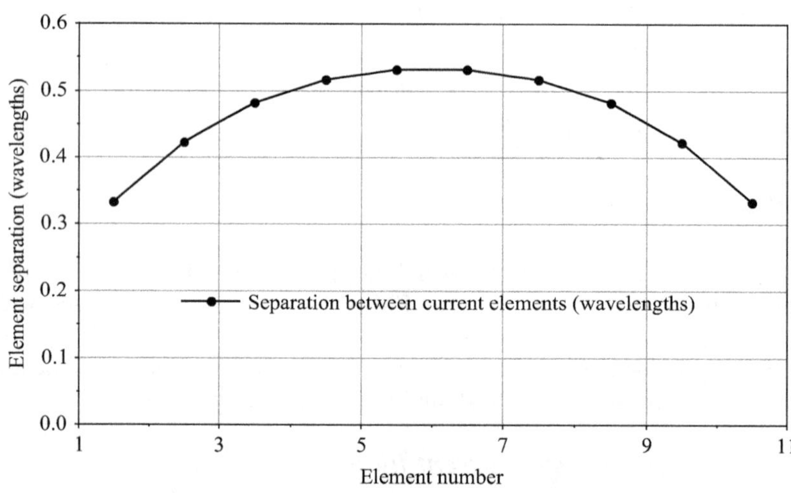

Figure 7.13 The separation between adjacent current elements in the fitting model of the $\cos^2(\pi z/L)$ *current*

sources and their separation for $P = 0$. The patterns in Figure 7.14 are graphically indistinguishable down to −80 dB where a difference occurs in the near axial direction. The nine FM sources are found to be isotropic and of zero phase in Figure 7.15 with their spacings varying from 0.43 to 0.72 wavelengths. It was

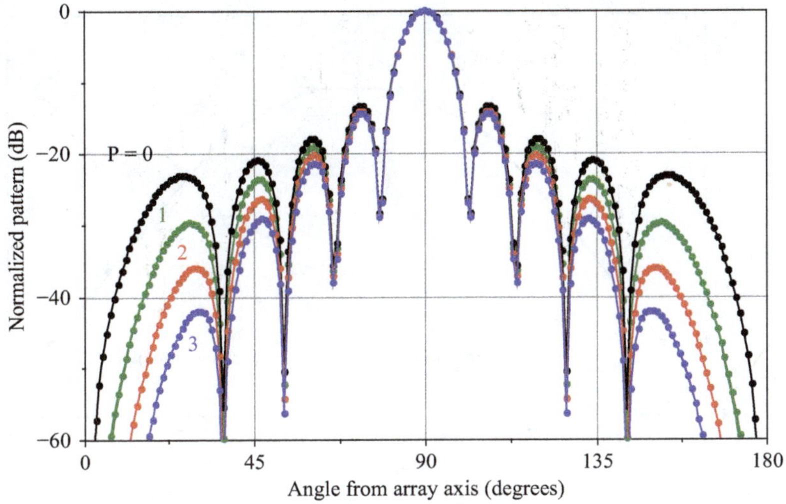

Figure 7.14 The generating model used here are the patterns of a uniform current filament 5 wavelengths and modified uniform current filaments for P = 0–3, long whose 11-term fitting models are shown by the dots

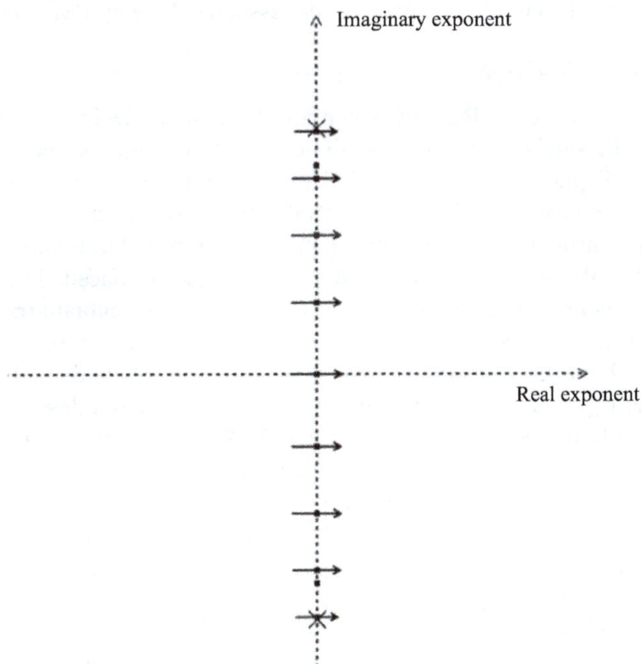

Figure 7.15 Source locations z_α for the P = 0 modified uniform current filament

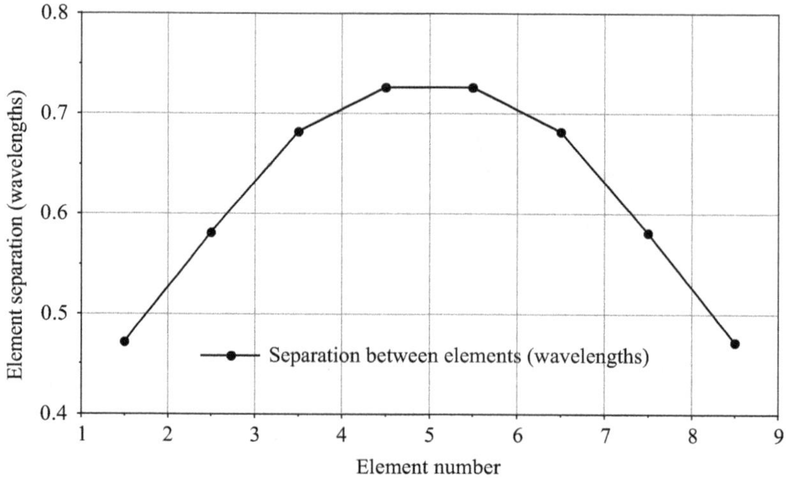

Figure 7.16 The separation between adjacent current elements of the $P = 0$ modified uniform current filament

found, however, that using $P \geq 4$ results in some $z_{a,rea;} \neq 0$, that is, these sources are not isotropic. From that perspective, the associated Prony FM is not realizable.

7.3.5 A Dolph–Chebyshev array

The normalized pattern of a 5-wavelength, 10-element Dolph–Chebyshev array having −26 dB sidelobes from an analytic model [13] and as obtained using the Prony FM is displayed in Figure 7.17. Results from the two approaches are graphically indistinguishable. The Prony model was developed using $N = 10$. The discrete-source array that produces this pattern is illustrated in Figure 7.18. For this example, all of the FM sources are pure real and equally spaced. The large X's on the vertical axis are 5 wavelengths apart, the length of the actual array.

The normalized pattern of a modified Dolph–Chebyshev array 7 wavelengths long having −20 dB sidelobe levels on the left and −40 dB sidelobe levels on the right as obtained using the synthesis approach described in [14] is compared with the Prony FM in Figure 7.19. These two results are again graphically indistinguishable with their current strengths in agreement to 4 or more digits. The Prony FM used here employs 15 elements and the pattern samples indicated by the solid circles of Figure 7.19. In contrast with the Dolph–Chebyshev array of Figures 7.17 and 7.18, however, the source plots in Figure 7.20 are not zero phase.

The somewhat unusual pattern of Figures 7.21 and 7.22 was synthesized also using the approach described in [14] and then modeled using the Prony discrete-source array. The agreement between the Prony FM and the original pattern in Figure 7.21 is comparable to the examples already presented with the GM pattern

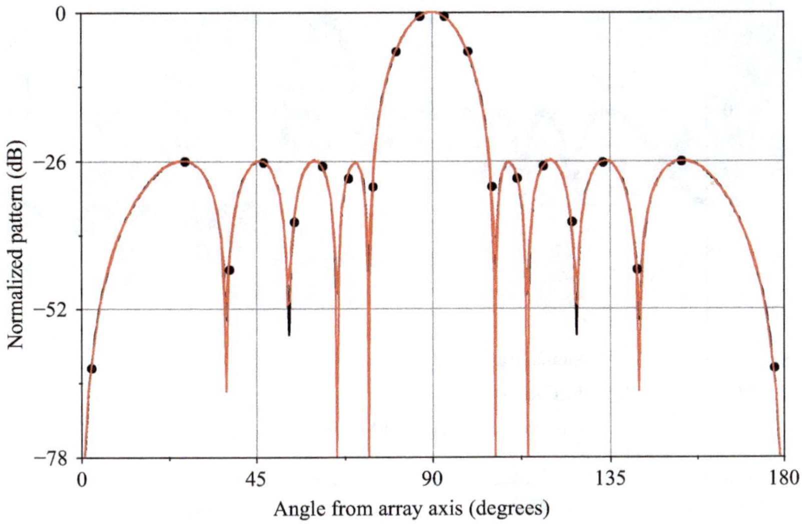

Figure 7.17 The radiation pattern of a 10-element Dolph–Chebyshev array 5 wavelengths long with −26 dB sidelobes as obtained using Prony synthesis

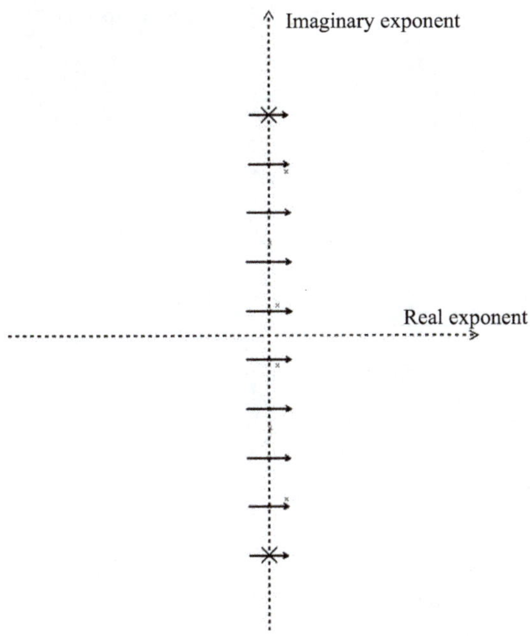

Figure 7.18 Source locations z_α for the 10-element Dolph–Chebyshev array 5 wavelengths long of Figure 7.17

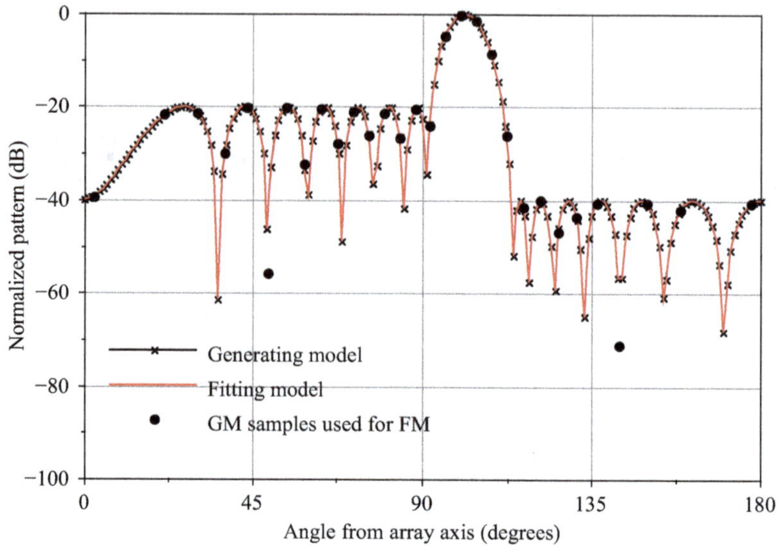

Figure 7.19 Prony model of a modified Dolph–Chebyshev array 7 wavelengths long having 15 current elements

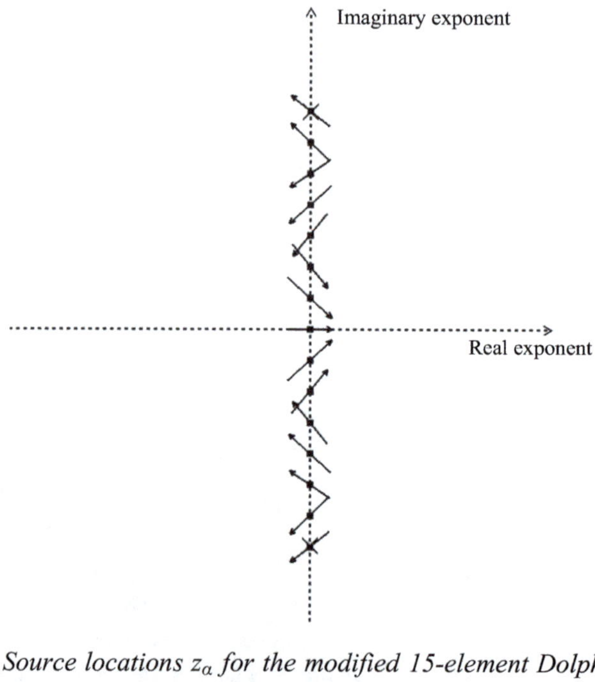

Figure 7.20 Source locations z_α for the modified 15-element Dolph–Chebyshev array 7 wavelengths long of Figure 7.19

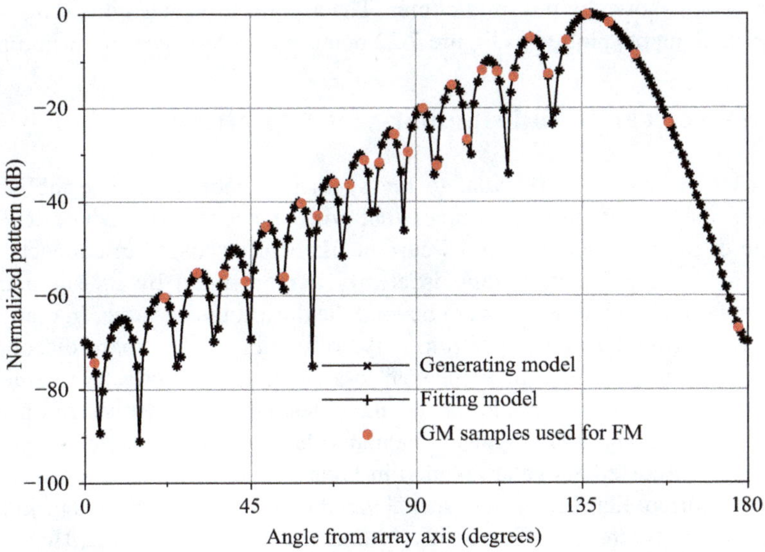

Figure 7.21 *Prony model for a 15-element array 7-wavelength that produces a pattern with lobes that successively increase by 5 dB from a minimum of −70 dB*

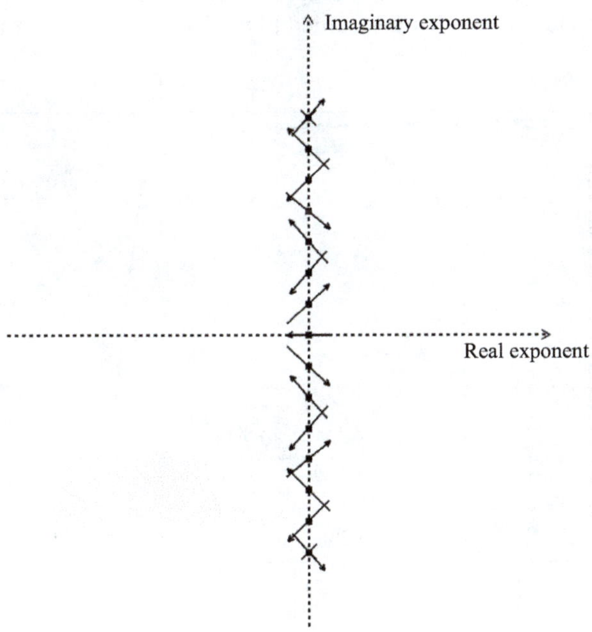

Figure 7.22 *Source locations z_α for the array that produces the pattern of Figure 7.21*

samples again shown by the solid circles. The array is 7 wavelengths long with the 15 current elements plotted in Figure 7.22 being non-zero phase as for Figure 7.20.

7.4 Source ranks and singular-value spectra

The rank of a pattern is relevant to the synthesis problem, as it establishes the minimum number of discrete sources that will be needed to produce it. For the example of the modified sinusoidal current filament discussed in connection with Figures 7.1–7.3, the pattern rank is clearly demonstrated by its singular-value spectrum falling to a numerical zero beyond the third singular value. In many cases, however, the singular-value spectrum may not exhibit such a sharp discontinuity. Then the pattern rank may be somewhat less clear, with its effective value determined by the data precision and the accuracy desired of the synthesized pattern.

As a concluding result, typical singular-value spectra for some of the source distributions discussed above are plotted in Figure 7.23. Three of the spectra are for continuous current filaments whose spatial variation is uniform, sinusoidal, and cosine squared. Their spectra roll off smoothly with increasing singular value. However, the spectrum for the 10-element Dolph–Chebyshev array drops abruptly to the equivalent of a numerical zero when the singular-value number exceeds the number of current elements. The smallest effective singular value of a pattern indicates the number of current elements that are likely to be needed to produce that pattern to some prescribed precision. Pattern spectra thus provide a useful tool for determining the number of discrete sources that are needed to produce a given pattern.

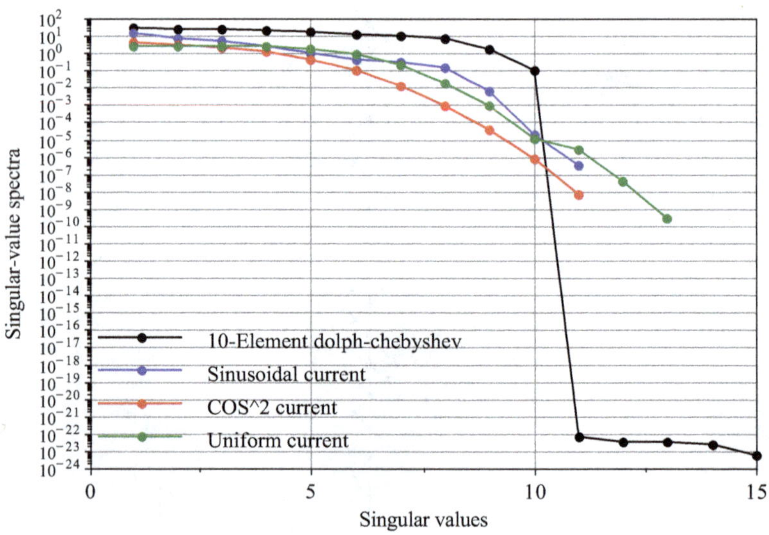

Figure 7.23 Singular-value spectra for the patterns of four different source distributions

7.5 Concluding comments

The use of Prony's method to synthesize discrete-source arrays that replicate the radiation patterns of continuous source distributions such as apertures and filamentary currents and prescribed patterns has been described and demonstrated with several examples. A potential problem in using the approach is that the matrix whose solution is needed to obtain the parameters of the Prony FM may become ill-conditioned. In order to minimize and avoid this possibility, the results presented here have been obtained using 24-digit computer precision.

The synthesized arrays are considered realizable only if their sources are isotropic in that they have numerically insignificant real exponents in the Prony FM, $\sum_{\alpha=1}^{N} S_\alpha e^{kz_\alpha \cos(\theta)}$, that is, $z_\alpha = z_{\alpha,real} + iz_{\alpha,imag} \approx iz_{\alpha,imag}$ Not all patterns and source distributions are found to be realizable, something that cannot be determined without having to perform the computations needed to use the procedure. However, for patterns and source distributions where a realizable discrete-source array can be determined, Prony synthesis offers a useful tool. It is also worth emphasizing that Prony synthesis can yield arrays having non-uniform spacing, a useful feature when grating lobes might otherwise become a problem. Whether a modified version of Prony synthesis might be constrained so as to produce only realizable discrete-source arrays seems worth further investigation.

References

[1] S. A. Schelkunoff, "A Mathematical Theory of Linear Arrays," *Bell System Technical Journal*, Vol. 22, pp. 80–107, 1943.

[2] C. L. Dolph, "A Current Distribution for Broadside Arrays Which Optimizes the Relationship Between Beam Width and Side-lobe Level," *Proceedings of the IRE*, Vol. 34, No. 7, pp. 335–348, 1946.

[3] P. M. Woodward, "A Method of Calculating the Field Over a Plane Aperture Required to Produce a Given Polar Diagram," *Journal Institute of Electrical Engineering, (London), Pt. III A*, Vol. 93, pp. 1554–1558, 1946.

[4] T. T. Taylor, "Design of Line-Source Antennas for Narrow Beamwidth and Low Sidelobes," *IRE Transactions on Antennas and Propagation*, Vol. 7, pp. 16–28, 1955.

[5] Robert S. Elliott, "On Discretizing Continuous Aperture Distributions," *IEEE Transactions on Antennas and Propagation*, Vol. AP-25, No. 5, pp. 617–621, 1977.

[6] O. M. Bucci, M. D'Urso, and T. Isernia, "Optimal Synthesis of Difference Patterns Subject to Arbitrary Sidelobe Bounds by Using Arbitrary Array Antennas," *Microwaves, Antennas and Propagation, IEE Proceedings*, Vol. 152, No. 3, pp. 129–137, 2005.

[7] Y. Liu, Z. Nie, and Q. H. Liu, "Reducing the Number of Elements in a Linear Antenna Array by the Matrix Pencil Method," *IEEE Transactions on Antennas and Propagation*, Vol. 56, No. 9, pp. 2955–2962, 2008.

[8] M. Garcia-Vigueras, J. L. Gomez-Tornero, G. Goussetis, A. R. Weily, and Y. J. Guo, "Efficient Synthesis of 1-D Fabry-Perot Antennas With Low Sidelobe Levels," *IEEE Antennas and Wireless Propagation Letters*, Vol. 11, pp. 869–872, 2012.

[9] E. K. Miller and D. L. Lager, "Radiation Field Analysis and Synthesis Using Prony's Method," *Electronics Letters*, Vol. 14, pp. 180–182, 1978.

[10] E. K. Miller, "Natural-Mode Methods in Frequency- and Time-Domain Analysis," in *Theoretical Methods for Determining the Interaction of Electromagnetic Waves with Structures*, Sijthoff and Noordhoff, pp. 173–212, 1981.

[11] E. K. Miller, "The Incremental Far Field and Degrees of Freedom of the Sinusoidal Current Filament," *IEEE Antennas and Propagation Society Magazine*, Vol. 49, No. 4, pp. 13–21, 2007.

[12] E. V. Jull, *"Radiation from Apertures," in Antenna Handbook*, Y. T. Lo and S. W. Lee (eds.), Van Nostrand Reinhold Company, New York, 1988.

[13] C. Balanis, *Antenna Theory: Analysis and Design*, Harper and Row, New York, 1982.

[14] E. K. Miller, "Synthesizing Linear-Array Patterns Via Matrix Computational of Element Currents," accepted for publication in *IEEE Antennas and Propagation Society Magazine*, 2013.

Chapter 8

Designing discrete arrays using Prony's method to synthesize exponentiated radiation patterns[1]

Abstract

This chapter discusses the rather novel idea of synthesizing patterns that result from exponentiating, or raising to an integer power M, the pattern of a Dolph-Chebyshev (DC) array. The exponentiated patterns retain a constant number of lobes whose amplitudes decrease in dB by an amount proportional to M. The main lobe becomes successively more narrow while the number of array elements proportionally increases. This procedure essentially creates what can be viewed as a new category of the DC array. Singular-value distributions are shown for the patterns of an exponentiated DC and various other patterns. Analytical expressions are developed for a 4-element DC array for comparison with the Prony-derived patterns. The effects of noisy element strengths on exponentiated patterns is also examined.

8.1 Introduction

Prony's method was demonstrated in Chapter 7 to provide a new way of designing a discrete-source array to produce radiation patterns that have specified kinds of sidelobe levels or nulls, or that reproduce the patterns of continuous source distributions. Various applications were discussed there to show the utility of the Prony-based approach. In this chapter, the application of Prony's method is to synthesize discrete-source arrays that generate a given pattern raised to an integer power. Among other possibilities, this procedure leads to a new class of Dolph–Chebyshev (DC) patterns.

8.2 Exponentiating a pattern

The use of Prony fitting models to develop discrete-source arrays to replicate the patterns of various source distributions and prescribed patterns was

[1]Adapted from Edmund K. Miller, A Collection of Presentations in Computational Electromagnetics: "*Model-Based Parameter Estimation*," pp. MBPE 0-95, and "*Pattern Synthesis*," pp. PS 0-56, Amazon ISBN 97963774707434, 2023.

presented in Chapter 7. The possibility of synthesizing discrete arrays that radiate a given pattern raised to a specified power, that is, an exponentiated pattern, is investigated in the following. This approach provides a way to reduce the sidelobe levels and the main beamwidths of a given pattern. It also offers the possibility of extending the performance capabilities of interesting patterns in a straightforward way.

Even if a discrete-source array could produce an exponentiated pattern that has some desirable properties, this would not be a useful concept unless that array can be synthesized. While such an array might not even be unique, all that is required is that it produce a pattern that matches the exponentiated pattern to some desired accuracy. Use of Prony's method for that purpose is developed here.

The basic idea can be best illustrated using a DC array. A good test case is provided by a 10-element array whose pattern is given in (8.1a) [1].

$$\begin{aligned} P_{10}(\theta) = & \ 2.798cos(D) + 2.496cos(3D) + 1.974cos(5D) \\ & +1.357cos(7D) + cos(9D) \end{aligned} \tag{8.1a}$$

where

$$D = [(\pi d/\lambda)\cos\ \theta] \tag{8.1b}$$

with d the element spacing and θ measured from the array axis. The uniform sidelobe level for this array is −26 dB as shown in [2] and repeated here as the $M = 1$ plot of Figure 8.1. Also included in Figure 8.1 are the patterns obtained by raising (8.1a) to integer powers of 2, 3, and 4 where these resulting exponentiated patterns are plotted. An element spacing of 0.5 wavelengths was used or an $M = 1$

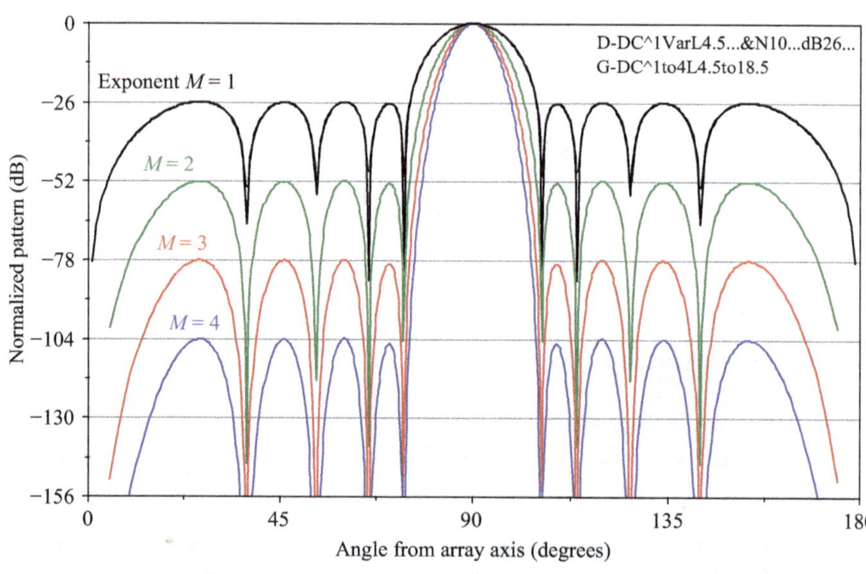

Figure 8.1 Patterns for the array of (8.1) for exponentiation factors of M = 1, 2, 3, and 4

array length of 4.5 wavelengths. Aside from the sidelobes adjacent to the main beam that deviate by a slightly increasing amount with each increase in M, the sidelobe levels of the exponentiated patterns are −52, −78 and −104 dB as required. At this point, these patterns exist only mathematically, as no array that would produce them has been derived.

One way to develop such an array is to employ the synthesis procedure described in [3]. In that approach, the element currents are obtained by inverting a matrix whose coefficients are the contributions made by each current element (initially having amplitudes of unity) to each of the pattern maxima. The inverted matrix is then used to multiply a right-hand-side vector whose entries are the amplitudes specified for the pattern maxima. Since the pattern maxima may move somewhat as the currents change, this approach usually requires a few iterations. Implementing this procedure for the same 10-element array, the results plotted in Figure 8.2 are obtained. The required sidelobe levels are indeed produced as displayed on the graph. Except for this, however, there is no other similarity between the patterns of Figures 8.1 and 8.2. Whereas the main-lobe width in Figure 8.1 decreases with decreasing sidelobe level, the main-beam width in Figure 8.2 increases, resulting in a not very desirable design. The patterns in Figure 8.2 indicate that the array length will have to be systematically increased beyond 4.5 wavelengths with increasing values of M to produce patterns like those in Figure 8.1.

As an alternative, Prony's method as described in [2] was explored to attempt synthesizing arrays that might produce the patterns of Figure 8.1. Samples of the pattern to be synthesized are equated to the Prony series as

$$P_{SYN}(\theta) = \sum_{\alpha=1}^{N} S_\alpha e^{kz_\alpha \cos\theta} \tag{8.2}$$

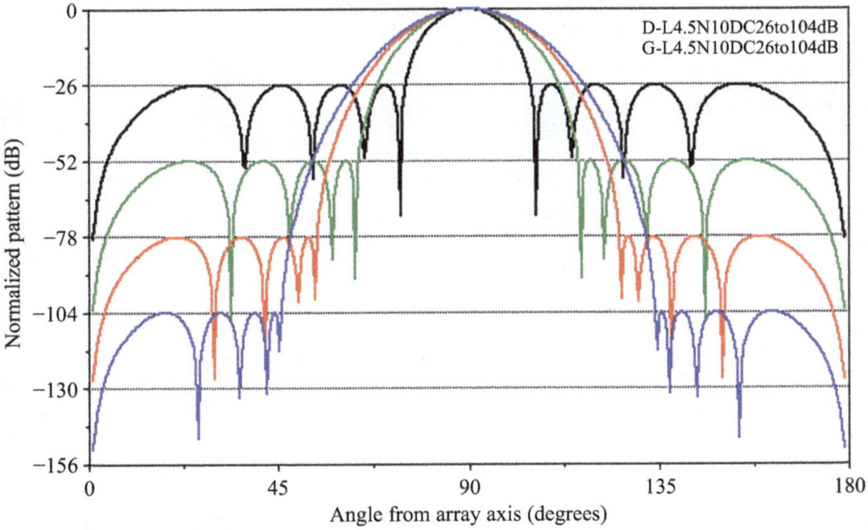

Figure 8.2 The pattern for the array of (8.1) and for successive 26 dB steps of that pattern as obtained using the matrix-synthesis approach [3]

where the Prony parameters are S_α, the strength of current source α, and $z_\alpha = z_{\alpha,real} + iz_{\alpha,imag}$ is its position in complex space. While the imaginary part of the exponent in (2) is thus the location in real space of the source, the real part accounts for the directivity D_α that source might have the variation in angle of which is given by

$$D_\alpha = e^{kz_{\alpha,real} \cos\theta} \tag{8.3}$$

When using Prony's method, it has been observed that the observation variable must be sampled in uniform steps. For the pattern application, this requires sampling in uniform steps of $\cos\theta$ over whatever angular window is used. The reason for this restriction is that the sequence of exponential terms in (8.2) is required to form a polynomial. The rationale for uniform sampling in $\cos\theta$ is illustrated in Figure 8.3 where the far field of a 20-wavelength, uniform-current filament is plotted. The plotting variable in the left half is $\cos\theta$ and the right half is θ. The pattern lobe widths are uniform in the former and vary in width by about a factor of 5 in the latter.

It is convenient to initiate the Prony approach for pattern applications by using a FM having a number of terms on the order of two times the size in wavelengths of the source being analyzed. The original and Prony patterns can be compared and if they do not agree acceptably, one or more terms are successively added to the FM. The computation is repeated until their comparison is satisfactory or it is obvious that convergence is not occurring.

Unless otherwise indicated, the $\cos\theta$ angle window used for the results that follow was usually −0.999 to +0.999. Upon applying Prony's method to the exponentiated patterns of Figure 8.1, the results in Figure 8.4 were obtained. The Prony patterns are graphically indistinguishable from the analytical patterns.

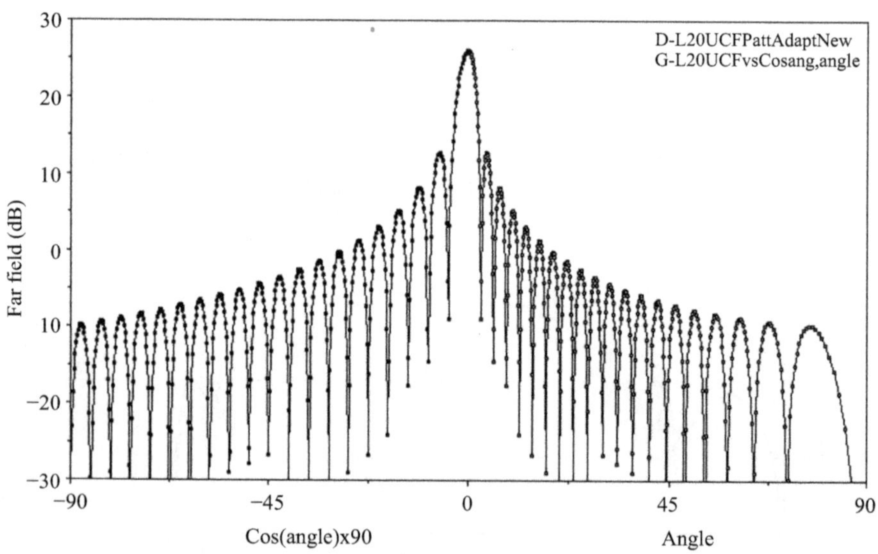

Figure 8.3 The far field of a 20-wavelength uniform-current filament plotted as a function of the cosine of the observation angle on the left as a function of the observation angle on the right

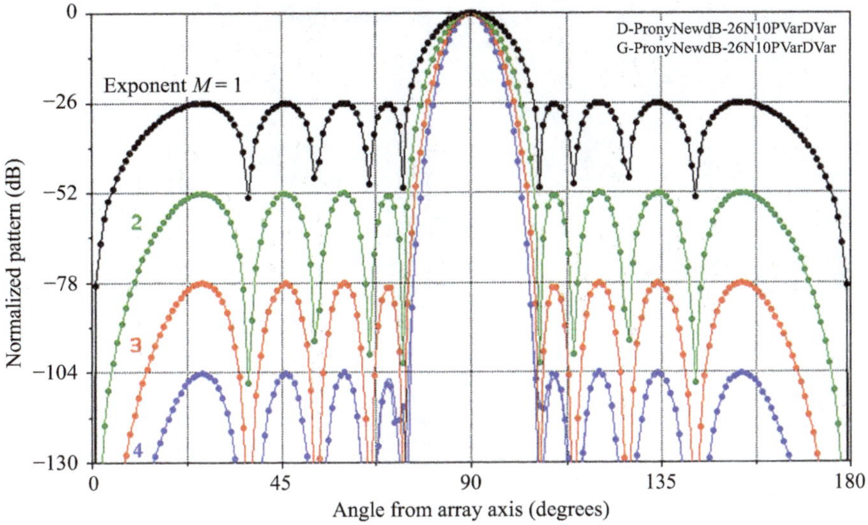

Figure 8.4 The radiation pattern of the 4.5 wavelength, 10-element Dolph–Chebyshev array having −26 dB sidelobes for M = 1 and for exponents of 2, 3, and 4, where the analytic patterns are solid lines and the Prony model results are shown by the dots

These results were developed by increasing N from a starting value of 10 until the Prony patterns matched those of the analytical patterns to within 0.1 dB. The converged Prony array widths for the −26, −52, −76, and −104 dB sidelobe patterns are 4.5, 9, 13.5, and 18 wavelengths long. Their respective numerically significant current elements numbered 10, 19, 28, and 37 spaced 0.5 wavelengths apart. As might be expected, an increasing array length and number of elements are required to produce the lower-level sidelobes that occur with each increase in M while retaining the 9-lobe structure of the starting array.

Plotted in Figures 8.5–8.7 on the complex exponent plane of (8.2) are the geometries of the Prony-derived arrays for the exponentiated patterns of Figure 8.4. The large X's in each plot are 4.5 wavelengths apart to show the physical extent of the $M = 1$ array. The arrow lengths depict the source strengths on a three-decade log scale, with their angles from the real axis showing the source phase. The arrows are arranged with their centers located at $z_\alpha = z_{\alpha,real} + iz_{\alpha,imag}$. The source phases are all zero in each case and their locations are in real space, that is, they are isotropic radiators. It is interesting that while the array lengths increase with M, the basic pattern shape remains the same with the sidelobes decreasing uniformly by 26 dB.

These computations were repeated for an increased element spacing of 0.83333 wavelengths and array length of 7.5 wavelengths to produce the patterns in Figure 8.8. The agreement between the analytical and Prony patterns is similar to that of Figure 8.3. An obvious difference between the 4.5 and 7.5 wavelength results is that that there are now 15 distinct lobes in the larger array compared with

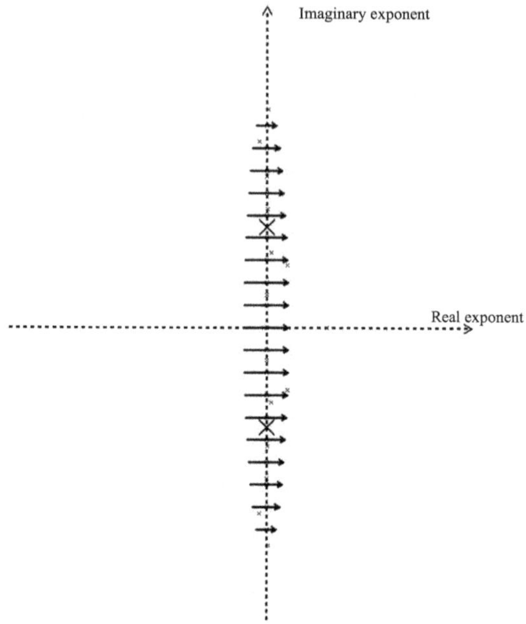

Figure 8.5 The discrete-source array of Figure 8.3 for an exponentiation factor M = 2

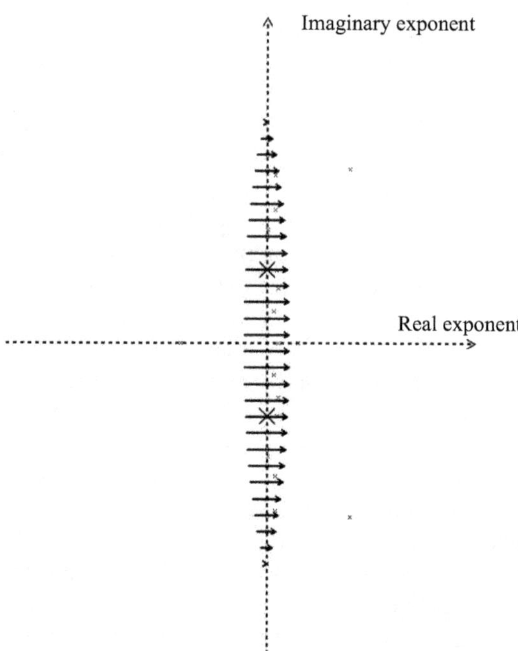

Figure 8.6 The discrete-source array of Figure 8.3 for an exponentiation factor M = 3

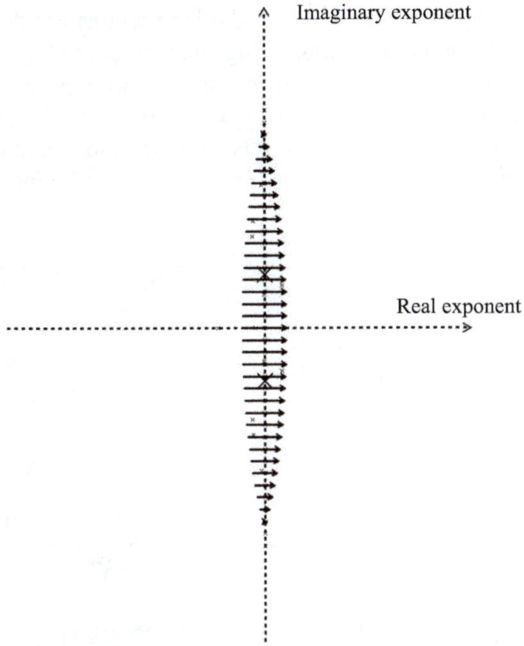

Figure 8.7 The discrete-source array of Figure 8.3 for an exponentiation factor M = 4

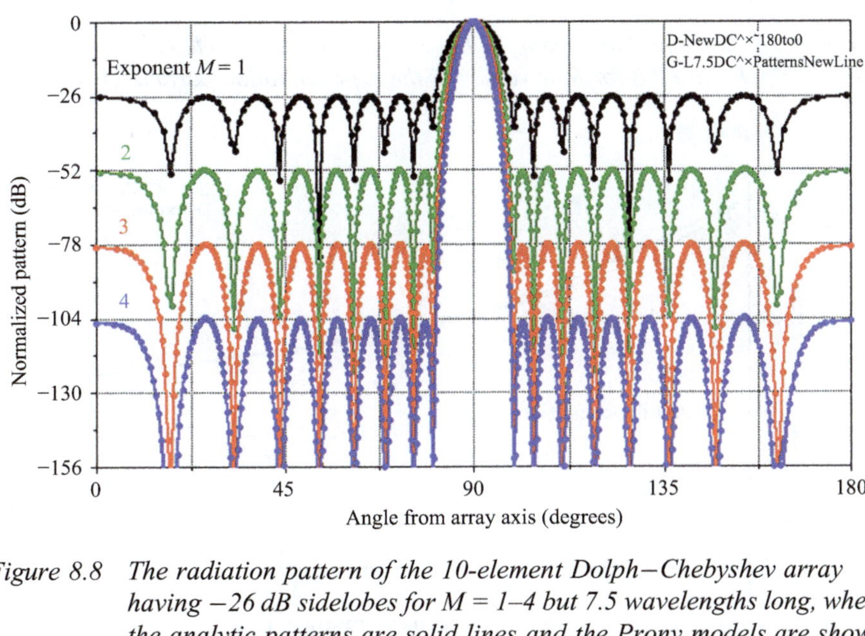

Figure 8.8 The radiation pattern of the 10-element Dolph–Chebyshev array having −26 dB sidelobes for M = 1–4 but 7.5 wavelengths long, where the analytic patterns are solid lines and the Prony models are shown by the dots

the 9 lobes of the original version. All of the lobe widths are decreased proportionately, with the result for the main lobe being summarized in Figure 8.9. The explicit variation of the 3 dB beamwidth as determined from Figure 8.9 is plotted as a function of the exponentiation factor in Figure 8.10 where the main beamwidth is nearly halved as M increases from 1 to 4. It is worth mentioning that in order to obtain converged results for $M = 4$, the compute precision had to be increased to 36 digits.

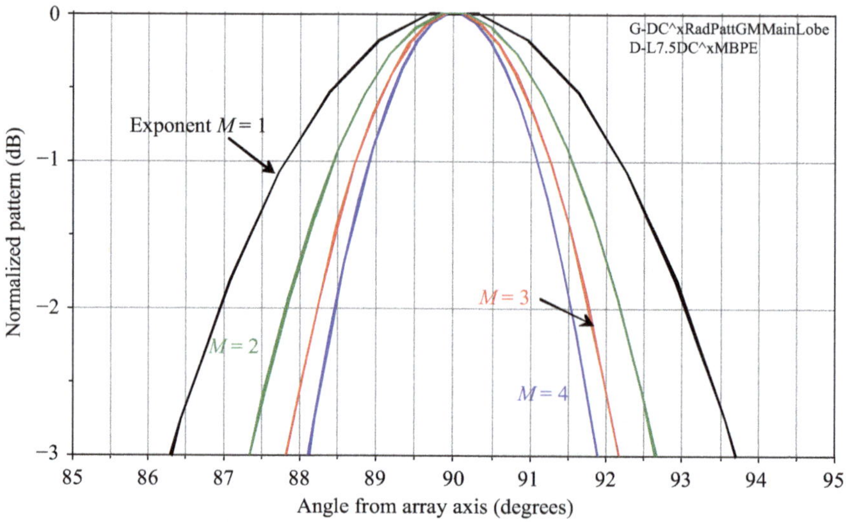

Figure 8.9 The main-lobe beam for the 7.5-wavelength, 10-element array of Figure 8.8 for four values of the exponentiation parameter

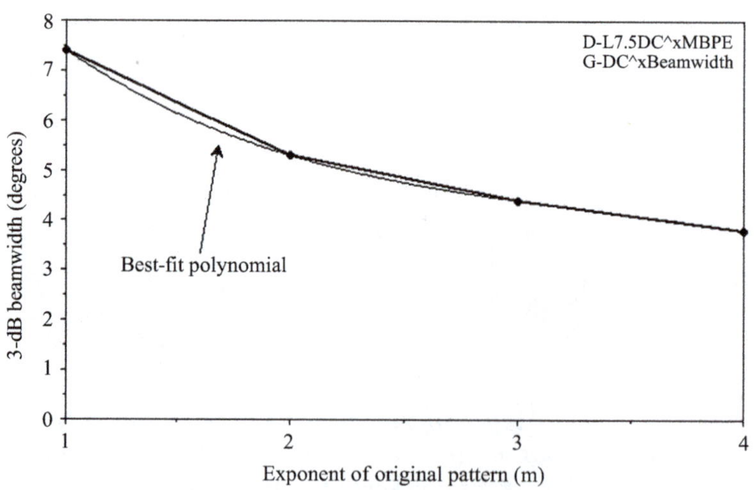

Figure 8.10 The 3-dB beamwidth of the main lobe for the patterns of Figure 8.9

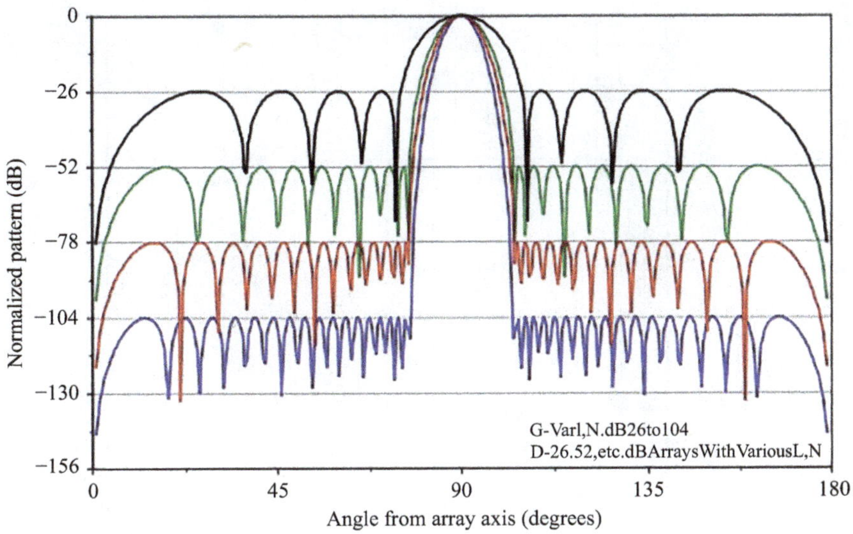

Figure 8.11 The radiation pattern of a 10-element Dolph–Chebyshev array 4.5 wavelengths long for the exponentiation parameter varying from 1 to 4 as obtained using a matrix solution for the element currents [3]

Although the number of sidelobes remains the same for the exponentiated patterns in Figures 8.4 and 8.8, their widths decrease slightly with increasing M. The lobe count would normally be expected to grow in proportion to an increase in array length as is exhibited for the $L = 4.5$ wavelength case in Figures 8.5–8.7. This behavior is demonstrated by the matrix-based synthesis procedure [3] that produced the patterns in Figure 8.11 where the parameters derived from the Prony patterns of Figure 8.3 were used. The array lengths for the Figure 8.11 patterns are 4.5, 9, 13.5, and 18 wavelengths and the corresponding element numbers are 10, 19, 28, and 37. In contrast with the Prony patterns in Figures 8.4 and 8.8, the number of sidelobes in Figure 8.11 increases with each decrease in sidelobe level and corresponding increase in array length. That the Prony-synthesized patterns maintain a fixed number of sidelobes while the array size increases is interesting when compared with the usual DC array.

The parameters of the discrete-source model whose patterns are shown in Figure 8.8 are plotted in Figure 8.12. The horizontal axis displays the spatial locations of the currents in wavelengths, while the vertical axis shows their real amplitudes with the exponentiation factor, M, a parameter. Their imaginary amplitudes are numerically negligible, having maximum values on the order of 10^{-6} of the real amplitudes and so are not shown. The wide dynamic range of the element currents implies the increasing difficulty that would arise in the

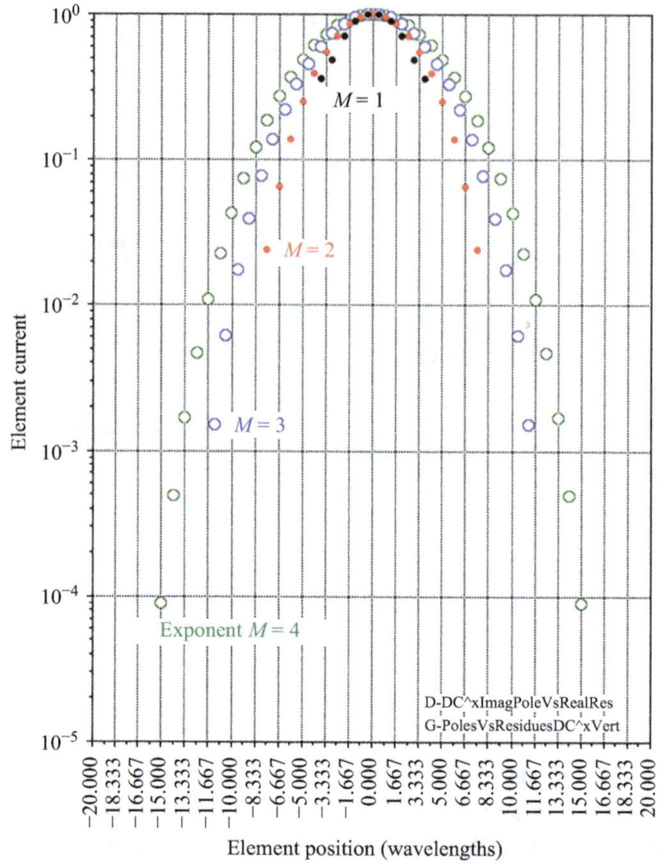

Figure 8.12 The source strength as a function of element position in wavelengths for the patterns of Figure 8. The number of sources is 10, 19, 28, and 37 for M varying from 1 to 4

physical realization of an array having extremely low sidelobes. The element spacings for all four arrays are uniform, which shows that a "standard" DC-type formula could be written for each pattern, that is, for an array having uniform spacing.

The width of the discrete-source array is plotted as a function of the exponentiation factor in Figure 8.13, where this data is a straight line of the form

$$W = 7.5 \times M \tag{8.4}$$

W of a discrete-source array.

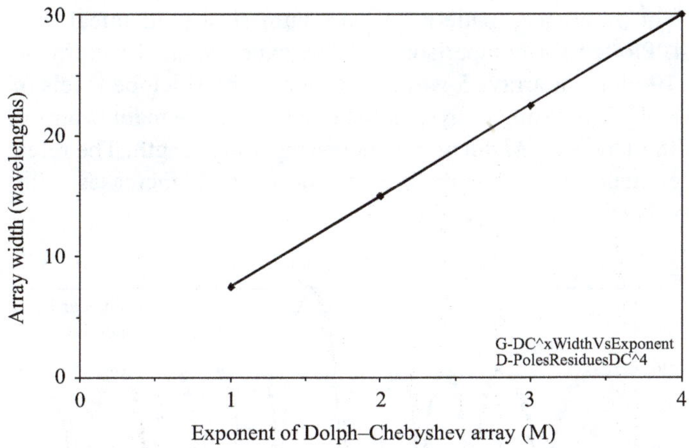

Figure 8.13 The exponentiated array width as a function of the exponentiation parameter M for the 7.5-wavelength array of Figure 8.8

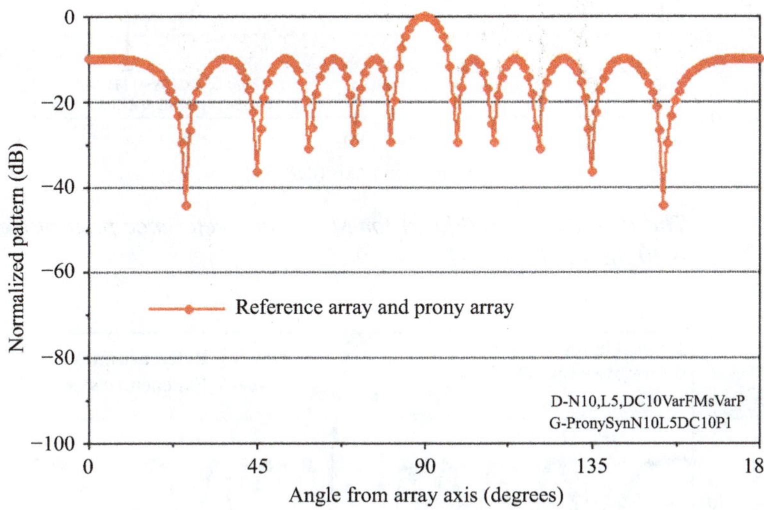

Figure 8.14 The pattern of a 10-element, 5-wavelength Dolph–Chebyshev array [4]

8.3 Comparison of Prony and standard Dolph–Chebyshev arrays for a given sidelobe range

A formula for the pattern of a 10-element DC array 5 wavelengths long having −10 dB sidelobes [4] was used to develop exponentiated patterns with $M = 2–4$ for which the Prony-model patterns are plotted in Figures 8.15–8.17. The Prony and "standard," or reference, pattern [4] are numerically identical for $M = 1$ in Figure 8.14. Plotted for comparison with the exponentiated patterns are reference patterns of 10-element arrays 5-wavelength long with sidelobe levels of −20, −30, and −40 dB [4]. The Prony arrays, on the one hand, have main beams that decrease in width with increasing M due to the increasing array length. The reference arrays, on the other hand, exhibit main beams whose width increases with decreasing sidelobe levels (Figures 8.15 and 8.16).

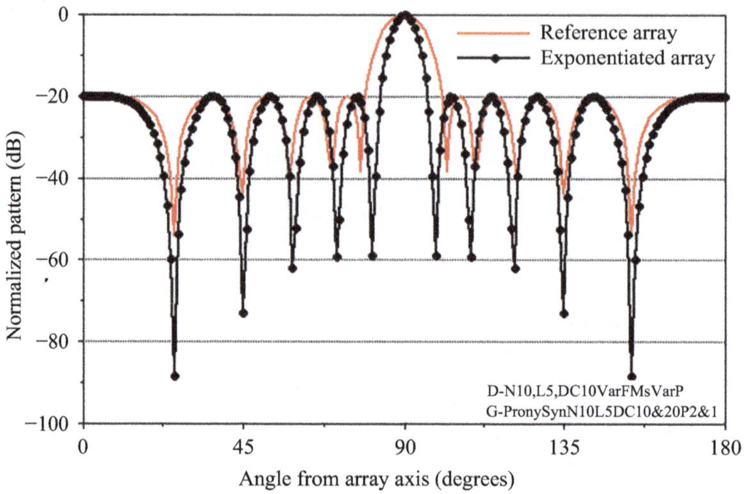

Figure 8.15 The Prony pattern (black) for M = 2 and reference pattern (red) for −20 dB sidelobes [4]

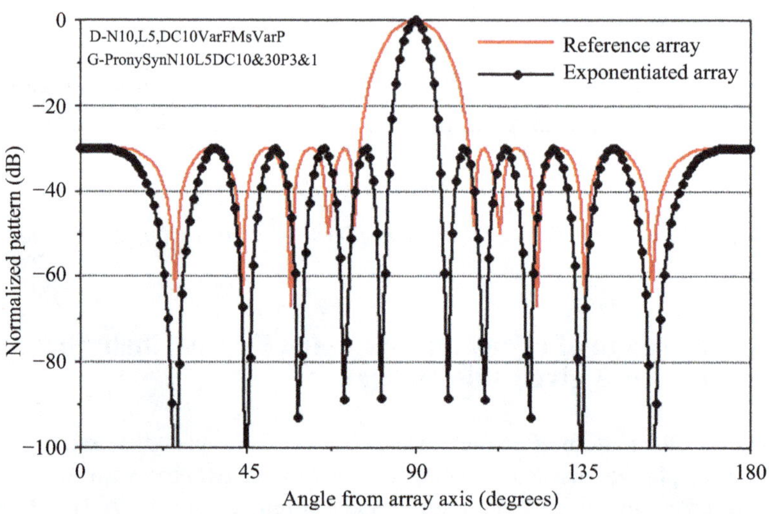

Figure 8.16 The Prony pattern (black) for M = 3 and reference pattern (red) for −30 dB sidelobes [4]

The element separations for these Prony patterns plus an additional Prony pattern (not shown here) for $M = 5$ are plotted in Figure 8.17. The element spacings are equal and uniform for $M = 1$, 2, and 3, but for $M = 4$ and 5, the element spacings are equal only near the end portions of each array. These element spacings increase substantially over their center regions. Thus, while all of these exponentiated patterns have uniform sidelobe levels, the $M = 4$ and 5 cases have element separations that are no longer equal and uniform, an unexpected result (Figure 8.18).

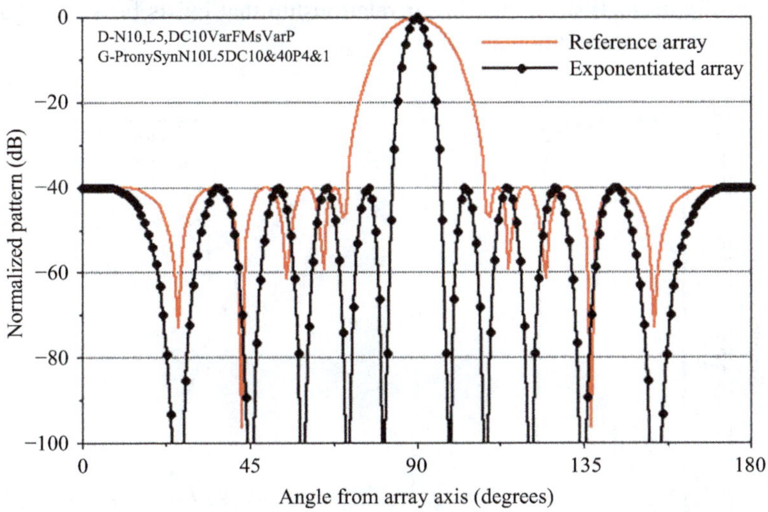

Figure 8.17 The Prony pattern (black) for M = 4 and reference pattern (red) for −40 dB sidelobes [4]

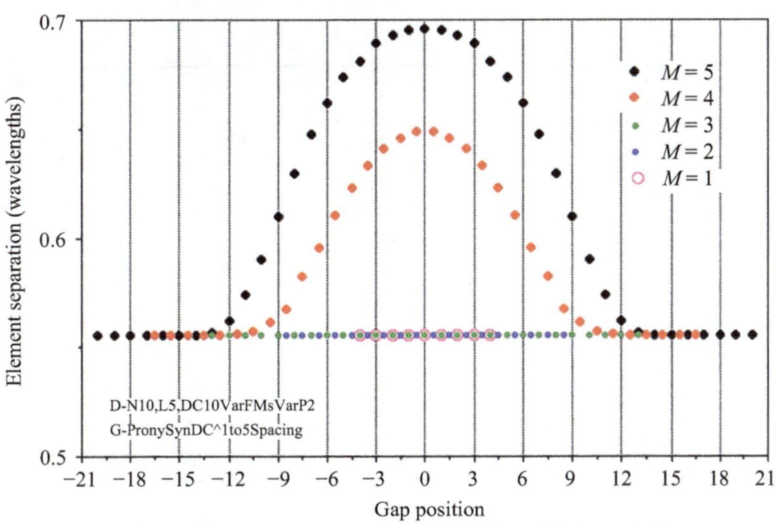

Figure 8.18 The element separations for the Prony models for exponentiation parameter M varying from 1 to 5

8.4 Singular-value spectra of various source distributions

The singular values of radiation patterns can be useful. The singular-value computations shown here were done in 24-digit precision, so that the noise "floors" or zeros for these spectra are on the order of 10^{-24} as displayed in Figure 8.19 for the 7.5-wavelength array patterns of Figure 8.8. The number of array elements for the various values of $M = 1$ is clearly indicated by the non-zero singular values of its spectrum. Figure 8.20 shows the linear relationship that exists between M and the highest-order, non-zero singular value, $SV_{\max}$.

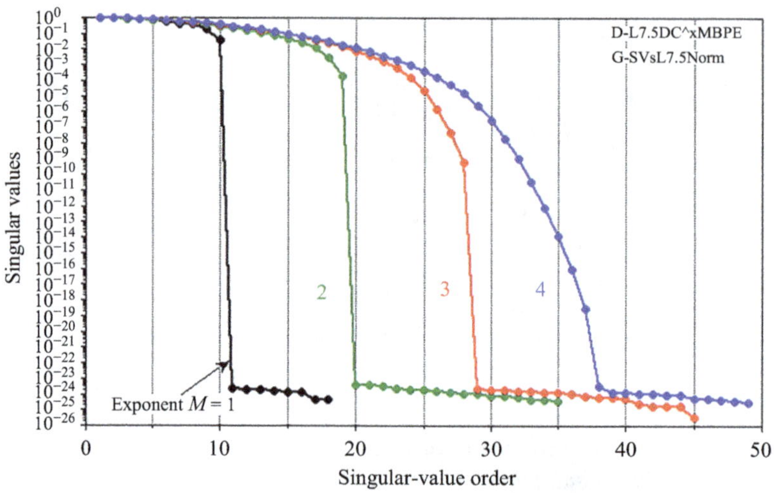

Figure 8.19 Singular-value spectra for the four patterns plotted in Figure 8.8

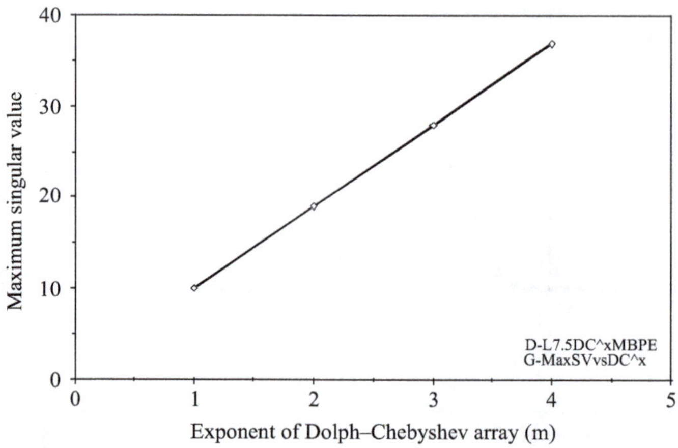

Figure 8.20 The variation of the maximum numerically significant singular values of the patterns of Figure 8.7 with the exponentiation parameter M

The normalized singular-value spectra of patterns computed from (8.1) for a DC array length of 10 wavelengths are plotted in Figure 8.21 with the window width in $\cos\theta$ a parameter. The singular values fall off at an increasing rate as the observation window shrinks. In spite of that, using even the narrowest ± 0.1 window, a Prony array whose coefficients matched those of (8.1) was obtained. However, it would generally be advisable to use an observation window that yields a singular-value spectrum that falls off minimally as for the ± 0.999 window of Figure 8.21.

Normalized singular-value spectra are plotted in Figure 8.22 for three continuous sources 5-wavelengths long where singular values for the 7.5-wavelength DC array are

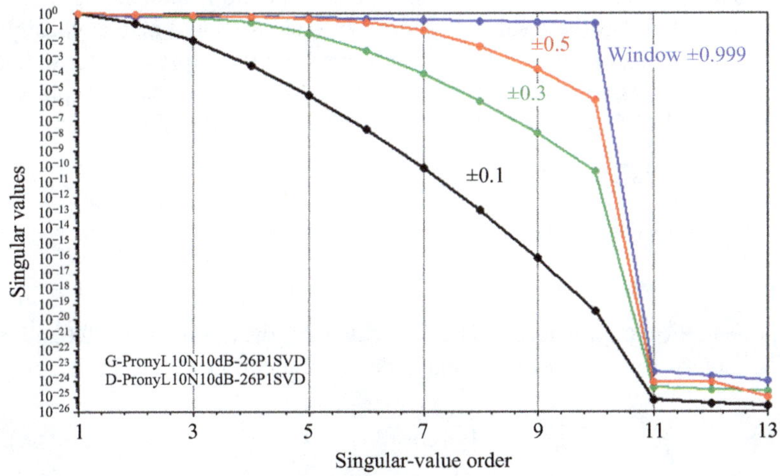

Figure 8.21 Singular-value spectra for a Dolph–Chebyshev array 10 wavelengths long for various values of the cos observation window

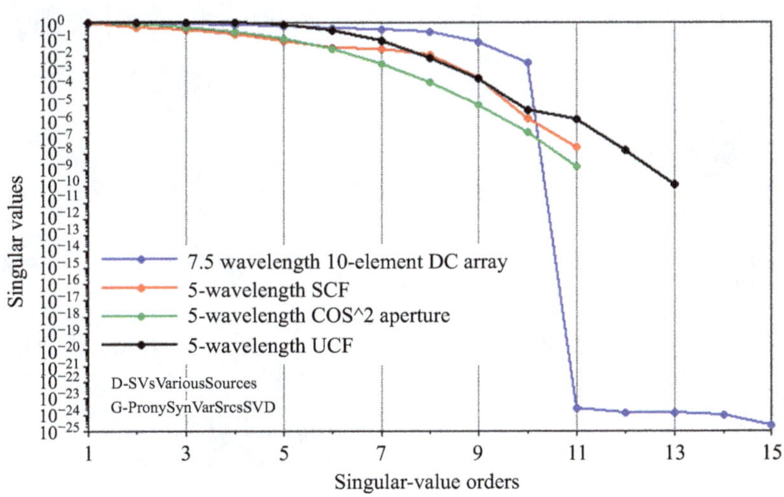

Figure 8.22 Singular-value spectra for a 7.5-wavelength, Dolph–Chebyshev array, and for a sinusoidal current filament, an aperture whose field varies as a cosine squared and a uniform current filament, all 5-wavelengths long

included for comparison. In contrast with the latter plot whose spectrum falls abruptly to the numerical noise level beyond the 10th singular value, the continuous spectra continue to decline smoothly at values well above the noise floor.

8.5 An example of analytically determining the parameters for an array that produces the exponentiated pattern

Various arrays were numerically determined using Prony's method and exponentiated patterns in the previous discussion. When the exponentiation parameter is an integer, it might be expected that the resulting pattern expressions should be analytically determinable. This is because all of the product terms from the expression for the original pattern can be expressed from multiple-angle formulas. As a simple example, consider the pattern of a DC 4-element array having −10 dB sidelobes given by [4]

$$P_4 = A_1\, cos(u) + A_2\, cos(3u) \tag{8.5a}$$

whose exponentiated pattern is then

$$P^M = [A\cos(u) + A_2\cos(3u)]^M \tag{8.5b}$$

for exponentiation parameter M. The element strengths are $A_1 = 0.8794$ and $A_2 = 1$.

The pattern for $M = 2$ can be expanded analytically to yield [5]

$$P_4^2 = \frac{A_1^2 + A_2^2}{2} + \left(\frac{A_1^2}{2} + A_1A_2\right)\cos(2u) + A_1A_2\cos(4u) + \frac{A_2^2}{2}\cos(6u). \tag{8.6}$$

Similarly, the patterns for $M = 3$ and $M = 4$ can be expanded to give

$$P_4^3 = \frac{3}{4}\left[A_1^3 + A_1^2A_2 + 2A_1A_2^2\right]\cos(u) + \frac{1}{4}\left[A_1^3 + 6A_1^2A_2 + 3A_2^3\right]\cos(3u)$$
$$+\frac{3}{4}\left[A_1^2A_2 + A_1A_2^2\right]\cos(5u) + \frac{3}{4}A_1A_2^2(7u) + \frac{1}{4}A_2^3\cos(9u) \tag{8.7}$$

and

$$P_1^2 = \frac{3}{2}\left[\frac{1}{4}A_1^4 + \frac{1}{3}A_1^3A_2 + A_1^2A_2^2\frac{1}{4}A_2^4\right] + \frac{3}{2}\left[\frac{1}{3}A_1^4 + A_1^3A_2 + A_1^2A_2^2 + AA_2^3\right]\cos(2u)$$
$$+\frac{3}{2}\left[\frac{1}{12}A_1^2 + A_1^3A_2 + \frac{1}{2}A_1^2A_2^2 + A_1A_2^3\right]\cos(4u) + \frac{3}{2}\left[\frac{1}{3}A_1^3A_2 + A_1^2A_2^2 + \frac{1}{3}A_2^4\right]\cos(6u)$$
$$+\frac{3}{2}\left[\frac{1}{2}A_1^2A_2^2 + \frac{1}{3}A_1A_2^3\right]\cos(8u) + \frac{1}{2}A_1A_2^3\cos(10u) + \frac{1}{8}A_2^4\cos(12u) \tag{8.8}$$

The numbers of current elements in the above symmetric arrays are 4, 7, 10, and 13. Those having an odd number of elements have a center current, while those with even numbers of elements do not. The current elements of all of the above patterns have strengths determined by only the 2 parameters for $M = 1$.

Patterns for the above expressions are plotted in Figure 8.23, together with results obtained using Prony synthesis. As previously found, these independent results are graphically indistinguishable. As a further check, the element strengths given in (8.6)–(8.8), normalized to their edge-element values and included in Table 8.1, agree to four places with those computed from the Prony models.

The above exercise demonstrates that formulas for the exponentiated patterns can be obtained by analytically manipulating the formula for the base pattern. An alternative approach to obtain formulas for the exponentiated patterns is simply to

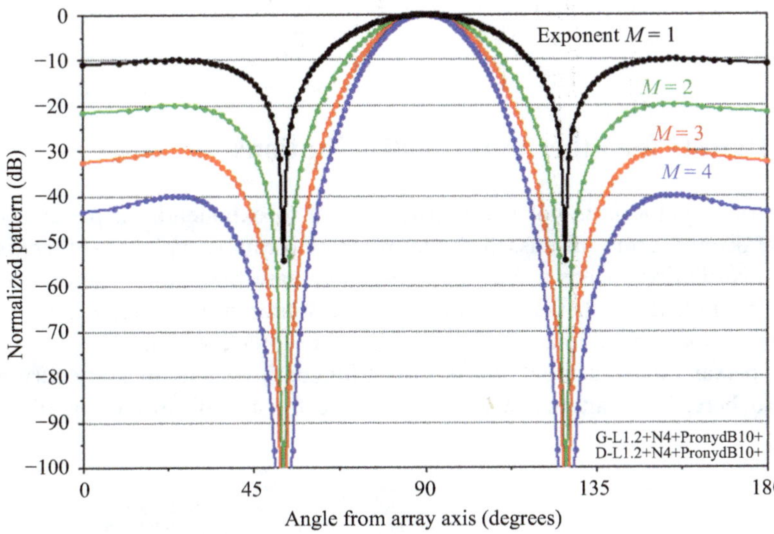

Figure 8.23 Analytical and Prony exponentiated patterns of a 2-wavelength, 4-element, Dolph–Chebyshev array

Table 8.1 Current element strengths for Figure 8.23

Element number	**Basic array 4 elements**	***M* = 27 elements**	***M* = 310 elements**	***M* = 413 elements**
1	0.8794	1.773	9.640	16.79
2	1	2.532	8.320	30.38
3		1.759	4.958	23.95
4		1	2.638	16.00
5			1	8.158
6				3.518
7				1

insert the computed residues and poles into the standard analytical form for the pattern of the DC array. Upon noting that the formulas for odd and even numbers of current elements in an array have the following forms:

$$P(\theta) = \sum_{n=1}^{N} a_n \cos[2(n-1)u], \text{ for } 2N \text{ elements} \tag{8.9a}$$

and

$$P(\theta) = \sum_{n=1}^{N+1} a_n \cos[2(n-1)u], \text{ for } 2N+1 \text{ elements} \tag{8.9b}$$

with

$$u = \frac{\pi d}{\lambda} \cos\theta \tag{8.9c}$$

8.6 Results for other current distributions

Prony-array synthesis is not necessarily limited to exponentiated patterns of the kinds of discrete arrays discussed above. Whether it can be applied to other kinds of source distributions needs to be established on a case-by-case basis. Also needing to be determined is whether a Prony discrete-source array is realizable, that is, whether or not its individual current elements are directive or not.

Two examples of using Prony synthesis for continuous filamentary currents are included here. The patterns in Figure 8.24 are for a uniform current filament 5

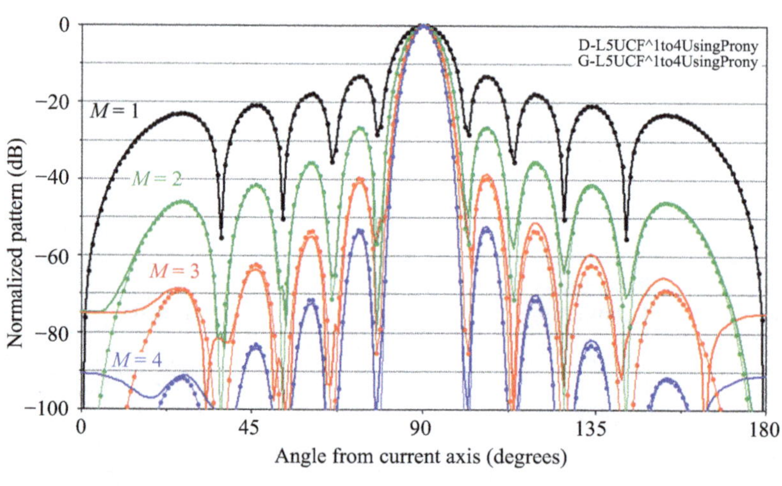

Figure 8.24 Exponentiated patterns for a uniform current filament

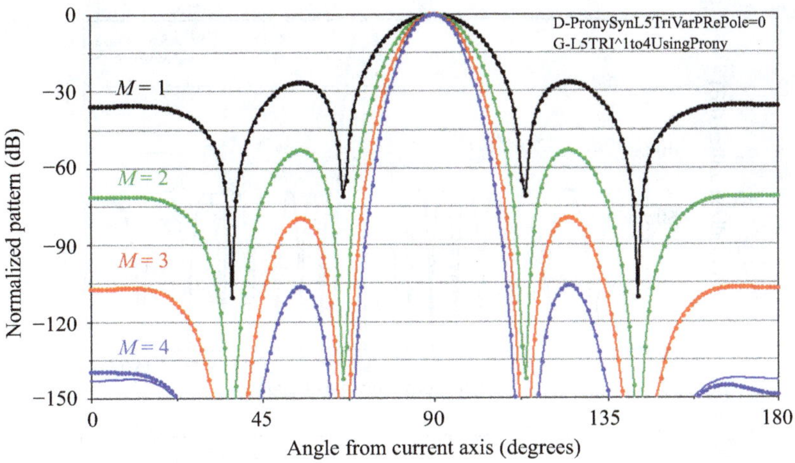

Figure 8.25 Exponentiated patterns for a triangular current filament

wavelengths long with $M = 1$–4. The dotted curves are the analytical patterns and the solid curves are the Prony patterns. For $M = 1$ and 2 they match within a dB or so down to about −65 dB. The Prony patterns diverge more from the actual patterns below −70 dB for $M = 3$ and 4, especially at end fire, but within ±45° of the main beam they exhibit maximum differences of 2 dB or so.

The results of Figure 8.25 are for a triangular current filament. Except for the $M = 4$ exponent where the sidelobe is below −135 dB, the Prony and actual patterns are graphically identical.

8.7 Noise effects

Because increasing the exponentiation parameter M decreases the sidelobe levels, the exponentiated pattern might be expected to become more sensitive to noise in the Prony discrete-array parameters. An example of the effect of noise in the amplitudes of the array strengths is illustrated in Figure 8.26. For a 10% noise level, the sidelobes in Figure 8.26(a) are increased by 12 or 13 dB on average, and the lobe structure effectively disappears. The effect of 1% noise increases the sidelobe levels in (b) and (c) to −60 to −70 dB from the noise-free levels of −78 and −104 dB. Many independent results to evaluate noise effects would be needed to fully assess the limitations that would result from realistic noise levels.

8.8 Concluding comments

Chapters 8 and 9 have outlined and demonstrated various applications of Prony's method to pattern synthesis. In Section 8.1, the primary emphasis was on

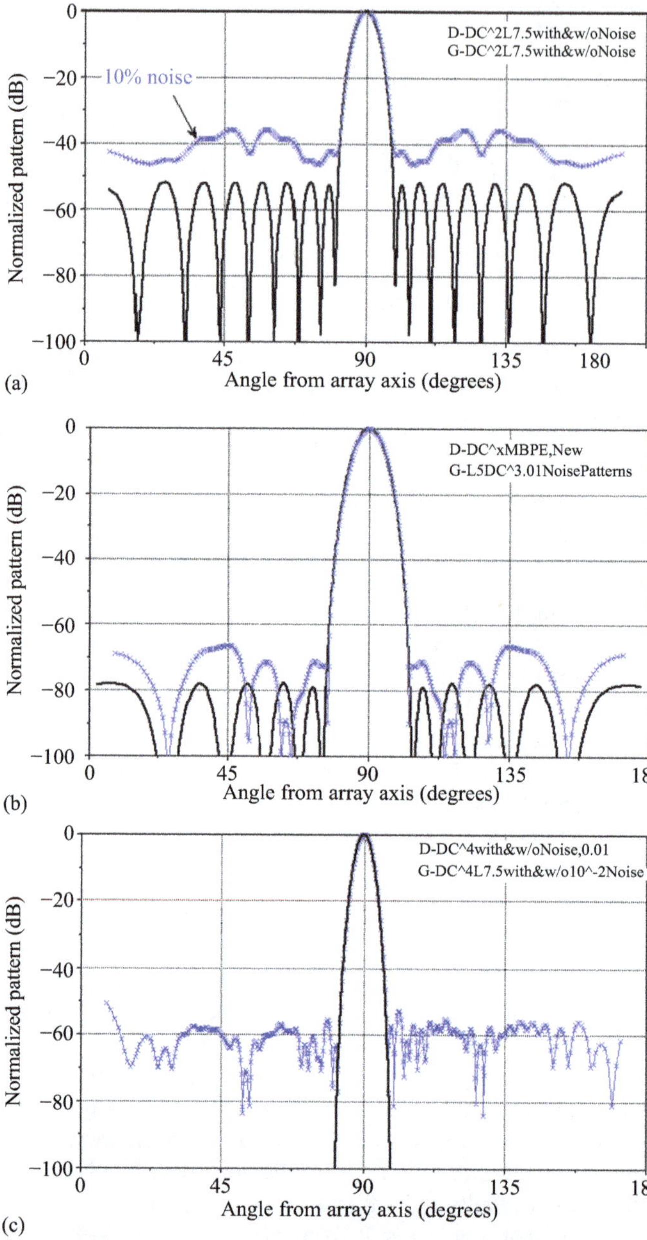

Figure 8.26 The patterns of 10-element, noise-free arrays having −26 dB sidelobes with M = 2, 3, and 4 (in black) and patterns where the source strengths are randomly varied by a maximum of 10% in (a) and 1% in (b) and (c) (in blue)

developing discrete arrays that produce the radiation patterns of continuous source distributions and prescribed radiation patterns. The fitting model involved in this process is a series of exponential terms of unknown amplitudes and exponents whose values are determined using Prony's method. Each exponential represents one current element, or source, in the Prony discrete-source array. If any of the numerically significant Prony sources has a complex exponent, this is equivalent to that source being non-isotropic, or directive. The Prony-derived array is considered non realizable in that case.

In Chapter 8, the focus has been on deriving discrete arrays that replicate radiation patterns that come from exponentiating a given pattern to integer powers of 2 or greater. For each increase in the exponentiation parameter, the sidelobe levels in dB decrease by the amount of their levels in the original array while the length of the synthesized array increases in length by that of the original array. When applied to DC arrays this procedure produces what can be regarded as a new class of DC arrays having fewer sidelobes than a standard design of the same size. In some cases, the elements in the synthesized arrays are uniformly spaced, while in some others this is not the case. It was also shown that for a simple DC array the exponentiated pattern can be derived analytically the parameters of which match the Prony model to several decimal places.

While most attention in Chapter 8 has dealt with DC arrays some continuous source distributions were modeled as well. Also discussed was analysis of the patterns being modeled using singular-value decomposition. For discrete-source arrays, the singular-value spectra are found to fall off abruptly beyond the number of resolvable sources. For continuous sources, the spectra fall off more smoothly. In either case, such spectra are useful to determine how many discrete sources will be needed to produce the synthesized pattern.

Not all patterns can be successfully modeled with Prony's method. There are at least two different outcomes where this can occur. One possibility was mentioned above where it can happen that the Prony discrete-source array is not realizable an example of which is provided by the simple sinusoidal current. An example of the second is the pattern of the traveling-wave current for which a converged result cannot be derived. Finally, all of the cases above were performed using integer exponentiation parameters. Some results have been obtained using a non-integer parameter such as 1.5. They do not converge as well and require more discrete sources than when using a smaller integer parameter for the same array.

References

[1] C. Balanis, *Antenna Theory: Analysis and Design*, Harper and Row, New York, 1982.

[2] Edmund K. Miller, "Using Prony's Method to Synthesize Discrete Arrays for Prescribed Source Distributions and Exponentiated Patterns," *IEEE Antennas and Propagation Society Magazine*, 57 (1), February, pp. 147–163, 2015.

[3] Edmund K. Miller, "Synthesizing Linear-Array Patterns via Matrix Computation of Element Currents," *IEEE Antennas and Propagation Society Magazine*, 55 (5), October, pp. 85–96, 2013.
[4] A. Safaai-Jazi, "A New Formulation for the Design of Chebyshev Arrays," *IEEE Trans- Actions on Antennas and Propagation*, **AP-42**, Vol. 3, pp. 439–443, 1994.
[5] G. J. Burke, *Private Communication*, 2013.

Chapter 9

Using adaptive estimation to minimize the number of samples needed to develop a radiation or scattering pattern to a specified uncertainty[1]

Abstract

Obtaining far-field patterns in electromagnetics or acoustics, although generally not as computationally expensive as solving for the sources induced on an object, can nonetheless at times be a substantial part of the overall computer time required for some problems. This can especially be the case in determining the monostatic radar cross section of large objects, since the current distribution must be computed for each incidence angle, or when computing the radiation patterns of large reflector antennas using physical optics. In addition, when employing the point sampling and linear interpolation of the far field that is most often used to develop such patterns, it can be necessary to sample very finely in angle to avoid missing fine details such as nulls.

A procedure based on model-based parameter estimation is described here that reduces the number of samples needed while developing an easily computed and continuous representation of the pattern. It employs windowed, overlapping fitting models whose parameters are estimated from the sparsely sampled far-field values. The fitting models themselves employ either discrete-source approximations to the radiating currents or Fourier models of the far field. For the cases investigated, as few as 1.5–2 samples per far-field lobe are found to be sufficient to develop a radiation-pattern estimate that is accurate to 0.1 dB, and 2.5 samples per lobe for a simple scatterer. In general, however, the sampling density is not determined by the lobe count alone, but by the effective rank of the field over the observation window, which in turn is a function of both the aperture size and the spatial variation of the source distribution within that aperture.

9.1 Introduction

An important goal of all numerical modeling is that of minimizing the number of samples needed of the relevant observables and equations so as to minimize the

[1]Adapted from E. K. Miller (2002), "Using Adaptive Estimation to Minimize the Number of Samples Needed to Develop a Radiation or Scattering Pattern to a Specified Uncertainty," *Applied Computational Electromagnetics Society Journal*, Vol. 17, No. 3, pp. 176–186.

total computer operation count (OC)[2] (or the computer cost) while achieving a desired accuracy. Achieving a desired accuracy is equivalent to reducing the uncertainty in the computed results to a specified level. This topic is considered below in the context of computing radiation and scattering patterns in electromagnetics. The goal of particular interest here concerns minimizing the number of far-field pattern samples that are required to represent a radiation pattern and/or the number of incidence angles that are required to develop a monostatic backscatter radar cross section (RCS) pattern. The procedure used to do this is model-based parameter estimation (MBPE) [1], which has the added benefit of obtaining an estimate of the uncertainty in the pattern that is developed.

MBPE is a procedure that uses physically based fitting models (FMs) whose parameters are computed from samples of first-principles generating models (GMs) such as Maxwell's equations. Computation of a FM sample is trivial compared with one evaluated from a GM for large, complex problems, potentially requiring orders of magnitude less computing time. The FM makes possible development of an essentially continuous and accurate pattern as opposed to the pointwise result that is usually accepted when the cost of each GM evaluation is large.

9.2 Background

One of the most frequently encountered problems in electromagnetic field computations is that of determining a radiation or scattering pattern from a known current distribution over some surface. For antenna applications, usually a single current distribution is of interest, while for RCS computations, a new current distribution occurs for each incidence angle of a plane-wave exciting field. In either case, the far field is usually needed at enough points to develop a smooth, in effect continuous, approximation of the overall pattern in one or more planes. The required number of radiation-pattern samples and resulting OC are proportional to the maximum body dimension, L, in the plane in which the pattern is being computed. Furthermore, the number of current samples on the surface, S, is itself proportional to the body's surface area, that is, $S \propto L^2$. As L, and therefore S, increases, the far-field computation can become a significant component of the cost in obtaining radiation and scattering patterns. Thus, reducing the number of angle samples could be effective for reducing the overall computer cost of obtaining the radiation pattern(s) of a large antenna or RCS of a large scatterer.

Some previous work by the author [2,3] and others [4,5] has described an approach that uses MBPE to decrease the number of samples that are needed to determine a far-field pattern. The work presented in [2,3] is briefly summarized and extended here, especially with respect to how the errors in the pattern can be estimated and controlled and how the pattern itself is modeled. Further examples of efficient pattern computation can be found in the work of Bucci and his various

[2]For convenience a list of acronyms used in this chapter follows the references.

collaborators who have developed signal-processing-like techniques for computing far-field patterns [see for example 6–8].

9.3 Choosing the quantity to model

Estimating a far-field pattern using MBPE requires choosing the kind of FM that is to be used and to what observable that FM is to be applied. Two obvious choices for the relevant observable are to model: (1) the source distribution, whether a current or aperture field; or (2) the far-field pattern itself. Furthermore, there are at least two ways in which the current distribution might be modeled. These various choices are briefly summarized below.

9.3.1 Modeling a current or aperture distribution

The two approaches to be discussed here for modeling a source distribution can best be described as discrete-source approximations (DSAs). The current over the surface of the object under consideration or field over an aperture is replaced by a linear array of discrete, or point, sources aligned in space as described below. (Other point-source geometries might also be used; attention is limited here to a linear geometry.) The parameters of the DSA model are obtained by fitting it to far-field samples obtained from the usual integration of the actual current distribution. The DSA is then used to approximate the pattern between these samples to thereby obtain a continuous pattern estimate without requiring additional current integrations.

There are two different candidate DSA geometries that might be considered. In the first, shown in Figure 9.1(a), the DSA axis is fixed and a sequence of angle windows are rotated about this axis over the complete range of observation angles of interest. In the second, shown in Figure 9.1(b), the DSA axis is itself rotated to be perpendicular to the angle that defines the midpoint of each angle window used for the successive DSA computations. Note that *L'* varies in proportion to the length of the object as seen from the center of the observation window when using the approach of Figure 9.1(b), whereas *L* is fixed at the maximum linear dimension of the object in the observation plane for approach 1a.

9.3.1.1 Using a Prony model

Prony's method (or its equivalent) can be used for the DSA computation, using as an FM [6]

$$F(\theta) = \sum_{\alpha=1}^{N} S_\alpha e^{ikd_\alpha \cos\theta} \tag{9.1}$$

which involves *N* point sources of strengths *S* located at positions *d* along the axis of the DSA array, with θ the angle to the far field measured from that axis. Since there are 2*N* unknown parameters in (9.1), the Prony DSA (called hereafter the Prony FM) requires 2*N* far-field samples of the actual pattern. Furthermore, these

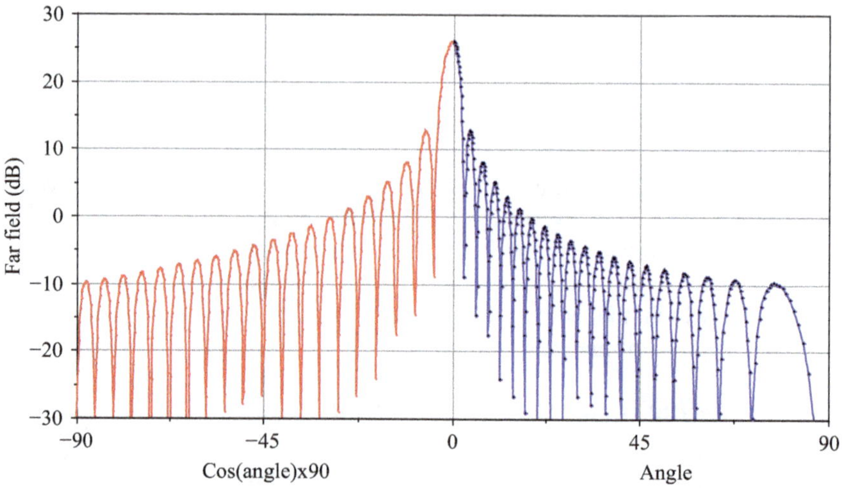

Figure 9.1 This plot of the far field of a uniform current filament demonstrates the equal spacing of the sidelobes as a function of the cosine of the observation angle (the left-hand plot) as opposed to the sidelobe variation as a function of the observation angle itself (the right-hand plot)

$2N$ samples need to be spaced uniformly in $\cos\theta$, a feature that makes the Prony FM less suitable for an adaptive-sampling approach such as described next. A possible advantage, however, of finding the locations of the discrete sources is the possibility of imaging the source whose far field is being sampled. A previously shown graph is again included here in Figure 9.1 to emphasize the distinct difference between angle sampling in θ and sampling in $\cos\theta$.

9.3.1.2 Using a specified discrete-source approximation

The specified DSA, or Residue FM is the same as (9.1) except that, since the source locations are now α specified, only the N source strengths S, or residues, are unknowns.

Samples of the actual far field are used only to obtain the discrete-source strengths. These pattern samples can be arbitrarily located in angle and only N are needed, two distinct advantages over the Prony model. These sources are equally spaced along the array axis, with numbers 1 and N located at the ends of the aperture L using the configuration shown in Figure 9.2(a). As the number of sources used in a particular Residue FM is increased, the source spacing is therefore systematically decreased in proportion to the number of sources that are used.

9.3.2 Modeling the pattern

An alternative to using a DSA for the pattern estimation is to model the pattern itself, using a Fourier series, an approach denoted here as the Fourier FM. In this

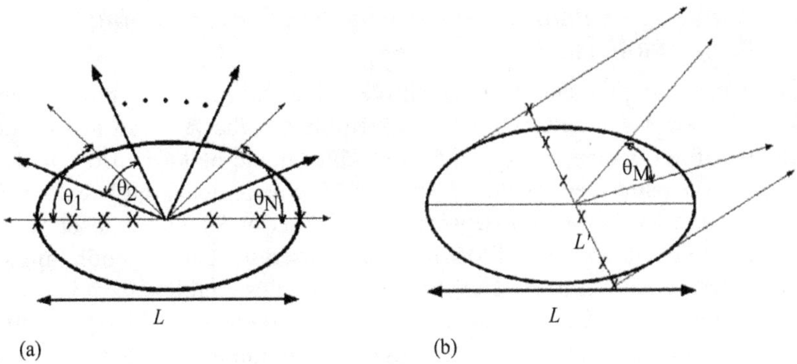

Figure 9.2 *Two ways of implementing the discrete-source array fitting model. In (a) the sampling window (denoted by alternately light and heavy lines) rotates in angle about the long axis of the object while the DSA FM axis remains fixed along which the discrete sources are located. In (b), on the other hand, the sampling aperture L′ is rotated with respect to the long axis of the object and the sampling window is bisected by a line perpendicular to it. For the Prony FM, an N-term FM requires a minimum of 2N field samples. Both the source locations (indicated by the X′s) and their strengths are determined by sampling the far field as a function of cosθ, where generally the sources will be non-uniformly spaced. Alternatively, for the Residue FM used to obtain most of the results presented here, the configuration (a) was used with N sources uniformly spaced along the array axis, thus requiring only N field samples for computation of their strengths.*

case, the FM is developed as [2]

$$F(\theta) = \sum_{\alpha=S}^{F} R_{\alpha} e^{i\alpha\theta} + \sum_{\alpha=S'}^{F} R'_{-\alpha} e^{-i\alpha\theta} \tag{9.2}$$

where F is set to Int($L + 1$), with Int(X) denoting the value of X rounded off to the nearest lower integer. Also, $2F - S - S' = N$ with N the total number of α terms in the FM, $S \leq S' \leq S + 1$, and R and R' are the amplitudes of the positive and negative modes respectively. As for the Residue FM, N far-field samples are also required to quantify the parameters of the Fourier FM. Note that the Fourier approach yields estimates of the far-field Fourier components directly. This is in contrast with either the Residue FM where the source strengths, and for the Prony FM the source locations as well, are the parameters being computed. Ultimately, however, the differences between the Prony and Residue FMs and the Fourier FM are rather slight, differing most in the former having an observation-angle dependence given by $e^{ikd\cos\theta}$ while the latter depends on $e^{i\alpha\theta}$.

9.3.3 Adaptive estimation using windowed, overlapping fitting models

To be effectively implemented, any adaptive numerical procedure requires an error estimate. The error estimate is used to determine whether a given FM is accurate enough, and if not, where in angle a new sample should be located for the specific case of adaptive pattern estimation. Various FM arrangements might be considered for error estimation. For example, with an initial set of S angles chosen to sample the GM, a single fitting model FM(S) might be used to span the entire angle window of interest. A subset of $S-1$ of these GM samples could be used to obtain the $S-1$ parameters of FM($S-1$). These two FMs could then be sampled more finely in angle, by a factor of 10 or so, than was used for the initial GM sampling, with the difference between them serving as an error measure. If the error measure exceeds the allowable specified uncertainty, GM sample $S+1$ would be added at the angle where the difference between FM($S-1$) and FM(S) is greatest. Using this new GM sample, the parameters of FM($S+1$) could then be computed and the difference between FM(S) and FM($S+1$) obtained. The process would be systematically continued until the maximum difference between FM($S+k-1$) and FM($S+k$) satisfies the uncertainty specification, with $S+k$ being the total number of GM samples required. The starting number of GM samples to be used is somewhat arbitrary, but could be chosen to be proportional to the number of pattern lobes expected over the observation window based on the object size L.

Using a single FM to cover an entire pattern is generally not the best approach, however, partly due to the growing cost of the FM computation itself as the order, or number of parameters, of the FM is increased. More important is the fact that the condition number of the data matrix needing solution for the FM parameters may increase unacceptably. This is especially the case where the FM brackets a large number of pattern lobes. Instead, as an alternative, some number of windowed, lower-order (i.e., fewer parameters) overlapping FMs is used. Each FM is arranged to share two or more GM samples with its neighbors, as is illustrated conceptually in Figure 9.3 where there are a total of N FMs. After the parameters of these overlapping FMs have been computed, the FMs are evaluated as finely in angle as desired and the differences between them are computed. If the maximum difference exceeds a specified error, a new GM sample then is added where the maximum difference occurs. The respective parameters of the affected FMs (i.e., those whose windows contain the new sample angle) are then updated, with the process of computing FM differences and new GM samples continuing until the specified uncertainty is satisfied over the entire pattern window.

An appropriate choice for the number of FMs for a given problem might require some experimentation. If it is desired to keep the order of all FMs below some specified value, then based on the computer experiments done in getting the results presented below, each FM should have an angle window that spans just a few pattern lobes, say three or four at most. However, it may happen that adding a new GM sample to an FM causes its order to exceed some specified limit. By dividing such an FM into two lower-order, overlapping ones, the problem of excess FM order can be avoided and the initial number of FMs will not be so important.

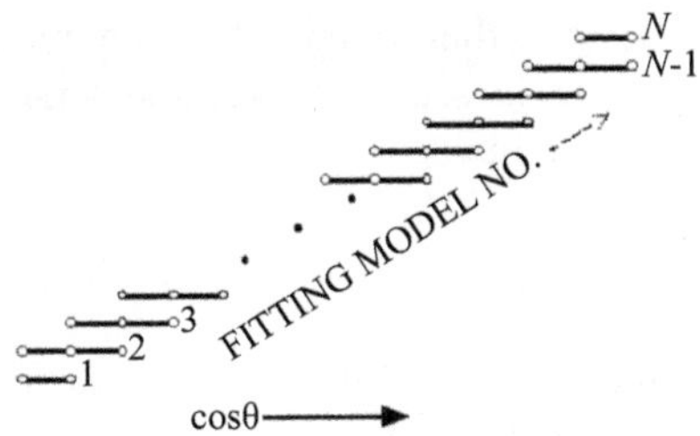

Figure 9.3 *A conceptual illustration of the use of overlapping fitting models for adaptive sampling of a far-field pattern. The horizontal lines indicate the angular extent of each FM with the open circles showing where the pattern (or GM) is initially sampled. In this particular case, all interior FMs begin with three GM samples while those at either end use only two, for a total of N + 1 starting samples. Additional GM samples are systematically located where the maximum difference occurs between two overlapping FMs until a specified convergence criterion has been satisfied. Only two FMs overlap in any of the observation windows here, but other overlap arrangements could be used as well.*

9.3.4 *Specification of the fitting-model error and computing the final pattern estimate*

In contrast to specifying a fixed FM difference, or fitting-model error (FME), between overlapping FMs as was done in [2,3], the maximum FME can also be scaled relative to how the magnitudes of the far-field samples vary. For example, as the magnitude decreases, the FME might be proportionately increased to accommodate the fact that sidelobe maxima may not be needed to the same accuracy as the main lobe. In the results to follow, the FME at the observation angle θ is required to satisfy

$$FME(\theta) = A_1 + A_2\left[\left|GM(\theta_{\max}) - \frac{1}{M}\sum_{\alpha=1}^{M} FM_\alpha(\theta)\right|\right] \le FME_{\max} \tag{9.3}$$

where A_1 and A_2 are specified parameters, $GM(\theta_{\max})$ is the maximum value determined for the pattern being modeled, $FM_\alpha^{(\theta)}$ is the value of the α' th FM at angle θ where a total of M FMs overlap and $FME_{\max}$ is the maximum acceptable error. The parameter A_1 determines the *FME* in the vicinity of the peak(s) of the GM. The parameter A_2 increases the allowable *FME* in proportion to the decrease in the FM values relative to $GM(\theta_{\max})$. For the results that follow, the nominal values for A_1 and A_2 were 0.1 and 0.05 dB, respectively, with $FME_{\max} = 3$ dB. There are many possible variations that might be used for setting the acceptable error or uncertainty, among which would be relaxing the desired accuracy in certain angular regions or increasing it in others consistent with the requirements of a particular application.

The final pattern estimate is then obtained as the average of the overlapping FMs as the observation angle is scanned over the angle pattern. Thus, for M overlapping FMs at angle θ, we have

$$F_{ave,M}(\theta) = \frac{1}{M}[F_i(\theta) + F_{i+1}(\theta) + \cdots + F_{i+M-1}(\theta)] \tag{9.4}$$

For the results presented below, only $M = 2$ was used as only 2 FMs are overlapped in the examples that are presented.

Note that in addition to controlling the adaptation process, the error specification provides a measure of the uncertainty in the average FM values of (9.4). Since (9.3) gives the maximum variation permitted between the FMs that overlap at a given angle, it is speculated that

$$F_{ave,M\pm}(\theta) = F_{ave,M}(\theta) \pm \frac{1}{2}FME(\theta) \tag{9.5}$$

will yield realistic upper and lower error-bound estimates for the average pattern values. That (9.5) does indeed provide a realistic error bound relative to the FM estimates of GM is demonstrated in the following examples.

It is worth emphasizing that, in contrast to using an entire-domain basis for the far field such as cylindrical- or spherical-wave expansions, the windowed approach employed here with discrete point sources can be applied to smaller, or limited, angular sectors with no computational penalty. Also observe that in implementing the Residue model, angle sampling is done in terms of cosθ rather than θ. This is because the pattern lobes tend to be distributed uniformly in terms of the former variable.

To summarize then, the following steps are involved in the MBPE modeling of radiation and scattering patterns:

1) The quantity to be modeled, that is, the source distribution or pattern itself, is first selected.
2) The total angle window over which the pattern is to be estimated is specified.
3) The number of pattern lobes, L, that are anticipated over the angle window of interest is estimated, taking into account the aperture or source size in.
4) The number, N, of initial FMs is then selected. Choosing a value for N between $L/2$ and $2L/3$, with L expressed in wavelengths, provides a reasonable starting point, noting that the smaller the value of N that is used the larger will be the required FM order for a given pattern.
5) The configuration of the FM overlap then needs to be selected. This can be fairly flexible. Arranging each window to be half overlapped with its nearest neighbors seems to be a good choice. Note that using half-overlapped FM*s* leads naturally to a minimum of 3 initial GM samples per FM window as shown in Figure 9.3.
6) Finally, the kind of FM error to be used and its numerical parameters, as in (9.3), must be specified.

9.4 Using Residue, Fourier, and Prony fitting models for a uniform current filament

In previous work [2,3], the Fourier FM approach was described and some initial results presented to demonstrate that the far-field pattern could be estimated to an amplitude uncertainty of 0.01 (about 0.05 dB) using only ~ 3–3.5 samples per lobe. For comparison, results are obtained for a uniform current filament using the Prony, Residue, and the Fourier FMs, in all cases using the arrangement shown in Figure 9.2(a). Following this, several results obtained using only the alternate Residue FM are presented together with an illustration of how the model performance depends on the uncertainty specification. The particular patterns used for testing the MBPE approach here were chosen as being representative of the kinds of patterns encountered in typical applications, as well as their having closed-form analytical expressions, except for the last example of the random-source array.

The far-field, $F_{UCF}(\theta)$, of a uniform current filament (or uniform aperture) is proportional to [7]

$$F_{UCF}(\theta) = L\frac{\sin[\pi L \sin\,\theta]}{\pi L \sin\,\theta} \tag{9.6}$$

where the filament length L is expressed in wavelengths and the observation angle θ is measured from the filament axis. A total of $N = 13$ FMs was used for an $L = 20$-wavelength UCF, arranged as shown in Figure 9.2, with the parameters of (9.3) given by $A_1 = 0.1$ and $A_2 = 0.05$ dB. Thus, all FMs initially have three GM samples except for those on either end, which have only two. This results in an FM sharing two samples with each of its nearest neighbors except for those on the end, which overlap with one adjacent FM and a total of 15 original GM samples.

The Residue and Fourier FMs were adaptively applied to GM samples of (9.6) to yield the results of Figures (9.4) and (9.5) where the patterns are normalized to a maximum of 0 dB. The adaptive procedure outlined above was employed while maintaining a fixed number of FMs in the adaptation process. The final number of GM samples required for convergence, indicated on the plots as solid circles, was 35 and 55, respectively, for the Residue and Fourier FMs. Note that the original GM samples are not always coincident with the average of the FM results since the sample angles used for the two can be slightly different and that only GM samples above −100 dB are plotted. The Prony FM of (9.1) with $N = 22$ with the number of GM samples 44 produced the results of Figure 9.6. Because the Residue FM provided the best overall results for this test in terms of flexibility and computational cost, it was used for the remaining examples that follow.

The upper and lower error-bound estimates for the pattern peaks, using (9.5), and the actual pattern from (9.6) are plotted for the Residue FM in Figure 9.7. The actual pattern is seen to lie between the upper- and lower-bound peaks throughout the entire window. In Figure 9.8, the average values of the overlapping FMs from (9.4) are compared with the actual pattern, where 34 of the 35 GM samples used for

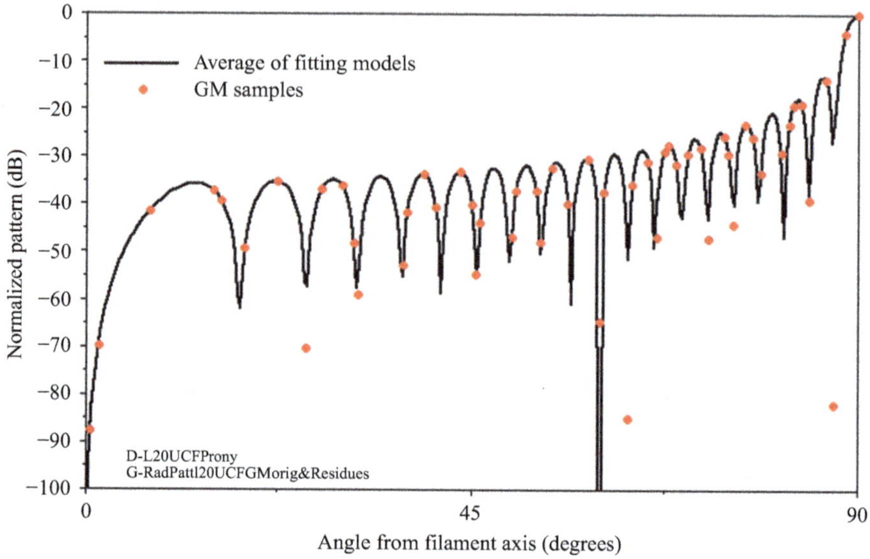

Figure 9.4 The 20-wavelength UCF pattern obtained with the Residue FM and using 35 GM samples

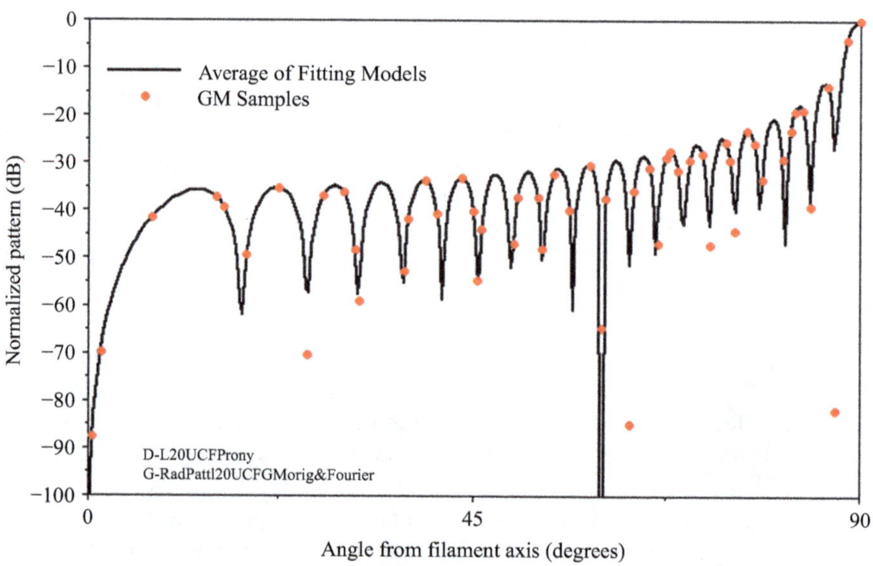

Figure 9.5 The 20-wavelength UCF pattern obtained with the Fourier FM and using 55 samples

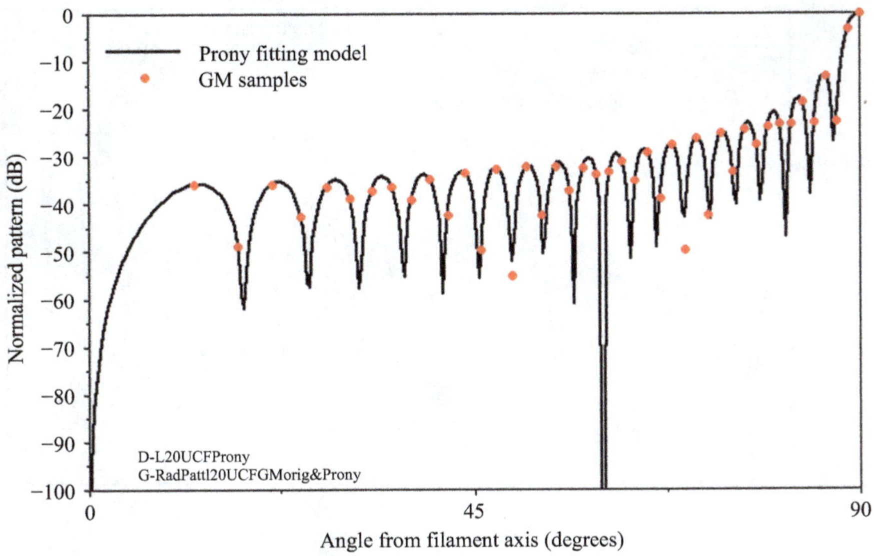

Figure 9.6 *The 20-wavelength UCF pattern obtained with the Prony FM and using 44 samples*

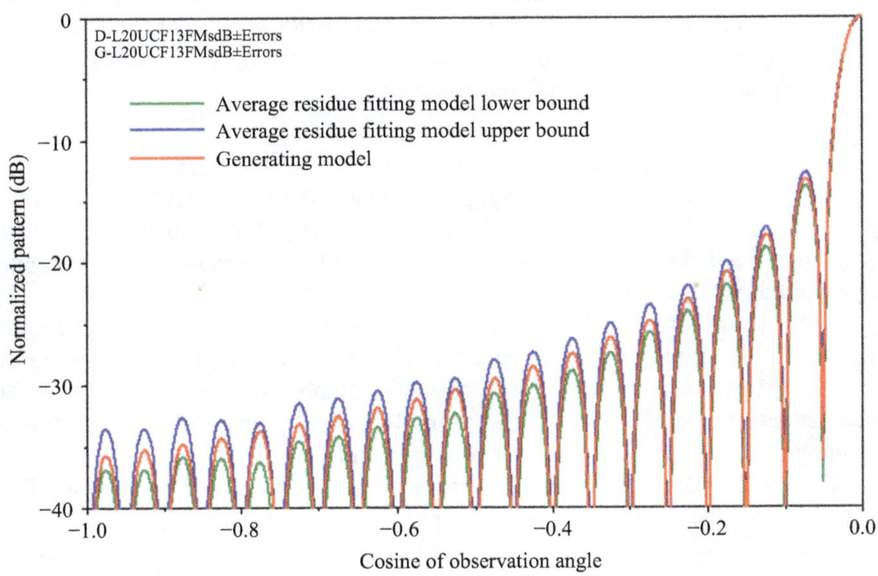

Figure 9.7 *One quadrant of the normalized radiation pattern for the field of an $L = 20$ (wavelengths) UCF as obtained using the adaptive Residue FM. The lower- and upper-bound curves estimated from (9.5) are shown together with the actual pattern from (9.6). The pattern is cutoff at -40 dB to show the difference between the various plots more clearly.*

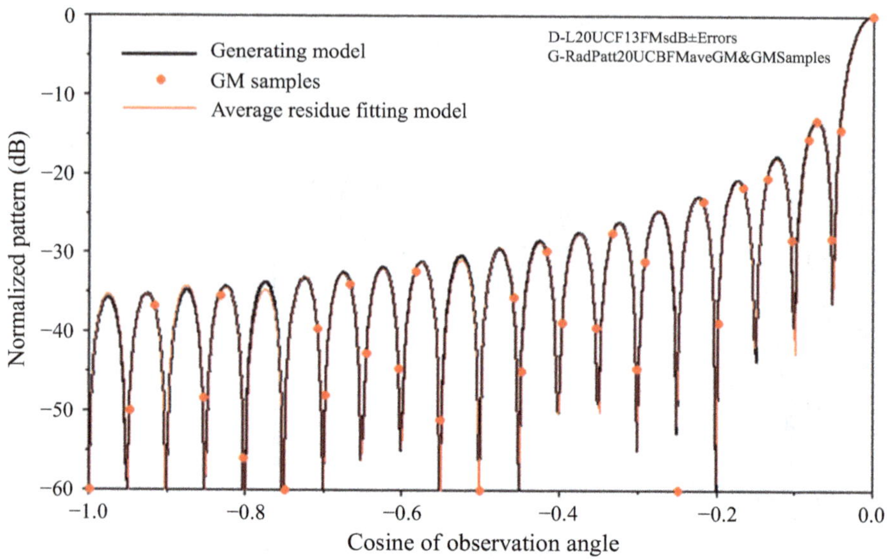

Figure 9.8 One quadrant of the normalized, actual field of the 20-wavelength UCF compared with average Residue FM. The circles indicate 34 of the 35 GM samples computed from (9.6) (one is below −60 dB), that were used in the estimation process. Because the cosine angle values used to obtain the average and actual patterns do not always precisely match those used to obtain the original GM samples, some of the latter do not coincide with the pattern plots.

computing the final FM parameters are also shown. The maximum difference between the actual pattern and average Residue FM is consistent with the error specification of (9.4) and the numerical values used for its parameters. With 35 samples needed to estimate a pattern having 20 lobes or maxima, 1.75 samples are required per lobe for the UCF.

The behavior of three different error measures for the $L = 20$ UCF is presented in Figure 9.9 as a function of the number of GM samples used for computing the Residue FM parameters. The upper plot on this graph shows the maximum difference between all pairs of the 13 finely sampled, overlapping FMs as a function of the number of GM samples used for their computation.

The middle plot exhibits the difference between all pairs of overlapping FMs averaged over the angles they both sample. The bottom plot displays the angle-averaged difference between the average FM and the actual pattern, or the GM. Of these three, only the first two would be available in actual application since the GM samples needed to compute the last error measure would be limited to those available up to that point in the adaptation process. It is useful to include the latter error measure, however, as the difference between the average FM and the actual pattern provides a reality check on the MBPE performance.

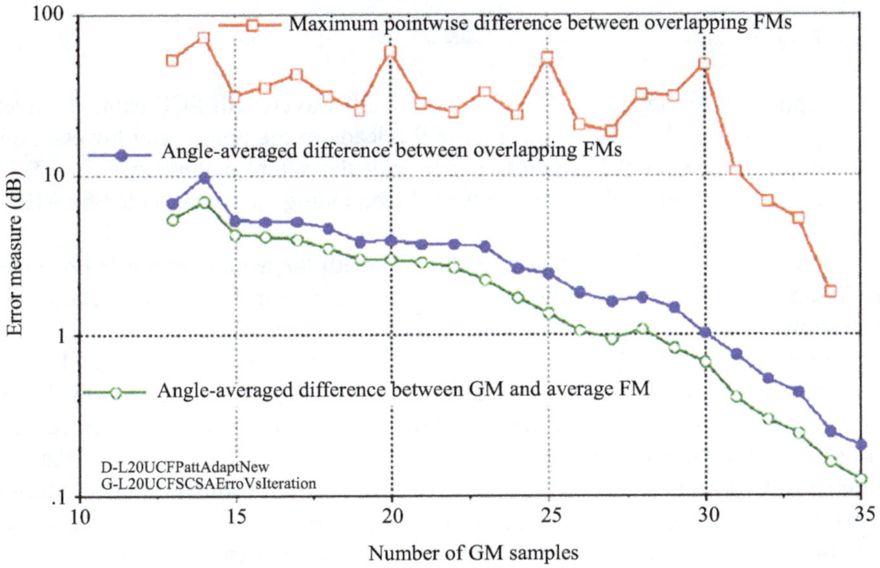

Figure 9.9 *Three error measures discussed in the text for the UCF are shown as a function of the number of GM samples used up to that point in the adaptation process over an observation window* $-90°$ *to 0, with the parameters in (9.3) given by* $A_1 = 0.1$ *dB and* $A_2 = 0.05$ *dB. The open squares denote the maximum pointwise difference between all pairs of overlapping FMs with the addition of a new GM sample. The closed and open circles exhibit the angle-averaged difference between the overlapping FMs and between the average FM and the GM itself. The fact that the latter is smaller than the former shows that the FM difference provides a conservative measure of the accuracy of average FM.*

Observe that the difference between the overlapping FMs for this example always exceeds the difference between the average FM and the GM. This shows that the FM differences provide a conservative, or high-side, estimate of the error or uncertainty in the average FM relative to the true GM values. Note also that the maximum FM difference does not always decline monotonically. This is because updating two FMs can at times increase the maximum differences with their overlapping neighbors.

9.5 Applying the Residue fitting model to other sources

9.5.1 The far radiated field of the sinusoidal current filament

The normalized far field of a center-fed sinusoidal current filament (SCF) L wavelengths long is proportional to [7]

$$F_{SCF}(\theta) = \frac{\cos(\pi L \cos\theta) - \cos \pi L}{\sin\theta}. \tag{9.7}$$

Application of adaptive MBPE to an $L = 20$-wavelength SCF using 13 overlapping Residue FMs as shown in Figure 9.2 leads to the upper- and lower-bound estimates for the far field shown together with the actual pattern in Figure 9.10. Again, the actual field values lie between the bounding curves provided by MBPE adaptive sampling.

A comparison of the average Residue FM with the actual pattern is presented in Figure 9.11 where most of the 40 GM samples used for the FM computation are also shown (some fall below −20 dB). The actual pattern and the average Residue FM curves are essentially graphically indistinguishable on the scale used. The 20-wavelength SCF has only 10 lobes rather than the 20 lobes of the 20-wavelength UCF over the same −1 to cosθ interval, but requires 6 more unknowns to achieve the same estimation uncertainty, or 4 samples per lobe. This indicates that the number of GM samples needed to achieve a given pattern-estimation uncertainty is sensitive not only to the lobe structure of the pattern itself but also to the spatial variation of the source distribution that produces that pattern.

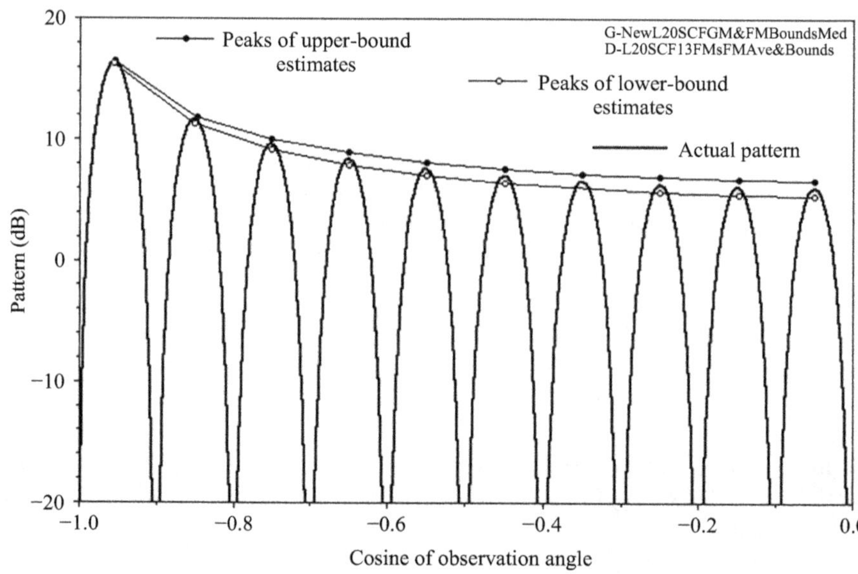

Figure 9.10 One quadrant of the normalized radiation pattern for the field of an $L = 20$ (wavelengths) SCF as obtained using the Residue FM. The lower- and upper-bound peaks estimated from (9.5) are shown together with the actual field from (9.7). The pattern is plotted over a limited dB range to show the difference between the various plots more clearly.

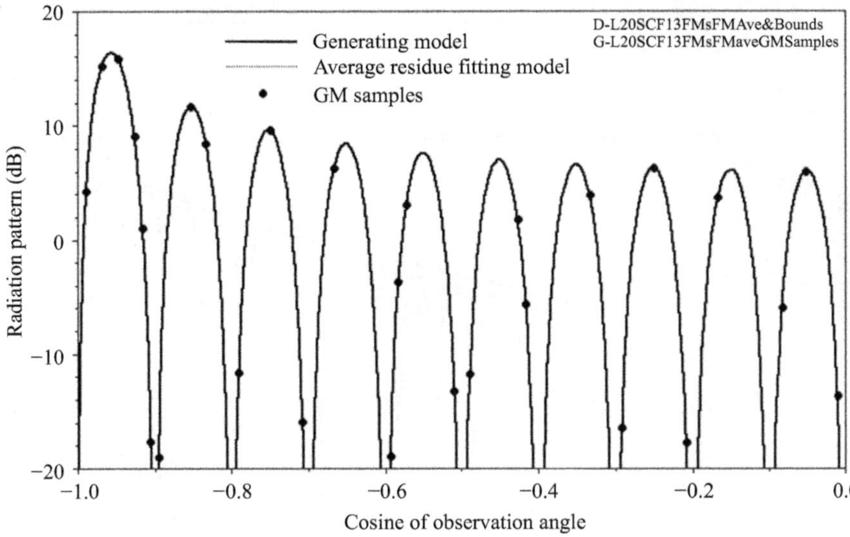

Figure 9.11 One quadrant of the actual field of the 20-wavelength SCF compared with the average Residue FM. The circles indicate 28 of the 40 GM samples, computed from (9.7), that were used in the estimation process

9.5.2 *The far-field scattered field from a thin, circular cylinder*

The GM here is the approximate far, scattered field of a thin, circular cylinder L wavelengths long, which is proportional to [8]

$$RCS \propto F^2_{CYL}(\theta) = \left[\cos\ \theta_i \frac{\sin(2\pi L \sin\ \theta_i)}{2\pi L \sin\ \theta_i}\right] \tag{9.8}$$

for which MBPE adaptive sampling produces the results shown in Figures 9.12 and 9.13 for $L = 10$ wavelengths with 7 FMs being used. In contrast with an L-wavelength SCF, where the number of lobes over a -1 to 0 interval in $\cos\theta$ is $\sim L/2$ and the UCF which has $\sim L$ lobes over that same interval, the L-wavelength scatterer has $\sim 2L$ lobes, 20 in this example. As for the previous cases, the average Residue FM, as obtained from 48 GM samples, and a finely sampled GM plot for the scatterer are graphically indistinguishable. The number of samples required per lobe for the 10-wavelength cylinder scatterer is thus 2.4, greater than the 1.75 needed for the UCF even though their patterns are quite similar, as can be seen by comparing Figures 9.4 and 9.12.

9.5.3 *The far radiated field of a randomized array of point sources*

The last example considered here is a linear array of 21, isotropic point sources having random amplitudes varying between -1 and $+1$ and located randomly

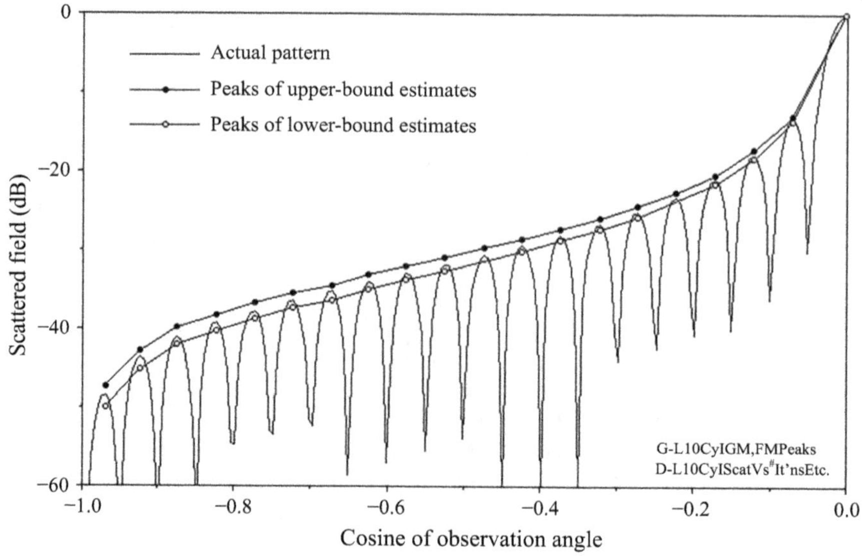

Figure 9.12 One quadrant of the normalized scattering pattern for the field of an L = 10 wavelengths thin cylinder as obtained using the Residue FM. The lower- and upper-bound peak estimates of the far field from (9.5) are shown together with the actual field from (9.8).

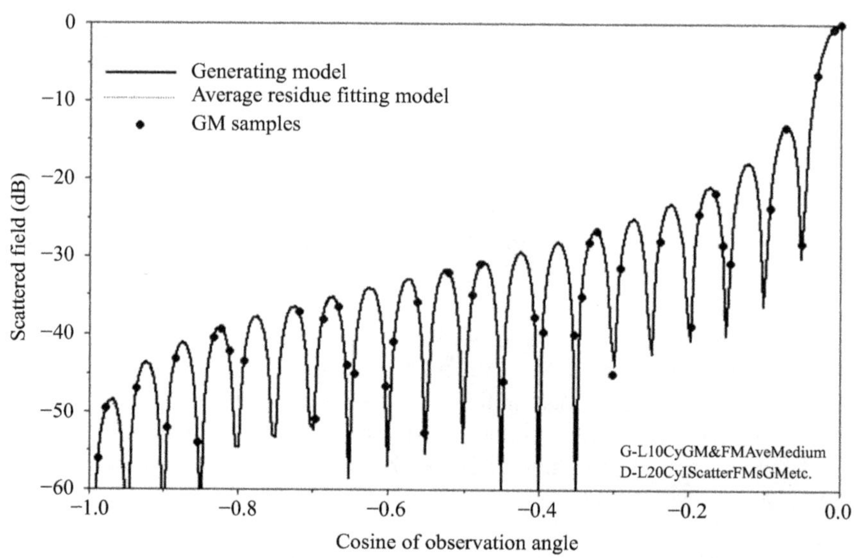

Figure 9.13 One quadrant of the actual scattered field of the 10-wavelength cylinder compared with the average Residue FM. The circles indicate some of the 48 GM samples, computed from (9.8) that were used in the estimation process.

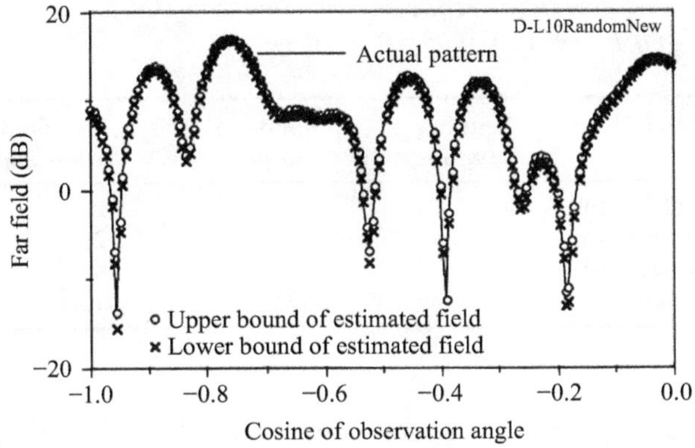

Figure 9.14 One quadrant of the radiation pattern for the field of 21 isotropic point sources having random amplitudes between −1 and + 1 and randomly located along an L = 10 wavelengths linear array together with upper-bound and lower-bound estimates

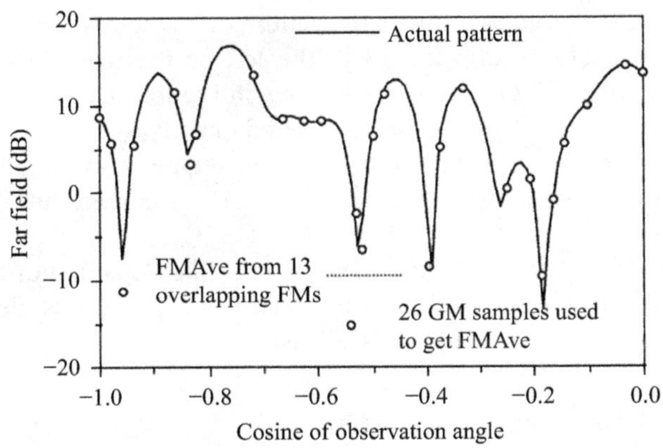

Figure 9.15 The actual pattern of 21 isotropic sources having random amplitudes and located at random positions along a 10-wavelength aperture compared with the average Residue FM. The circles indicate the 26 GM samples used in the estimation process.

within a 10-wavelength aperture. Results for cosθ varying from −1 to 0 are shown in Figures 9.14 and 9.15, again using 13 FMs. A total of 26 GM samples is needed to achieve the same specified estimation error as used for the previous cases. Since there are on the order of 8.5 maxima in the pattern, this works out to about 3 samples per lobe, midway between that required for the scatterer and the SCF. The number of samples needed per lobe or per wavelength of aperture to achieve the same specified estimation accuracy (as given by (9.3) with $A1$ and $A2$ 0.1 dB and

Table 9.1 Far-field sampling required over a 90° window for sources tested using A_1 = 0.1 and A_2 = 0.05 dB in (9.3)

Source type tested	Number of samples	Far-field samples/lobe	Far-field samples/ wavelength of aperture
20-WL sinusoidal current	40	4	2
10-WL randomized array	26	3	2.6
10-WL cylinder scatterer	48	2.4	4.8
20-WL uniform current	35	1.75	1.75

0.05 dB, respectively) for the various sources considered here is summarized in Table 9.1.

9.6 Displaying the GM sampling sequence for various arrays

9.6.1 A modified Dolph–Chebyshev array

The Residue FM was used to estimate the pattern of a Dolph–Chebyshev type array having uniform sidelobes of −20 dB to the left of the main lobe and −40 dB sidelobes to the right [9] with the results shown in Figures 9.16 and 9.17. A total of 10 FMs were used and 22 GM samples were adaptively added for a total of 32 to yield the pattern in Figure 9.16. The distribution sequence of the added GM samples as a function of cosθ is plotted in Figure 9.17 to demonstrate the adaptation process.

In order to provide an alternate, more pattern-like depiction of the FM uncertainty, a pattern error associated with the adaptation process is also included in Figure 9.16, measured in dB, as computed using

$$P_{error}(\theta) = 10 \times Log\left[\left(FM_{ave,real}(\theta) - GM_{real}(\theta)\right)^2 + \left(FM_{ave,imag}(\theta) - GM_{imag}(\theta)\right)^2\right] \tag{9.9}$$

The error pattern is about 15 dB below the skirt of leftmost sidelobe to more than 50 dB below the main lobe and is generally 40 dB or more below the rest of the GM pattern. A plot like this of $P_{error}(\theta)$ represents a noise floor for the estimated pattern.

It is interesting to note that when the same pattern was estimated using 13 FMs, a total of 36 GM samples were required employing the same FME parameters. Reducing the number of FMs to 7 resulted in 30 GM samples being required. Further study about how the number of FMs and the aperture size might be related could be useful.

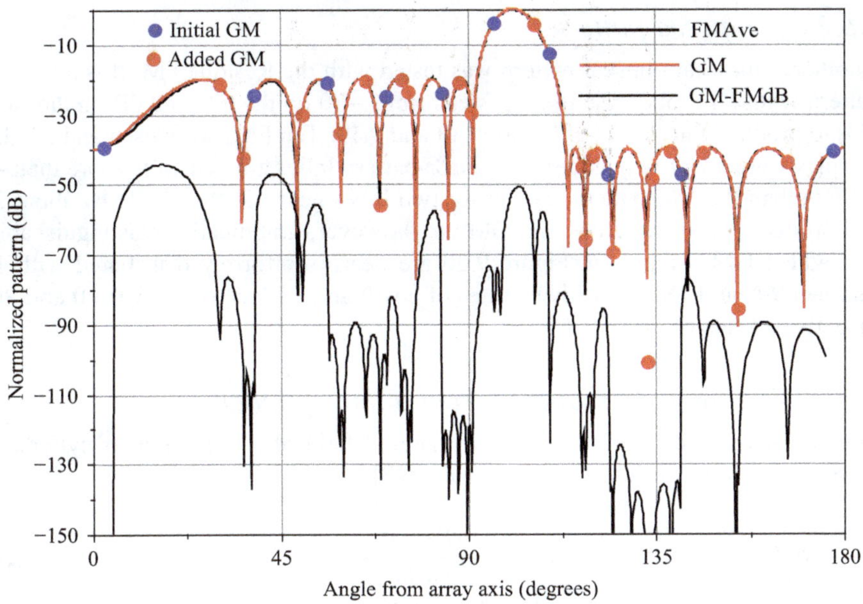

Figure 9.16 *The average FM and GM patterns with the error pattern of (9.9) for the modified Dolph–Chebyshev pattern, where the initial GM samples are indicated by the solid blue circles and the added samples by the red*

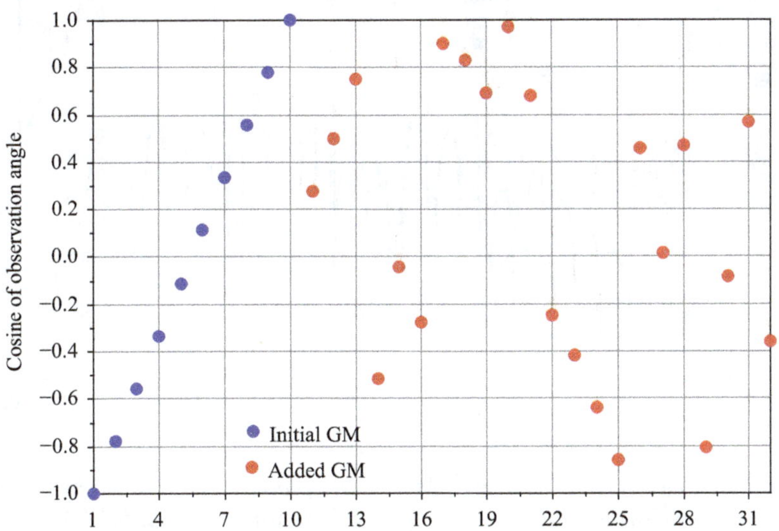

Figure 9.17 *Sequence of GM samples showing their locations in the cosine of the observation angle for a modified Dolph–Chebyshev array using 10 FMs, where the initial GM samples are indicated by the solid blue circles and the added samples by the red*

9.6.2 A tapered pattern

Another somewhat unusual pattern was tested with the Residue FM. It is a tapered pattern whose 15 lobes rise in 5 dB steps from −70 on the left to 0 dB on the right [13] to produce the results of Figure 9.18 and 9.19. Ten FMs were used and 24 GM samples were adaptively added. The error-pattern lobes are generally more than 40 dB less than the GM pattern, except for two lobes near 65° that differ by about 20 and 30 dB. The average FM and GM are, however, graphically indistinguishable. The added GM samples in Figure 9.20 are nearly uniformly distributed with 11 between 90° and 180° (positive values of cosθ) and 13 located between 0 and 90° (negative values of cosθ).

9.6.3 The far-field scattered from a thin cylinder

The GM here is the approximate far, scattered field of a thin, circular cylinder L wavelengths long, which is proportional to [8]

$$RCS \propto F^2_{CYL}(\theta) = \left[\cos\theta_i \frac{\sin(2\pi L \sin\theta_i)}{2\pi L \sin\theta_i}\right] \tag{9.10}$$

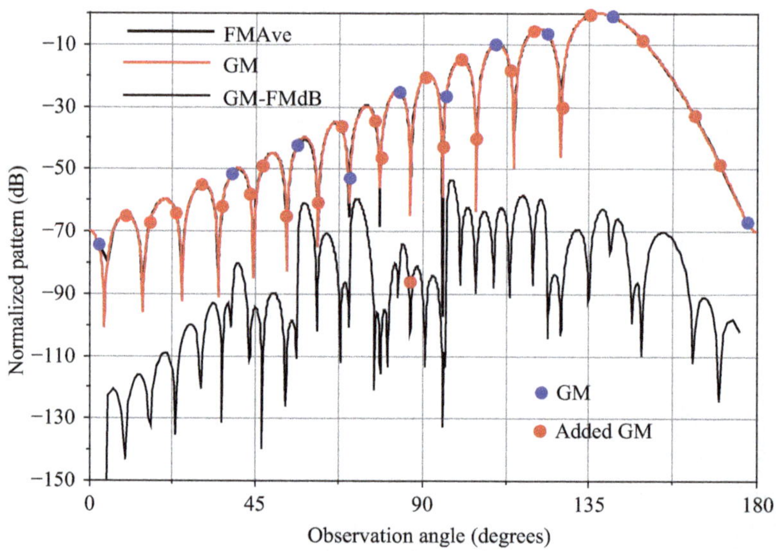

Figure 9.18 The tapered average FM and GM patterns with the error pattern for a 7-wavelength array using 10 FMs where the initial GM samples are indicated by the solid blue circles and the added samples by the red

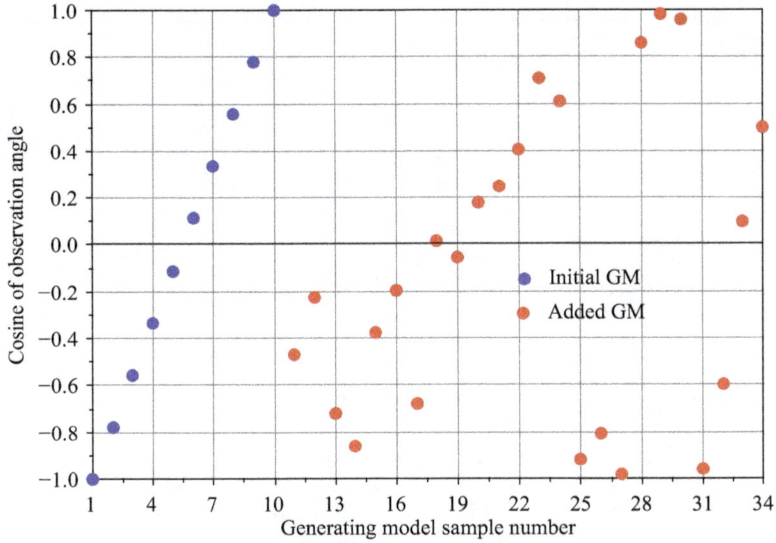

Figure 9.19 Sequence of GM samples showing their locations in the cosine of the observation angle for a 20-wavelength UCF using 10 FMs where the initial GM samples are indicated by the solid blue circles and the added samples by the red

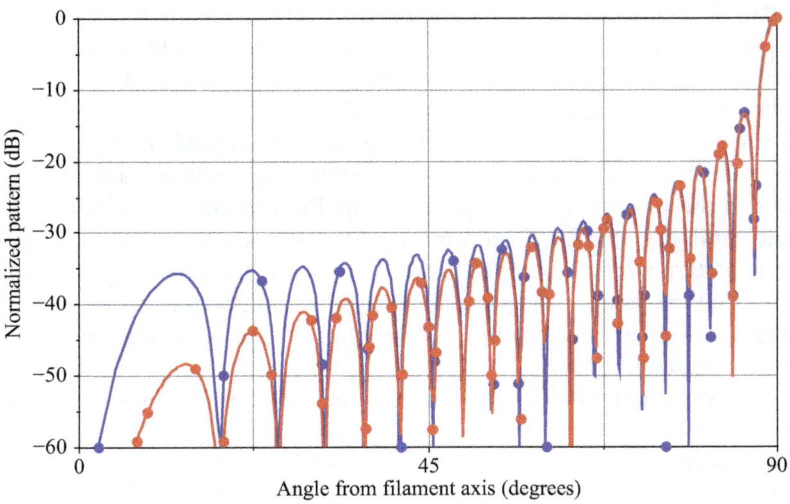

Figure 9.20 The GM and average FM patterns for the 20-wavelength UCF (the upper curve) and the 10-wavelength cylinder (the lower curve)

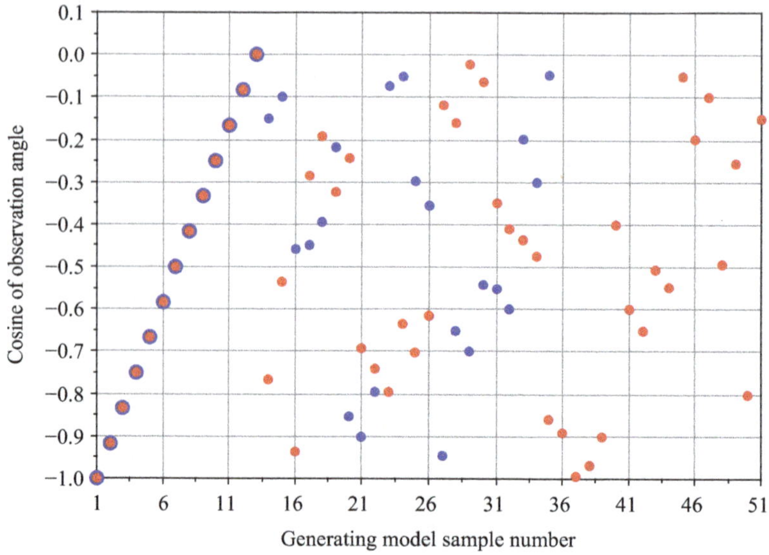

Figure 9.21 Sequences of GM samples showing their locations in cosθ for the 20-wavelength UCF (the blue smaller solid circles) and the 10-wavelength cylinder (in red) with both using 10 FMs

with the incidence angle θ_i measured from broadside to the cylinder axis. Adaptive sampling of (9.10) using a Residue FM produces the results shown in Figures 9.20 and 9.21 for L = 10 wavelengths and using 13 FMs. In contrast with the UCF which has ~ L lobes over a 90° window extending from 0 to 90°, the L-wavelength scatterer has ~ $2L$ lobes. As exhibited in Figure 9.20, the 10-wavelength cylinder and the 20-wavelength UCF patterns are very similar, except that the former that falls off more rapidly near end-on incidence. In spite of their similarity, the UCF required on 35 GM samples for estimation of its radiation pattern while 51 samples were required to estimate the cylinder scatting pattern using the same *FME* parameters. As for the previous cases, the average Residue FM and a finely sampled GM plot for the scatterer are graphically indistinguishable.

The distribution of GM samples for the UCF and cylinder is plotted in Figure 9.21. It is interesting to observe that between cosθ = −0.6 and −1.0, while 11 GM samples are used for the UCF, there are 21 samples for the cylinder. The number of samples required per lobe for the scattering pattern of the 10-wavelength cylinder is 2.55. This is significantly greater than the 1.75 needed for the UCF in spite of their having the same number of lobes.

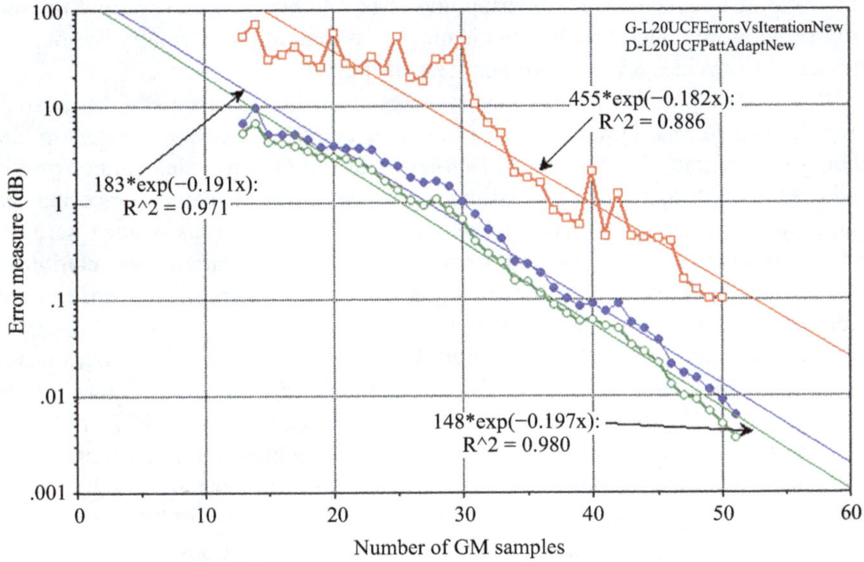

Figure 9.22 *The three error measures shown in Figure 9.6 are plotted here for the 20-wavelength UCF but with A1 = 0.05 and A2 = 0 in (9.3) to obtain an estimate that is both more accurate and remains constant over the dynamic range covered by the pattern sampled in a −90° to 0 observation window. A best-fit straight line is computed for all three, where R^2 is the square of the correlation coefficient. All three best-fit lines are seen to be nearly parallel on this log-linear plot, being well described as exponential functions of the number of GM samples Since an UCF L-wavelength in extent produces L lobes per 90° in its pattern, this sampling density translates to about 2 samples per lobe as well. As can be deduced from Table 9.1, it is clear that the far-field sampling density depends on more than just the lobe count in the pattern.*

9.7 Concluding observations concerning radiation-pattern estimation sampling and uncertainty

The preceding examples demonstrate that adaptive sampling of radiation and scattering patterns using MBPE with DSA FMs can be effective in not only reducing the number of samples needed to obtain a reduced-order, continuous representation of a pattern but also in constraining the estimated pattern to satisfy an uncertainty specification. Some additional computations are included here to shed further light on the sampling requirements. The $L = 20$ UCF problem was repeated using $A_1 = 0.05$ and $A_2 = 0$ dB to reduce the estimation uncertainty to a smaller, and constant, value compared with the criterion used in obtaining the previous results. Using these coefficient values in the *FME* given in (9.3) results in a maximum acceptable difference between overlapping FMs, and hence, a maximum estimation error, of no more than 0.05 dB. This is much less than might normally be sought in practice but provides a more stringent test of the MBPE procedure.

It should be noted that if the GM samples are of limited accuracy, for example being derived from a numerical first-principles model for a complex problem, then seeking an accuracy in the pattern estimate that the GM samples cannot provide might result in "stagnating" the estimation process, that is, convergence may not occur. But when using an analytical expression for a pattern, as is done here, this will not be a problem. It is also worth noting that the variational relationship between the far fields and the sources that produce them results in errors in the latter not translating into comparable errors in the former.

Upon running the $L = 20$-wavelength UCF problem using these new values for the FME coefficients and plotting the same error measures as for Figure 9.5, the results shown in Figure 9.22 are obtained. Also included in Figure 9.22 are lines of the form $A\exp(-Bx)$ where x is the number of GM samples and A and B are best-fit parameters. It is interesting to see that the three best-fit lines are nearly parallel, with all decreasing exponentially as a function of the number of GM samples. This behavior is similar to the convergence of numerical solutions for various wire geometries as the number of unknowns in a moment-method solution is increased. A 0.1 dB in the angle-averaged FM difference and the difference between the average Residue FM and the actual pattern is achieved using about 37 GM samples.

This computation was repeated for UCFs having lengths of $L = 10$, 15, and 25 wavelengths, yielding the results of in Figure 9.23 where, for clarity, only the FM–FM differences are plotted. The data for each of these UCF lengths are found to be fit comparably well by exponentials having different slopes. This is an expected result since the number of GM samples needed to achieve a given uncertainty for a specified source distribution is expected to be related to the number of pattern lobes, which are, in turn, proportional to the aperture size. When the latter effect is removed by replotting the data of Figure 9.23 as a function of the number of GM samples per wavelength of aperture, the results shown in Figure 9.24 are obtained. The best-fit lines for the various UCF lengths are nearly coincident, shifting to the left slightly with increasing L and exhibiting slopes that are within about 1.5%. A 0.1 dB angle-averaged FM difference is seen to require about 2 samples per wavelength of aperture.

How this sampling density might generalize to 2D or 3D source distributions needs to be considered. First note that the sampling density of 2 per wavelength of aperture for the UCF applies to a pattern symmetric about broadside where an angle-observation window of only 90° needs to be sampled. For a non-symmetric, but otherwise equivalent, linear source distribution, this would then work out to 4 samples per wavelength since a far-field window of 180° would then need to be sampled. For a true 2D source whose pattern has to be sampled over 360°, this would imply that 8 samples per wavelength of its maximum linear aperture are needed to achieve a comparable pattern accuracy of 0.1 dB. This measure would seem to hold in planar cuts for the radiation patterns of 3D source distributions as well.

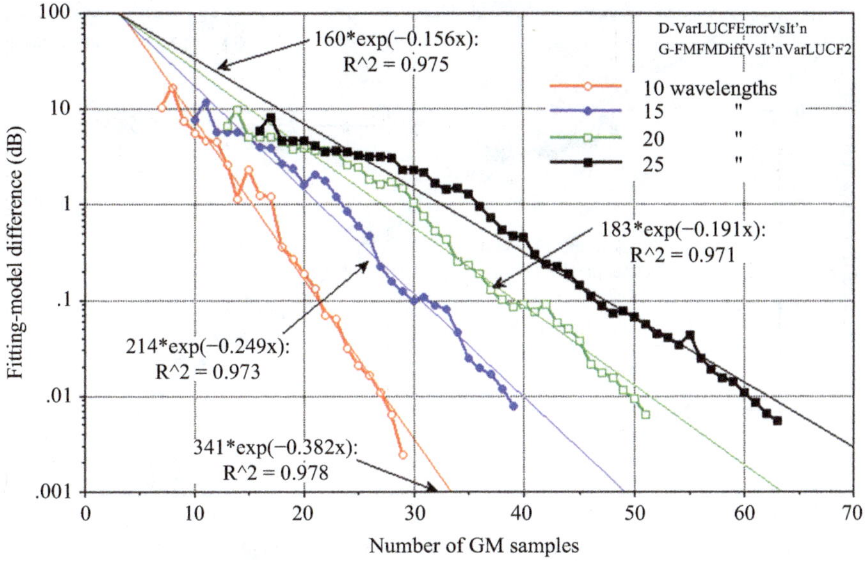

Figure 9.23 *The angle-averaged FM–FM differences for a UCF varying from 10 to 25 wavelengths using A1 = 0.05 and A2 = 0 in (9.3) yields the results shown here, again for a −90° to 0 observation window. Best-fit straight lines are also shown together with their respective correlation coefficients. The constants in the exponentials for the best-fit lines are seen to decrease with increasing filament length. The exponential decrease of the error measures as the number of GM samples increases is a very beneficial attribute of a numerical process.*

Results obtained for the cylinder scatterer indicate, however, that nearly 5 samples per wavelength of cylinder length are required for the same kind of pattern-estimation accuracy for a 90° sector of the scattered field to be estimated. Following the same line of reasoning as above for a general 2D scatterer, or for planar cuts of 3D objects, sampling over 360° can be estimated to require on the order of 20 samples per wavelength. One obvious question arises about why a greater sampling density is apparently needed for a scattering pattern as compared with a radiation pattern when the two are similar-appearing, as exhibited in Figures 9.4 and 9.9. Perhaps the answer is that for the latter situation a single current distribution produces the entire radiated field. For the former, however, the current distribution, being a function of the angle of incidence, changes with every viewing angle when determining the monostatic radar cross section.

It would also seem to follow then that the number of GM samples needed to achieve a specified estimation uncertainty must depend on both the aperture size and spatial variation of the source within it, a dependence that ultimately is exhibited by the pattern "rank" or the number of degrees of freedom of the pattern

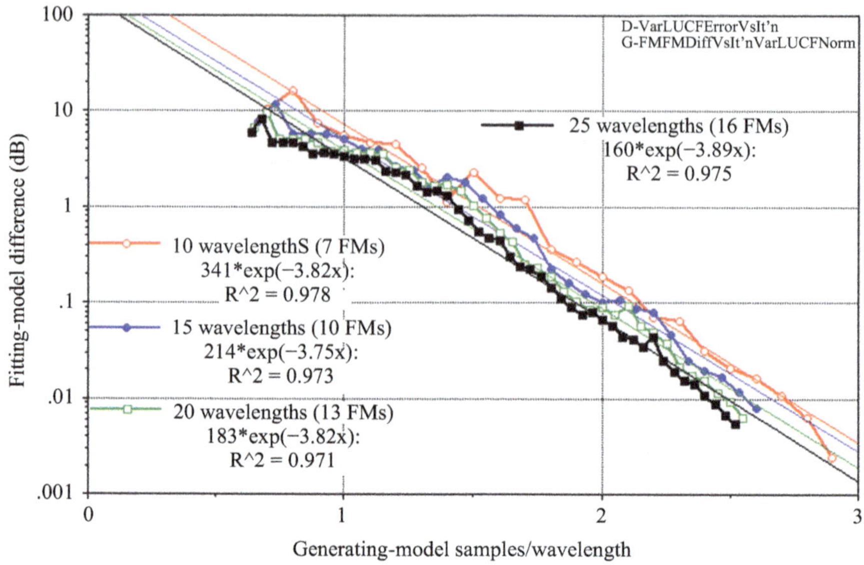

Figure 9.24 Upon replotting the results of Figure 9.23 but with the number of GM samples/wavelength as the independent variable, the best-fit angle-averaged FM–FM differences become nearly coincident. These results show that for this particular case, the number of GM samples needed for a specified angle-averaged FM difference in the pattern is linearly proportional to the filament length. They also show that the sampling density required for a given value of error measure decreases somewhat as the UCF increases in length.

or rank R, over the chosen observation window. As a specific example of how the source characteristics and aperture might influence the rank, consider an aperture L-wavelengths long having N point sources uniformly spaced within it.

For the simplest case of two point sources, as L is increased, the number of pattern lobes over 360° in an observation plane containing the sources will be of order $4L$. However, R as determined by eigenvalue analysis of the data matrix that arises when using Prony's method remains fixed at two when L exceeds 0.5, regardless of how large the aperture is made and how many lobes are in the pattern. However, if L is fixed and the number of point sources is systematically increased, R will increase proportionately until $N >= 2L$, whereupon R remains fixed at $\sim 2L$ since only about 2 source/wavelength can be resolved in the far field. Because it can model such point-source arrays, a Prony-based procedure can exploit the reduced rank of such special problems, but the DSA and a Fourier model of the far field are not as well suited for doing so. Generally speaking, with everything else being equal, the "best" pattern estimator would be one for which the number of GM samples can be reduced to as close to R as possible. More investigation is needed to settle these issues and to generalize the results beyond the simple cases considered here. Other approaches related to the problem of reducing the amount of computation needed to obtain reliable and accurate estimates of radiation and scattering patterns and measures of their complexity can be found in [10–12].

Acronyms

DSA	discrete-source approximation
FM	fitting model
FME	Fitting-model error
GM	generating model
L	object size in plane of far-field pattern
MBPE	model-based parameter estimation
OC	operation count
R	rank
RCS	radar cross section
S	number of currents on the surface of an object whose pattern is being obtained
SCF	sinusoidal current filament
SDSA	specified discrete-source approximation
UCF	uniform current filament

References

[1] E. K. Miller, "Model-Based Parameter Estimation in Electromagnetics: I–Background and Theoretical Development," *Applied Computational Electromagnetics Society Newsletter*, Vol. 10, No. 3, pp. 40–63, 1996, "II–Applications to EM Observables," Vol. 11, No. 1, pp. 35–56, 1995, "III–Applications to EM Integral Equations," *Applied Computational Electromagnetics Society Journal*, Vol. 10, No. 3, pp. 9–29, 1995. Change to APS xxx

[2] E. K. Miller, "Computing Radiation and Scattering Patterns Using Model-Based Parameter Estimation," *IEEE AP-S International Symposium*, Renaissance Waverly Hotel, Atlanta, GA, June 21–26, pp. 66–69, 1998.

[3] E. K. Miller, and T. K. Sarkar, "An Introduction to the Use of Model-Based Parameter Estimation in Electromagnetics," in *Review of Radio Science, 1999 URSI General Assembly*, pp. 139–174, 1999.

[4] R. J. Allard, D. H. Werner, J. S. Zmyslo, and P. L. Werner, "Spectral Domain Interpolation of Antenna Radiation Patterns Using Model-Based Parameter Estimation and Genetic Algorithms," *Proceedings of the 14th Annual Review of Progress in Applied Computational Electromagnetics (ACES)*, Vol. 2, pp. 964–971, 1998.

[5] D. H. Werner, and R. J. Allard, "The Simultaneous Interpolation of Antenna Radiation Patterns in Both the Spatial and Frequency Domains Using Model-Based Parameter Estimation," *IEEE Transactions on Antennas and Propagation*, Vol. 48, No. 3, pp. 383–392, 2000.
[6] R. W. Hamming, *Numerical Methods for Scientists and Engineers*, Dover Publications, Inc., New York, 1962.
[7] C. A. Balanis, *Antenna Theory: Analysis and Design*, Harper & Row, Publishers, New York, 1982.
[8] E. G. Knott, J. F. Shaeffer, and M. T. Tuley, *Radar Cross Section*, 2nd Edition, Artech House, Boston, MA, 1993.
[9] E. K. Miller, "Synthesizing Linear-Array Patterns via Matrix Computation of Element Currents," *IEEE Antennas and Propagation Society Magazine*, Vol. 55, No. 5, pp. 85–96, 2013.
[10] O. M. Bucci, and G. Franceschetti, "On the Degrees of Freedom of Scattered Fields," *IEEE Antennas and Propagation Society Transactions*, Vol. 37, No. 7, pp. 918–929, 1989.
[11] O. M. Bucci, C. Gennarelli, and C. Savarese, "Optimal Interpolation of Radiated Fields Over a Sphere," *IEEE Antennas and Propagation Society Transactions*, Vol. 39, No. 11, pp. 1633–1643, 1991.
[12] O. M. Bucci, C. Gennarelli, and C. Savarese, "Representation of Electromagnetic fields Over Arbitrary Surfaces by a Finite and Nonredundant Number of Samples," *IEEE Antennas and Propagation Society Transactions*, Vol. 46, No. 3, pp. 315–359, 1998.

Chapter 10

Modeling dipole arrays that produce synthesized patterns using NEC[1]

Abstract

Pattern synthesis usually involves determining the strengths of the currents in a given array so that it radiates a specified pattern. Demonstrating that this pattern can be produced by an actual array of physical elements is the next step in the synthesis problem. This entails determining the excitations needed for a physical array when mutual interactions are accounted for. These excitations can be either source voltages or the synthesized currents themselves, with both options available in NEC 4 (the Numerical Electromagnetics Code). While this is done here using NEC, the same approach could use any wire-antenna computer code. When the current option is chosen, the code first solves for the voltages that induce the required currents. This is a simple procedure that automatically includes the effect of mutual coupling among the array elements. The purpose of this chapter is to examine how well a numerical model of a given array will match the specified pattern. Several examples of this approach are included here to demonstrate the process.

10.1 Introduction

The pattern-synthesis problem involves determining the complex (possibly) values of the N current elements required to produce a desired radiation pattern from a specified array configuration using various analytical procedures [1–19]. The next step is then to design a physical antenna of N wire dipoles of appropriate lengths and radii and excited so as to carry the currents derived in the synthesis process. The most critical part of this step is determining the exciting voltages needed to induce the needed current values in the physical array while including the interaction effects among the antenna elements.

The Numerical Electromagnetics Code (NEC) includes two options for exciting a wire structure being modeled as an antenna. The more-often used option is to specify the input voltage(s) that excites it. An alternative option is to instead specify the input current(s), the approach used here for the synthesis problem, a feature offered in NEC version 4.2. It is worth emphasizing that this approach is compatible with any numerical antenna-array model. Unless stated otherwise, the NEC-array

[1]Published in *Progress in Electromagnetics Research C*, Vol. 121, 145–161, 2022.

models used for the results that follow consist of 7-segment, half-wavelength dipoles 10^{-3} wavelengths in radius. The dipoles are parallel to the z-axis with their centers at $z = 0$ as illustrated below and the patterns are computed in the x–y plane. The dipoles are equally spaced along the x-axis except for Section 10.3.2 where they are randomly located, and Sections 10.3.3 and 10.3.4 where the array elements lie in the x–y plane.

Observe that this discussion is not about pattern-synthesis methodology *per se* but instead addresses the question about whether a real antenna, that is, one that is modeled numerically using a validated computer code, will radiate the pattern of interest. For those who might want to investigate various pattern synthesis approaches, a representative list of relevant articles is included in [1–19]. A reviewer observed "the problem of synthesized arrays by actual antenna has already been discussed by many papers" and provided three examples. Of the three, only one of these [20] is relevant but the approach it presents for including mutual coupling requires three journal pages to describe. This is much more complicated than that offered by the NEC approach which uses only the synthesized currents as the input to the NEC model. A thorough review of the mutual-coupling problem can be found in [21]. The combination of the synthesis procedure in [22] with the current-input treatment included in NEC offers a straightforward way to synthesize non-uniform and nonlinear geometries.

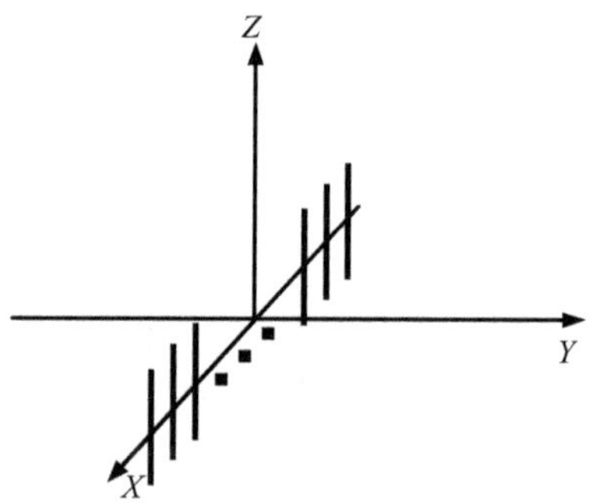

Array with equally spaced elements aligned along the X-axis.

10.2 Representative numerical results

In [20], a synthesis procedure was described for creating a variety of patterns based on iteratively deriving the element currents of a prescribed array that produce a specified sequence of associated pattern maxima. This iterative approach was demonstrated to converge to the desired pattern maxima for arrays having equal sidelobe levels such as the Dolph–Chebyshev (DC) array and various other sidelobe specifications, usually in fewer than a dozen iterations. The synthesis approach in [20] was demonstrated to provide results that conformed to a variety of specified patterns, some of which are examined here to test their compatibility with an actual physical array. Another procedure for synthesizing the element locations and currents based on Prony's method was described in [23] and in Chapters 7 and 8 of this book. An example developed from that approach is also included.

The pattern-synthesis results included in the following were obtained with 24-digit compute precision using Future Basic, a compiled language for Macintosh computers. The NEC computations were done using a Fortran version for the HP computer and performed in double precision. Note that the dipole excitation voltages are specific to the tangential-electric-field source model used in NEC.

Upon running NEC using the currents determined for some of the synthesized patterns and arrays discussed in [20], the excitation voltages for the array dipoles are obtained as are their associated patterns. The NEC-derived patterns and the original synthesized patterns from [20] are then plotted together to determine how closely the physical (NEC-modeled) array pattern agreed with them. The NEC excitation voltages are listed only for the first two examples since their specific values can depend somewhat on the source treatment used for a numerical model. The element currents are included for all of the examples for readers to perform similar experiments using their own models.

10.3 Dolph–Chebyshev patterns

10.3.1 The standard Dolph–Chebyshev array

The pattern of a DC array is one of the more popular examples of pattern synthesis. The patterns included in Figure 10.1 are for a 4-wavelength, 9-element array with −20 dB sidelobes. The pattern from the NEC model agrees to within a few tenths of a dB with the DC array of specified element currents except for the widest sidelobe where the difference is 0.76 dB. For reference purposes, the DC currents derived for the synthesized pattern and the excitation voltages needed for the NEC array are included in Table 10.1 for the DC array and in Table 10.2 for the randomized DC array. Tables 10.3–10.11 for the remaining 9 arrays include only the array currents.

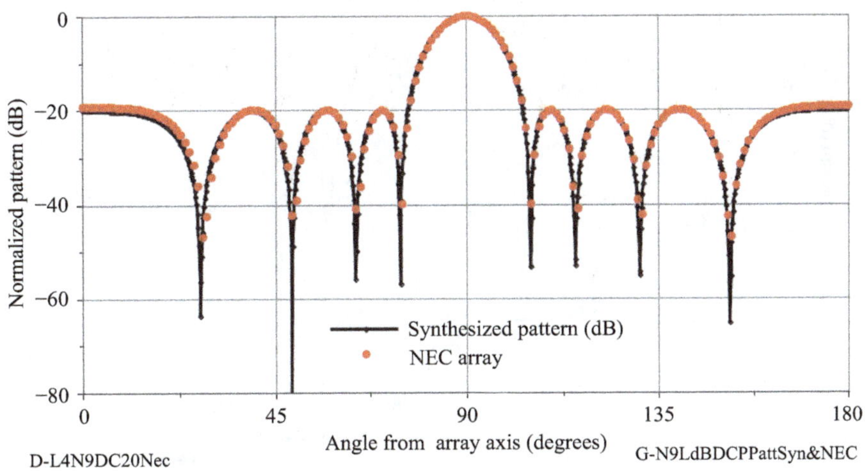

Figure 10.1 The synthesized and NEC-computed patterns for a 9-element, 4-wavelength DC array with −20 dB sidelobes

10.3.2 A randomized Dolph–Chebyshev array

The next variation on the −20 dB DC array is for a linear array 3.2 wavelengths long having randomly spaced dipoles along the *x*-axis as depicted in Figure 10.2(a).

Table 10.1 Exciting voltages and currents for the 9-element Dolph-Chebyshev array having −20 dB sidelobes of Figure 10.1

Element no.	V_{Real}	V_{Imag}	I_{Real}	I_{Imag}
1	7.20977E+1	3.21162E+1	1E+0	0E+0
2	4.99843E+1	−1.38425E+1	1.0231E+0	0E+0
3	8.32325E+1	2.38433E+1	1.3501E+0	0E+0
4	8.80578E+1	−2.71995E+0	1.5796E+0	0E+0
5	1.00128E+2	2.34736E+1	1.6622E+0	0E+0
6	8.80578E+1	−2.71995E+0	1.5796E+0	0E+0
7	8.32325E+1	2.38433E+1	1.3501E+0	0E+0
8	4.99843E+1	−1.38425E+1	1.0231E+0	0E+0
9	7.20977E+1	3.21162E+1	1E+0	0E+0

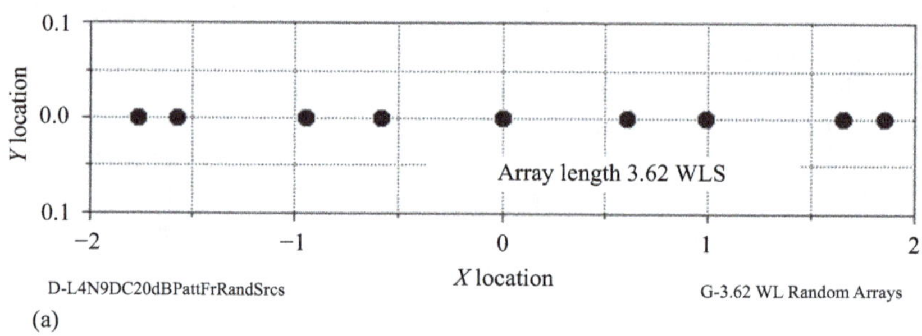

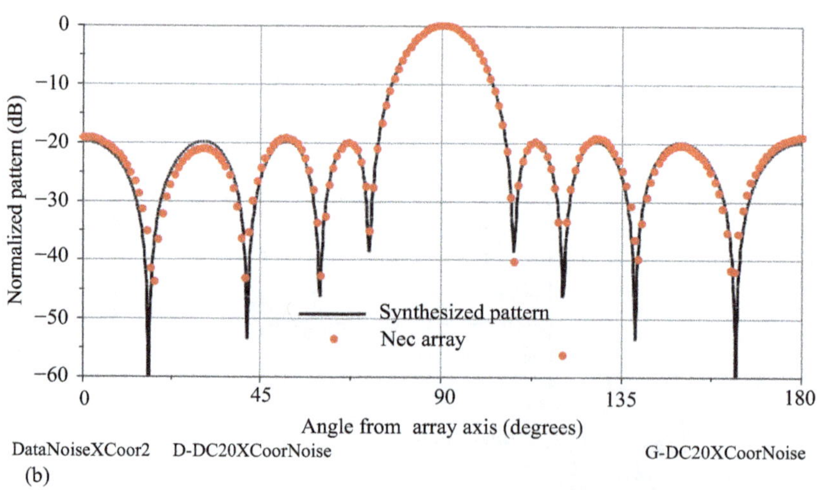

Figure 10.2 (a) The element locations and (b) the synthesized and NEC-computed patterns for a 9-element, 3.2-wavelength randomized array

Table 10.2 Exciting voltages and currents for the 9-element Dolph–Chebyshev randomized array having −20 dB sidelobes of Figure 10.2

Element no.	V_{Real}	V_{Imag}	I_{Real}	I_{Imag}
1	−5.95835E+00	−1.10614E+00	−3.34188E−02	4.61864E−06
2	−6.88189E+00	−1.99729E+00	−1.06344E−01	1.21018E−05
3	−7.32816E+00	−5.13839E+00	−1.40777E−01	9.62530E−06
4	−9.22876E+00	3.93987E+00	−1.20033E−01	5.24156E−06
5	−1.00203E+01	−1.04210E+01	−2.27637E−01	−3.97401E−07
6	−8.79976E+00	1.62886E+00	−1.32988E−01	−6.41862E−06
7	−6.49036E+00	−3.56334E+00	−1.21087E−01	−9.70153E−06
8	−6.22546E+00	−5.46955E+00	−1.29663E−01	−1.70477E−05
9	−4.57429E+00	1.18237E+00	1.19477E−02	1.97798E−06

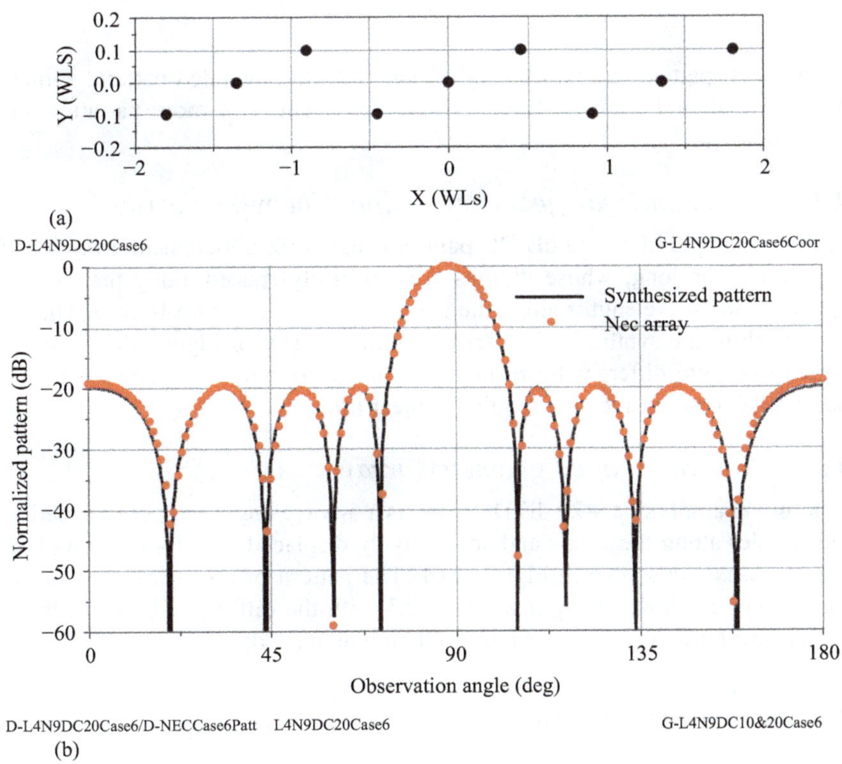

Figure 10.3 (a) The element locations and (b) the synthesized and NEC-computed patterns for a 9- element, 3.6-wavelength nonlinear array

Table 10.3 Currents for 9-element Dolph-Chebyshev nonlinear array having −20 dB sidelobes of Figure 10.3

Element no.	I_{Real}	I_{Imag}
1	−.09279187	−.0270467
2	−.07186295	.03204793
3	−.11254131	.069066712
4	−.11436718	−.03627464
5	−.16210607	−7.684294E−26
6	−.11436718	.03627464
7	−.11254131	−.06906671
8	−.071862950	−.03204793
9	−.092791872	.027046706

The radiation pattern (b) exhibits agreement with the intended pattern, somewhat similar to that of Figure 10.1 for the end lobes but with more variation in the interior lobes.

10.3.3 The nonlinear geometry Dolph–Chebyshev array

The third example of a −20 dB DC pattern is again for a 9-element array which is 3.6 wavelengths long, whose dipoles are uniformly spaced along the x-axis but displaced ±0.1 wavelengths along the y-axis as shown in Figure 10.3(a). The NEC-array match to the synthesized pattern is similar to that of Figure 10.1. This array geometry is quite different from the usual linear arrangement, showing that generating a DC pattern does not require a linear array.

10.3.4 The vee-shaped geometry array

The fourth example of a −20 dB DC array is a 5-wavelength long vee-shaped array of 11 dipoles along the x-axis and successively displaced at 0.05-wavelength steps along the y-axis is shown in Figure 10.4. The pattern of this array does not have nulls as deep as those of Figures 10.1–10.3 with the difference between the synthesized and NEC patterns near end fire being about 1 dB different.

10.4 A tapered pattern

A 15-element 7-wavelength array having a synthesized pattern with sidelobes tapering from −15 dB in −5 dB steps to −45 dB is presented in Figure 10.5. The NEC results and the synthesized patterns agree within a few tenths of a dB throughout the 45 dB range of the pattern.

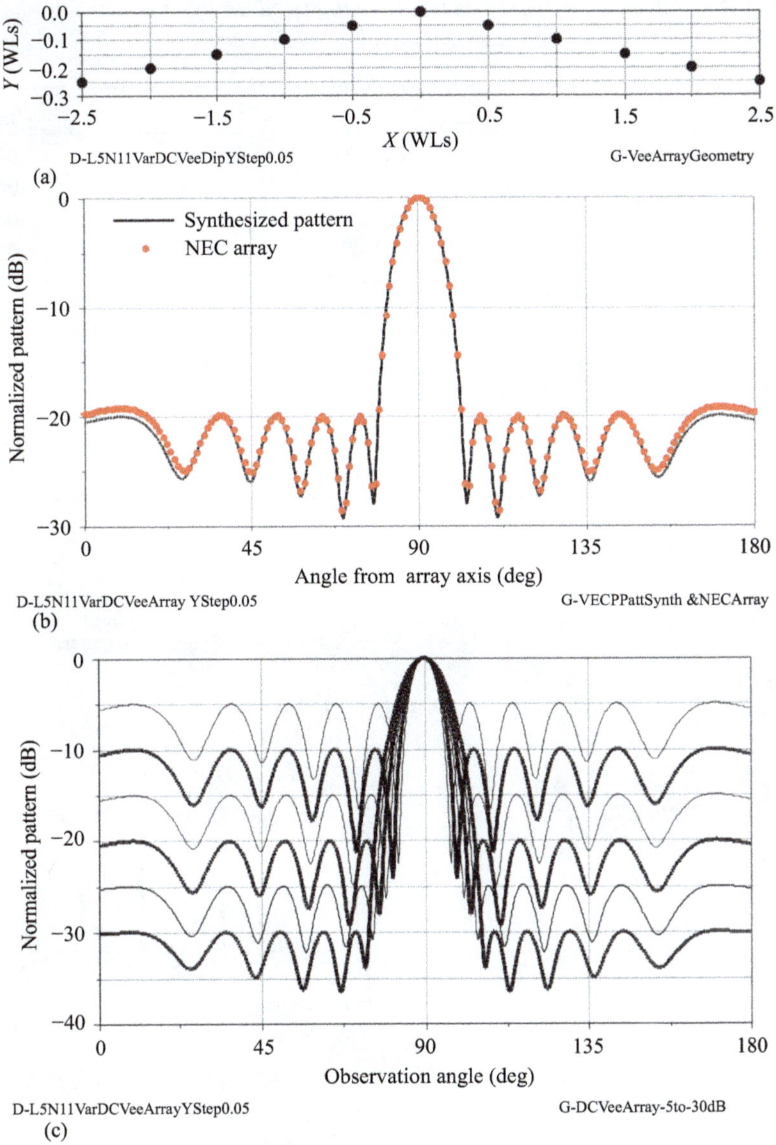

Figure 10.4 *(a) The element locations and (b) the synthesized and NEC-computed −20 dB patterns and (c) synthesized patterns ranging from −5 to −30 dB for an 11-element 5-wavelength vee-shaped array*

Table 10.4 Currents for the 11-element vee-shaped array Dolph-Chebyshev pattern having −20 dB sidelobes of Figure 10.4

Element no.	I_{Real}	I_{Imag}
1	.0178059	.0470000
2	.0138911	.0641140
3	.0541936	.0656606
4	.0858278	.0713842
5	.1191862	.0395169
6	.1315947	.0167981
7	.1191862	.0395169
8	.0858278	.0713842
9	.0541936	.0656606
10	.0138910	.0641140
11	.0178059	.0470000

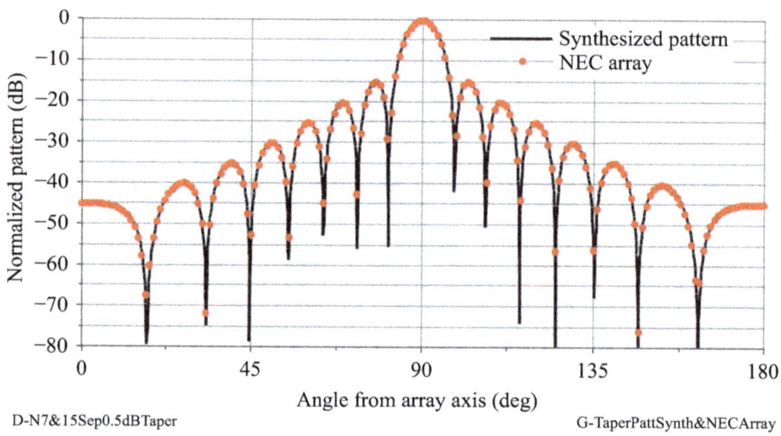

Figure 10.5 The synthesized and NEC tapered patterns for a 15-element, 7-wavelength array

10.5 An alternating lobe array

An array of 15 dipoles that produce an alternating lobe pattern of −20 and −40 dBs is illustrated in Figure 10.6. The agreement between the synthesized pattern and the NEC array is again within a few tenths of a dB.

Table 10.5 Currents for 15-element 7-wavelength array that radiates the symmetric tapered pattern of Figure 10.5

Element no.	I_{Real}	I_{Imag}
1	.0314320	−2.52774E−04
2	.0630510	−4.34806E−04
3	.0702429	−4.04356E−04
4	.0714669	−3.28602E−04
5	.0729315	−2.48783E−04
6	.0752856	−1.65582E−04
7	.0767993	−8.15574E−05
8	.0775808	6.10516E−24
9	.0767993	8.15574E−05
10	.0752856	1.65582E−04
11	.0729315	2.48783E−04
12	.0714669	3.28602E−04
13	.0702429	4.04356E−04
14	.0630510	4.34806E−04
15	.0314320	2.52774E−04

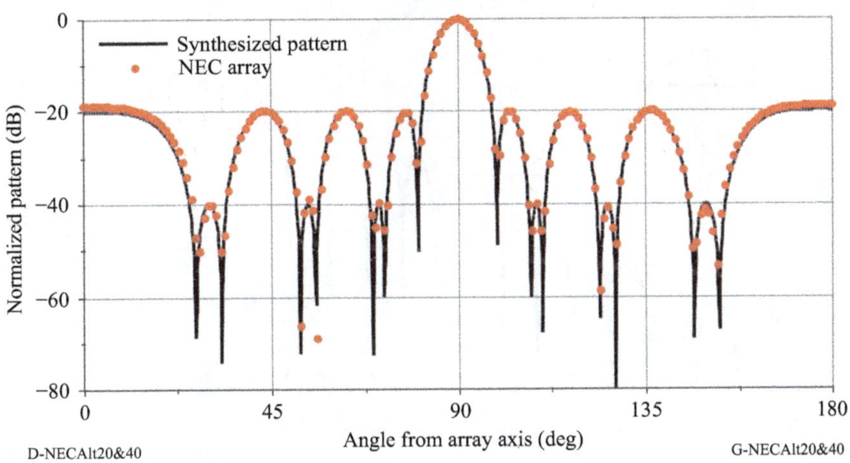

Figure 10.6 The synthesized and NEC alternating-lobe patterns for a 15-element, 7-wavelength array

10.6 A ramped pattern array

A synthesized, ramped pattern having a dynamic range of 70 dB with 15 lobes in 5 dB steps and its NEC counterpart is illustrated in Figure 10.7. The lower 3 lobes

Table 10.6 Currents for 15-element 7-wavelength array that produces the alternating-lobe pattern of Figure 10.6

Element no.	I_{Real}	I_{Imag}
1	.0438904	−7.42780E−04
2	.0396625	−5.65045E−04
3	.0537312	−6.34156E−04
4	.0677421	−6.40609E−04
5	.0803945	−5.79658E−04
6	.0904791	−4.27731E−04
7	.0969836	−2.28195E−04
8	.0542323	1.16675E−23
9	.0969836	2.28195E−04
10	.0904791	4.27731E−04
11	.0803945	5.79658E−04
12	.0677421	6.40609E−04
13	.0537312	6.34156E−04
14	.0396625	5.65045E−04
15	.0438904	7.42780E−04

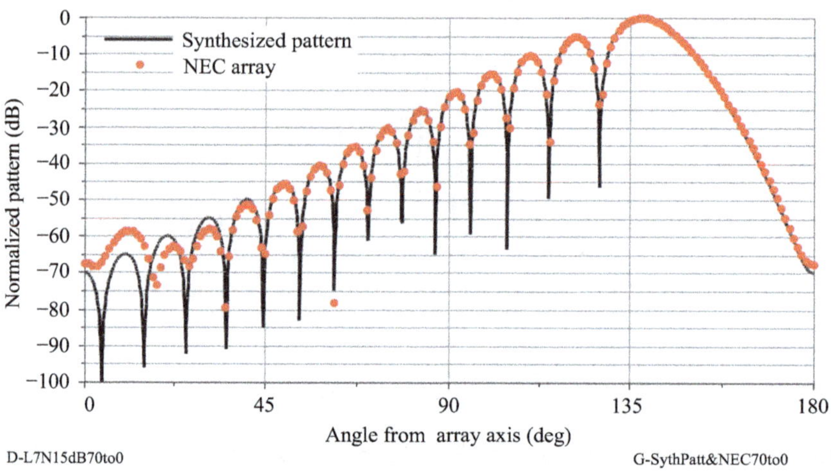

Figure 10.7 The synthesized and NEC ramped patterns for a 15-element, 7-wavelength array

exhibit differences as large as 5 dB but the agreement over the upper 50 dB of the patterns is graphically indistinguishable, even to the −70 dB minimum at 180°. It is not clear why the NEC result differs this much from the target pattern but seems likely due to the compute precision.

10.7 Two-level sidelobe array

Another 17-element, 8-wavelength array is demonstrated in Figure 10.8 for −20 and −40 dB left and right sidelobe levels. The NEC pattern agrees with the target result within a few tenths of a dB except for an approximate +2.5 dB error in both end-fire directions.

Table 10.7 Currents for 15-element 7-wavelength array that radiates the ramped pattern of Figure 10.7

Element no.	I_{Real}	I_{Imag}
1	1.43891E−02	−2.53967E−02
2	−5.10131E−02	−5.33530E−02
3	−6.11367E−02	7.20386E−02
4	9.04855E−02	2.41335E−02
5	−3.50013E−02	−7.84976E−02
6	−2.46251E−02	7.45970E−02
7	6.06884E−02	−4.19274E−02
8	−7.21317E−02	0.00000E+00
9	6.06884E−02	4.19274E−02
10	−2.46251E−02	−7.45970E−02
11	−3.50013E−02	7.84976E−02
12	9.04855E−02	−2.41335E−02
13	−6.11367E−02	−7.20386E−02
14	−5.10131E−02	5.33530E−02
15	1.43891E−02	2.53967E−02

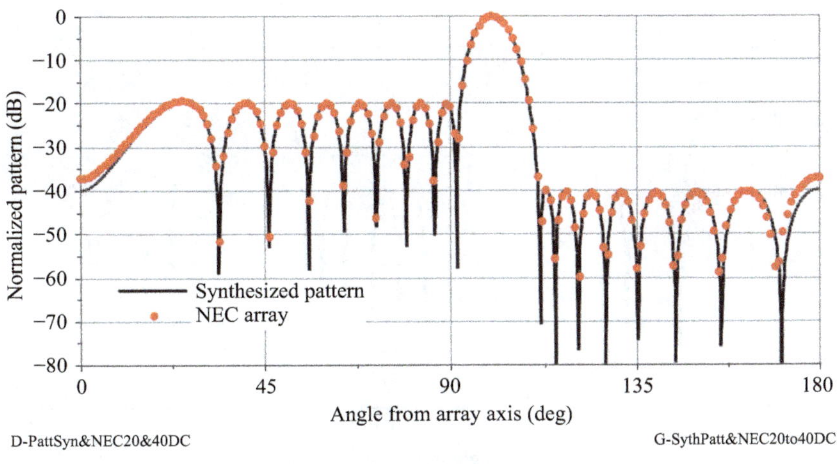

Figure 10.8 The synthesized and NEC patterns for a 17-element, 8-wavelength 2-level-type DC array

10.8 The two main-lobe array

The synthesized and NEC patterns of a 15-element, 7-wavelength array having two main lobes and −40 dB sidelobes are illustrated in Figure 10.9. There is again on the order of a few tenths maximum difference between the two results. The main lobes in the example are relatively wide, a feature that might be reduced by using an array with nonuniformly spaced elements.

Table 10.8 Currents for 17-element 8-wavelength array that radiates the two-level-type dc pattern of Figure 10.8

Element no.	I_{Real}	I_{Imag}
1	2.9735e−2	1.48663e−2
2	2.51053e−2	3.87608e−2
3	3.42868e−2	5.83771e−3
4	5.429e−2	−1.19063e−2
5	3.735e−2	−4.91706e−2
6	1.4209e−2	−7.0772e−2
7	−3.62175e−2	−7.37937e−2
8	−6.8423e−2	−4.60163e−2
9	−8.99515e−2	0e+0
10	−6.8423e−2	4.60163e−2
11	−3.62175e−2	7.37937e−2
12	1.4209e−2	7.0772e−2
13	3.735e−2	4.91706e−2
14	5.429e−2	1.19063e−2
15	3.42868e−2	−5.83771e−3
16	2.51053e−2	−3.87608e−2
17	2.9735e−2	−1.48663e−2

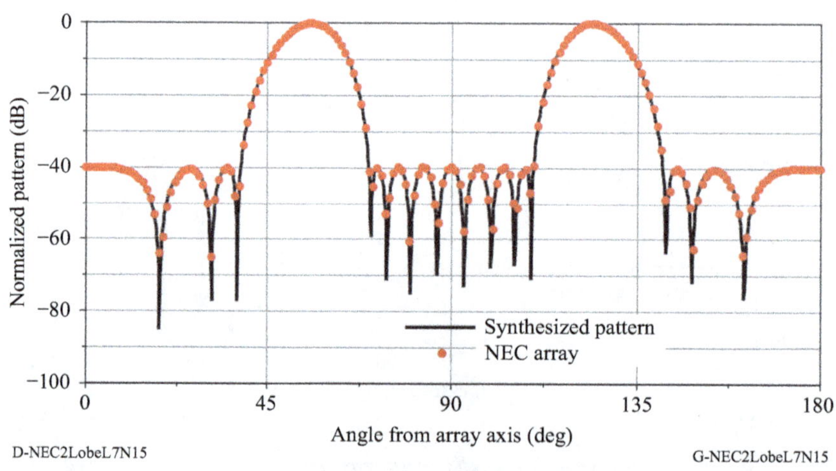

Figure 10.9 The synthesized and NEC two-beam DC patterns for a 15-element, 7-wavelength array

10.9 Scanned main lobe of a DC array

A DC array, having equal-level sidelobes, is well suited for angle scanning of its main beam. An example of this possibility is shown in Figure 10.10. A 17-element, 8-wavelength array with −20 dB sidelobes was synthesized with the main beam scanned to the sixth lobe position from the center using the approach described in [20]. Each of the 17 total lobes of the 17-element array can be made the main lobe to scan the pattern to 17 "look" angles. The NEC samples of this example match the synthesized pattern to the same degree of accuracy as found for most of the previous examples.

Table 10.9 Currents for 15-element 7-wavelength array that radiates the two-beam-type DC pattern of Figure 10.9

Element no.	I_{Real}	I_{Imag}
1	3.12939E−02	−5.34272E−04
2	−2.10927E−02	3.08985E−04
3	−7.54694E−02	9.19119E−04
4	8.91787E−02	−8.75262E−04
5	9.13544E−02	−6.70290E−04
6	−1.79067E−01	8.80241E−04
7	−4.21789E−02	1.03891E−04
8	2.21963E−01	0.00000E+00
9	−4.21789E−02	−1.03891E−04
10	−1.79067E−01	−8.80241E−04
11	9.13544E−02	6.70290E−04
12	8.91787E−02	8.75262E−04
13	−7.54694E−02	−9.19119E−04
14	−2.10927E−02	−3.08985E−04
15	3.12939E−02	5.34272E−04

Table 10.10 Currents for 17-element 8-wavelength array for one scan angle for the DC pattern of Figure 10.10

Element no.	I_{Real}	I_{Imag}
1	−.0539317	−.0381817
2	.0350317	−8.80232E−03
3	−.0196925	.0395528
4	−.0202375	−.0477967
5	.0561397	.0178807
6	−.0551144	.0341402
7	.0106420	−.0685240
8	.0469328	.0547769

(Continues)

Table 10.10 (Continued)

Element no.	I_{Real}	I_{Imag}
9	−.0730901	1.85599E−24
10	.0469328	−.0547769
11	.0106420	.0685240
12	−.0551144	−.0341402
13	.0561397	−.0178807
14	−.0202375	.0477967
15	−.0196925	−.0395528
16	.0350317	8.80232E−03
17	−.0539317	.0381817

Table 10.11 Real currents for a 9-element 4-wavelength array that radiates a −20 dB DC pattern and those for the P = 2, 3, and 4 patterns of Figure 10.11

Element no.	$P = 1$	$P = 2$	$P = 3$	$P = 4$
1	.500000	.250000	.125000	.0624999
2	.511550	.511550	.383662	.255775
3	.675150	.936833	.8988876	.730110
4	.790000	1.48074	1.76248	1.69915
5	.831350	2.09542	3.04966	3.44528
6	.789999	2.70729	4.78821	6.31920
7	.675149	3.23007	6.93649	10.7861
8	.511549	3.58256	9.37219	17.4791
9	.499999	3.87436	12.0227	27.1054
10		3.58256	14.3880	39.8125
11		3.23007	16.3597	55.1984
12		2.70729	17.6512	72.0278
13		2.09542	18.1002	88.4041
14		1.48074	17.6512	102.126
15		.936833	16.3597	109.961
16		.511549	14.3880	−2.87387E−26
17		.249999	12.0227	−4.19929E−09
18			9.37219	48.9678
19			6.93649	109.961
20			4.78821	102.126
21			3.04966	88.4041
22			1.76248	72.0278
23			.898887	55.1984
24			.383662	39.8125
25			.124999	27.1054

(Continues)

Table 10.11 (*Continued*)

Element no.	P = 1	P = 2	P = 3	P = 4
26				17.4791
27				10.7861
28				6.31920
29				3.44528
30				1.69915
31				.730110
32				.255775
33				.0625000

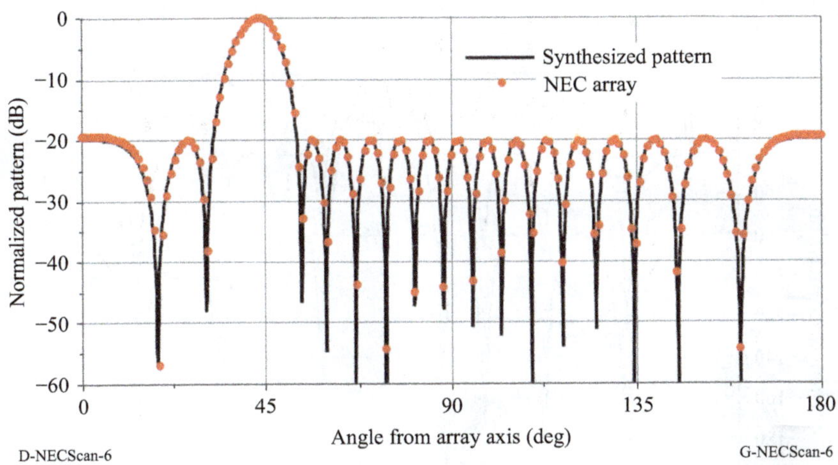

Figure 10.10 *The synthesized and NEC −20 dB DC scanned patterns for a 17-element, 8-wavelength array*

10.10 Exponentiated Dolph–Chebyshev array

A last comparison of synthesized- and NEC-array patterns is presented in Figure 10.11. The "$P = 1$" pattern is the standard 9-element, 4-wavelength DC −20 dB result previously shown in Figure 10.1. The patterns labeled "$P = 2$" and "$P = 3$" come from squaring and cubing the $P = 1$ pattern, thus creating sidelobes of −40 dB and −60 dB. How these two patterns are synthesized is described in [3]. Prony-method synthesis yields arrays 8 and 12 wavelengths long having 17 elements and 25 elements, respectively. The standard DC arrays of 8 and 12 wavelengths would have 17 and 25 lobes. The number of lobes in the $P = 2$ and $P = 3$ versions retain the 9-lobe

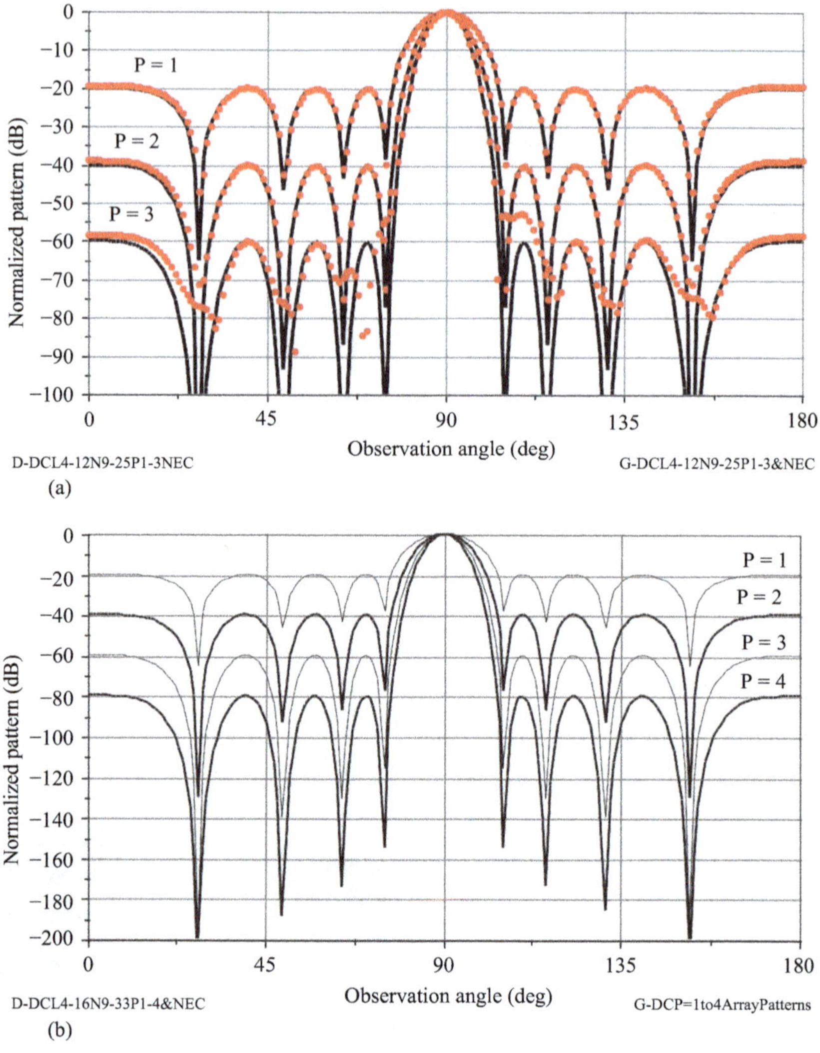

Figure 10.11 (a) The synthesized and NEC −20 dB DC-type patterns for a 9-element, 4-wavelength array for P = 1 and the P = 2 and P = 3 versions of that pattern and (b) synthesized versions for patterns with P ranging from 1 to 4

structure of the 4-wavelength array. However, the widths of the lobes are successively reduced and their null depths are increased. The NEC pattern for $P = 3$ differs substantially from the synthesized version, likely due to inadequate compute precision, while the agreement demonstrated with the other two patterns is similar to the other NEC results.

10.11 Summary

The output of the typical pattern-synthesis exercise is normally the individual currents needed to produce a desired radiation pattern from a specified array geometry. The array is typically comprised of radiating elements, such as dipole antennas, that are spaced closely enough that mutual interactions between them needs to be taken into account. Whether that pattern can be produced by a physical array requires the further step of determining the exciting voltages needed to produce the desired individual currents. The goal of this discussion has been to explore the latter step for a variety of antenna patterns using the NEC computer model. For reference purposes, the NEC voltages needed to match the target pattern were given for two cases in Tables 10.1 and 10.2 and the array element currents were included in Tables 10.1–10.11 for all of the synthesized patterns.

The NEC results were generally found to agree within a few tenths of a dB for the patterns for which they were developed. The two exceptions where the differences were greater occurred when the lobe maxima were at levels below −50 dB. The final step in demonstrating the practicality of producing a desired synthesized pattern would be implementation and testing an array design in actual hardware, a step beyond this study.

References

[1] S. A. Schelkunoff, “A Mathematical Theory of Linear Arrays,” *Bell System Technical Journal*, Vol. 22, pp. 80–107, 1943.

[2] C. L. Dolph, “A Current Distribution for Broadside Arrays Which Optimizes the Relationship between Beam Width and Side-Lobe Level,” *Proceedings of the IRE*, Vol. 34, No. 7, pp. 335–348, 1946.

[3] P. M. Woodward, “A Method of Calculating the Field Over a Plane Aperture Required to Produce a Given Polar Diagram,” *Journal Institute of Electrical Engineering*, Pt. IIIA, Vol. 93, pp. 1554–1558, 1946.

[4] T. T. Taylor, “Design of Line-Source Antennas for Narrow Beamwidth and Low Sidelobes,” *IRE Transactions on Antennas and Propagation*, Vol. 7, pp. 16–28, 1955.

[5] R. S. Elliott, “On Discretizing Continuous Aperture Distributions,” *IEEE Transactions on Antennas and Propagation*, Vol. 25, No. 5, pp. 617–621, 1977.

[6] R. S. Elliott, *Antenna Theory and Design*, Prentice-Hall, Englewood Cliffs, NJ, 1981.

[7] C. A. Balanis, *Antenna Theory: Analysis and Design*, Harper & Row, Publishers, New York, 1982.

[8] W. H. Kummer, “Basic Array Theory,” *Proceedings of the IEEE*, Vol. 80, No. 1, pp. 127–140, 1992.

[9] H.-P. Chang, T. K. Sarkar, and O. M. C. Pereira-Filho, “Antenna Pattern Synthesis Utilizing Spherical Bessel Functions,” *IEEE Transactions on Antennas and Propagation*, Vol. 48, No. 6, pp. 853– 859, 2000.

[10] M. Durr, A. Trastoy, and F. Ares, "Multiple-Pattern Linear Antenna Arrays with Single Pre-Fixed Amplitude Distributions: Modified Woodward-Lawson Synthesis," *Electronics Letters*, Vol. 36, No. 16, pp. 1345–1346, 2000.

[11] D. Marcano, and F. Duran, "Synthesis of Antenna Arrays Using Genetic Algorithms," *IEEE Antennas and Propagation Magazine*, Vol. 42, No. 3, pp. 12–20, 2000.

[12] K. L. Virga, and M. L. Taylor, "Transmit Patterns for Active Linear Arrays with Peak Amplitude and Radiated Voltage Distribution Constraints," *IEEE Transactions on Antennas and Propagation*, Vol. 49, No. 5, pp. 732–730, 2001.

[13] O. M. Bucci, M. D'Urso, and T. Isernia, "Optimal Synthesis of Difference Patterns Subject to Arbitrary Sidelobe Bounds by using Arbitrary Array Antennas," *IEE Proceedings on Microwaves, Antennas and Propagation*, Vol. 152, No. 3, pp. 129–137, 2005.

[14] R. Vescovo, "Consistency of Constraints on Nulls and on Dynamic Range Ratio in Pattern Synthesis for Antenna Arrays," *IEEE Transactions on Antennas and Propagation*, Vol. 55, No. 10, 2662–2670, 2007.

[15] Y. Liu, Z. Nie, and Q. H. Liu, "Reducing the Number of Elements in a Linear Antenna Array by the Matrix Pencil Method," *IEEE Transactions on Antennas and Propagation*, Vol. 56, No. 9, pp. 2955–2962, 2008.

[16] N. G. Gomez, J. J. Rodriguez, K. L. Melde, and K. M. McNeill, "Design of Low-Sidelobe Linear Arrays with High Aperture Efficiency and Interference Nulls," *IEEE Antennas and Wireless Propagation Letters*, Vol. 8, pp. 607–610, 2009.

[17] M. Comisso, and R. Vescovo, "Fast Iterative Method of Power Synthesis for Antenna Arrays," *IEEE Transactions on Antennas and Propagation*, Vol. 57, No. 7, pp. 1952–1962, 2009.

[18] A. M. H. Wong, and G. V. Eleftheriades, "Adaptation of Schelkunoff's Superdirective Antenna Theory for the Realization of Superoscillatory Antenna Arrays," *IEEE Antennas and Wireless Propagation Letters*, Vol. 9, pp. 315–318, 2010.

[19] P. S. Apostolov, "Linear Equidistant Antenna Array with Improved Selectivity," *IEEE Transactions on Antennas and Propagation*, Vol. 59, No. 10, pp. 3940–3943, 2011.

[20] T. Zhang, and W. Ser, "Robust Beampattern Synthesis for Antenna Arrays with Mutual Coupling Effect," *IEEE Transactions on Antennas and Propagation*, Vol. 59, No. 8, pp. 2889–2895, 2011.

[21] H. Singh, H. L. Sneha, and R. M. Jha, "Mutual Coupling in Phased Arrays: A Review," *International Journal of Antennas and Propagation*, Vol. 2013, No. 1, 23 pages, 2013.

[22] E. K. Miller, "Synthesizing Linear-Array Patterns via Matrix Computation of Element Currents," *IEEE Antennas and Propagation Society Magazine*, Vol. 55, No. 5, pp. 85–96, 2013.

[23] E. K. Miller, "Using Prony's Method to Synthesize Discrete Arrays for Prescribed Source Distributions and Exponentiated Patterns," *IEEE Antennas and Propagation Society Magazine*, Vol. 57, No. 1, pp. 147–163, 2015.

Chapter 11

Using model-based parameter estimation to assess the accuracy of numerical models[1]

Abstract

Determining the accuracy of the results produced by a numerical model remains an ongoing and important problem in computational electromagnetics, one for which there are relatively few options. Perhaps the most desirable and convincing approach is to perform a quantitative comparison between experimental data and the model output. Such a comparison, though undeniably useful, has the disadvantage that it embodies two different kinds of errors, these being the physical modeling error (PME) and the numerical modeling error (NME). The former arises from any differences that exist between the actual physical problem and its numerical representation, and the latter occurs because the numerical results actually obtained represent only an approximate solution to that numerical representation. Although the PME can normally be expected to be the larger of the two and also the more difficult to evaluate, quantitative knowledge of the NME itself is essential, estimation of which is the topic of this chapter.

There are two approaches now used to estimate the NME, one being the internal checks that can be done within a specific model or code (e.g., convergence tests, boundary conditions, power conservation), and the other the external checks that involve comparing the results of two or more different models. The approach discussed here is another kind of internal check, but different from the others in the sense that it seeks a "noise" level in the model output as a measure of the model accuracy using model-based parameter estimation.

The basic idea is to apply a low-order, physically motivated, fitting model (FM) to a sampled, generating model (GM), based on a first-principles formulation) to be tested. Were these GM samples known exactly, then some minimum number would in principle permit the continuous transfer function (TF) to be exactly reconstructed across the frequency range that they span if the FM exactly represents the GM behavior. In fact, however, the GM samples and the FM are only approximations, and were no more samples available the FM-based TF could possess no better accuracy either. However, by developing more samples from the GM, the

[1]Adapted and expanded from Using Model-Based Parameter Estimation to Estimate the Accuracy of Numerical Models, in *Proceedings of 12th Annual Review of Progress in Applied Computational Electromagnetics*, 1996, pp. 588–595.

additional information they provide can be used to estimate the accuracy of the TF derived from the FM and of the GM samples themselves. Furthermore, these extra GM samples might also be used to improve the accuracy of the FM estimate.

11.1 Introduction

However obtained, determining the quantitative accuracy of the numerical results developed for a particular electromagnetic (EM) problem is essential if such results are to be reliable. This is a problem of data validation. Another related, but broader, problem is that of establishing the accuracy that may be expected to be obtained from a given computer model when it is employed following accepted guidelines for its use. This is a problem of code or model validation. The latter problem is intrinsically open-ended and one in which one or just a few code developers might be involved, while the former is more specific and can involve relatively many more users of that code. Attention in the following is addressed to the problem of data validation.

The basic question that confronts anyone who has to deal with numerical results is how accurate they are relative to the requirements of the problem for which they have been obtained. This issue can be expressed in various ways, but ultimately reduces to a problem of associating error bars, or uncertainty bounds, with numerical data. Traditionally, the purpose of error bars is to indicate that the results are thought to have some numerical probability of lying within the band they span at some level of confidence. The question of accuracy, in other words, has a probabilistic, or statistical, context. While that degree of analytical formality and rigor could be useful in computational electromagnetics (CEM), it goes well beyond current practice and is not a necessary first step towards satisfying some minimally acceptable accuracy statement for computed results. In the discussion here, attention is limited instead to deriving a quantitative estimate of numerical solution, or data, accuracy. Systematic extension of the idea presented here, or some alternative, could lead to a more rigorous methodology for error assessment. In the following the kinds of data validation that can be used are discussed; how model-based parameter estimation (MBPE) models can provide an estimate of the error in the transfer function (TF) that data represents; and demonstrations of this procedure to a pole spectrum, the current induced on an infinite, circular cylinder and other selected examples. It is worth emphasizing that using MBPE for accuracy estimation requires more GM samples than those needed to simply compute the FMs.

11.2 Kinds of data validation

There are two fundamental ways by which data validation, or estimating the accuracy of EM observables, can be attempted. One is to compare the results in question with independent data, the expectation being that the true values are probably as close to these data as they are to each other, something that can be described as an external validation or check. This is a more convincing test when the independent data are experimental and come from the actual problem of interest, as the close agreement of two numerical results does not necessarily prove that they are correct and also does not shed any light on the physical modeling error (PME).

The other is to perform some test on the numerical results themselves, either to evaluate the conformity of the results relative to Maxwell's equations or their numerical consistency, as an internal validation or check. In the context of CEM, an internal check that derives from Maxwell's equations is most satisfactory, of which some examples are reciprocity, energy conservation, and boundary-field checks. An example of perhaps the most commonly used internal numerical check is a convergence test whereby the behavior of the numerical solution is examined as the number of unknowns is progressively increased. Unfortunately, the convergence of a numerical model is not always assured to occur, nor is convergence to a correct answer guaranteed either. Furthermore, a convergence test can be very expensive and even impractical for large problems.

A different kind of internal check using Maxwell's equations and model-based parameter estimation (MBPE) is discussed here. It is based on the observation that the various field quantities derived from a defining microscopic description, that is, Maxwell's equations, or a generating model (GM), must subsequently exhibit a known, reduced-order, macroscopic behavior. The specific behavior of interest for our purposes is the resonance-like structure of an EM frequency response, which is well described by a rational function [1]. As shown in the next section, knowledge that the TF has a simple analytical description, termed here a fitting model (FM), that can be exploited to test the numerical accuracy of the macroscopic field variables, and hence, of the GM that produced them. Note that determining the parameters of the FM requires some minimum number of GM samples while further additional GM samples are needed to estimate the GM accuracy using MBPE.

The approach taken here to estimate the accuracy of a numerical model is thus based on modeling the solution as a function of frequency. An alternative model-based approach might also be employed, where the error in the solution is the quantity being modeled rather than the value of the solution itself. This approach requires knowledge of how the error varies with the number of unknowns, X or equivalently with the spatial sample interval, h, since this permits combining solutions using different values of X, to obtain a solution estimate better than that for the largest X used. When the error varies as $\mathrm{A}h^n$ with n an integer, a behavior true of many numerical operations, a procedure known as Richardson extrapolation results [2]. Using the observation that an EM numerical modeling error (NME) for an integral-equation formulation varies as $\mathrm{A}e^{(-BX)}$ [3] similar approach can be used to improve the best result from a convergence test.

11.3 Modeling generating-model samples

It has been demonstrated [1] that a sequence of numerical TF samples, $S_i = S(f_i)$, computed from a CEM model (the GM) as a function of frequency f at sample frequencies f_i can be well approximated using a rational function [an FM] as a generalization of the pole series that alone describes the resonance behavior of the former. A P-term poles series can be expressed as the rational function

$$\mathrm{R}(\mathrm{P}-1,\mathrm{P},\mathrm{f}) = \mathrm{A}(\mathrm{P}-1,\mathrm{f})/\mathrm{B}(\mathrm{P},\mathrm{f}),$$

where

$$A(P-1,f) = n_0 + n_1 f + n_2 f^2 + \ldots + n_{P-1} f^{P-1},$$

and

$$B(P,f) = d_0 + d_1 f + d_2 f^2 + \ldots + d_P f^P,$$

that is, as a ratio of two polynomials of degrees $P - 1$ and P, respectively. However, as P increases, this model becomes less robust numerically. Furthermore, an EM TF is not completely described by a pole series, possibly having in addition a non-pole contribution. For these reasons, it was found that a series of overlapping, windowed, low-order rational functions having the form $R(P+1,P,f)$ is a better choice to model a wideband TF. This increase in the order of $A(P+1,f)$ relative to $B(P,f)$ or a rational function $R(P+1,P,f) = R(N,D,f)$ can also help to model the effects of poles that lie outside the frequency window modeled by the unknown, non-pole contribution where N and D represent the orders of the numerator and denominator polynomials that comprise the FM rational function.

Each windowed, FM is quantified by fitting it to $M = N + D + 1$ computed GM samples over the frequency range that it spans, from which its polynomial coefficients are then obtained as a solution to an Mth-order data matrix, assuming a square data matrix is used. The number of GM samples required is one less than the number of coefficients in $R(N,D,f)$ because one of the coefficients is set to unity, typically d_P. The minimum number of GM samples needed, were its TF to be a purely pole series, would be two per pole or resonance; otherwise, a non-unique FM will result. However, because of the factors just discussed, more than two samples per resonance may be required. Thus, if a FM is to span P poles, as many as $M \sim (2 \text{ to } 3)\times P$ GM samples may be needed. If the FM order is to be kept below a "reasonable" number, say 20, then the number of resonances spanned by an FM should be no more than 6, with numerator and denominator polynomial orders of at most 9 or 10.

Because the condition number (CN) of the data matrix grows with increasing M for an FM spanning a fixed frequency band, the numerical match or data fit (DF) achieved between that FM and the GM samples used for quantifying it will decrease commensurately. However, the model fit (MF) obtained between the FM and additional GM samples not originally used for its quantification will increase with increasing M until some maximum, M', is reached, beyond which MF will also decrease because of the increasing CN. The opportunity for estimating the accuracy, or error, of the GM samples lies in the range between some minimum value of M and $M' > M$. Some results obtained using a computer precision (CP), of 24 digits for a 15-pole spectrum having $s_i = \sqrt{i/20} + j{*}i$ ($j = \sqrt{-1}$) are included in Figure 11.1 to illustrate these points, where the $M = N + D + 1$ GM samples are evenly spaced between frequency end-points of 5.5 and 8.5.

The CN and DF as a function of N with D a parameter are shown in Figure 11.1(a) with all results displayed in digits, that is, where a CN of 10^X has the numerical value of X, etc. It can be seen that increasing N and/or D increases CN and decreases DF, the sum of which is furthermore about equal to CP. Thus, were a better fit to be sought

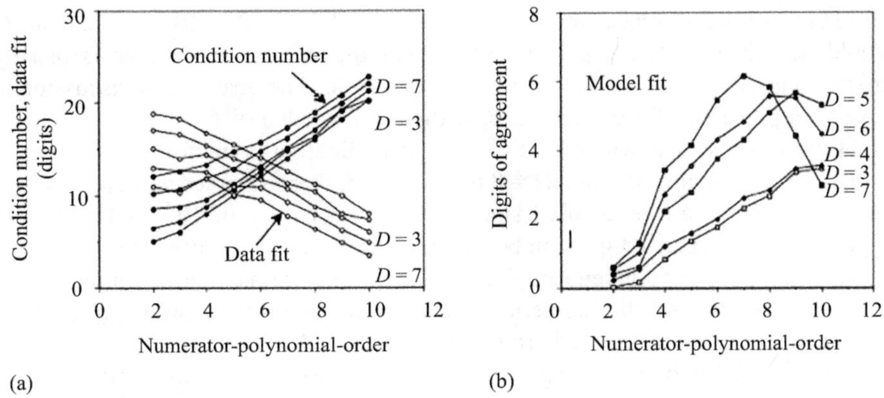

Figure 11.1 The condition number and data fit between the fitting model and the pole-series generating-model samples used to quantify it are presented in part (a), while in (b) the agreement between the fitting model and additional generating-model samples is presented, all results being shown in digits and as a function of the fitting-model numerator-polynomial order with the denominator-polynomial order, D, a parameter. The maximum model fit shown in (b) indicates that the smallest error detectable for this particular example would be about 10^{-6}. These calculations were performed in 24-digit computer precision.

between the original data and the FM, this might be accomplished by increasing CP. This possibility is not pursued further here but is an option to be considered.

The corresponding average match between the FM and additional GM samples, MF, is shown in digits in Figure 11.1(b). It can be seen that MF increases with both N and D, until the growing CN causes a downturn for the larger values of these parameters. By varying its parameters, it is possible to develop a FM from the original GM samples that produces an approximate 6-digit match to additional samples within the frequency range spanned by the FM. The significance of this result is further considered below.

11.4 Estimating the error in a transfer function

The accuracy of the GM samples can be expressed relative to an exact answer for the numerical representation of the physical problem being modeled, $E_i = E(f_i)$, as

$$A_i = [|(S_i - E_i)/E_i|] \tag{11.1}$$

or in a somewhat more convenient measure, digits of agreement using

$$B_i \equiv \log10[|(S_i - E_i)/E_i|]. \tag{11.2}$$

Thus, a 1% difference would yield $B_i = 2.0$, while for 2×10^{-3} agreement, B_i would be 2.7, etc. The *a priori* accuracy of the numerical samples is usually unknown as is E_i, so B_i (and A_i) is also not available. The goal of this discussion to develop a procedure for estimating B_i without knowledge of E_i.

First observe that when S_i is expressed in floating-point form, for example using an x-digit mantissa where $x = 8$ is often used, these samples normally contain more digits than can be justified since a relative accuracy of 10^{-8} is usually not expected. These excess digits can be regarded as numerical "noise" that represents an accuracy floor for a given sample. Assuming no systematic error source affects the computed samples, the numerical noise should be random and the S_i samples should be randomly distributed around the unknown E_i values.

The possibility of estimating B_i is based on this assumed random variation of S_i relative to E_i by comparing GM and FM samples. The idea is to oversample across the frequency range of interest, F_{GM}, the GM whose output is to be tested, using a subset of these data samples, S_{DF}, to develop an FM that spans part or all of that bandwidth, F_{FM}. This FM is then used to compute an average DF of $\underline{B}_{DF}$ digits to the S_{DF} samples and an average MF of $\underline{B}_{MF}$ to the remaining S_{MF} samples over F_{FM}. The value of $\underline{B}_{MF}$ relative to $\underline{B}_{DF}$ provides a consistency measure between the two data sets and the FM. This is so because all of the data comes from the same GM and so S_{DF} and S_{MF} must be internally consistent to an amount determined by the average GM accuracy, $\underline{B}_{GM}$. The value of $\underline{B}_{DF}$ for the data set S_{DF} will be only as accurate as permitted by the CN of the data matrix and CP used as illustrated in Figure 11.1(a). It is, however, relatively independent of $\underline{B}_{GM}$ and can actually exceed it since $\underline{B}_{DF}$ provides only a measure of how well the FM can match a set of numbers whose data matrix has a CN when the computations are performed to a specified CP. However, the value of $\underline{B}_{MF}$ for the data set S_{MF} will be limited in accuracy to the lesser of $\underline{B}_{GM}$ and $\underline{B}_{DF}$ because the FM can match the data set S_{MF} no better than it matches the data set S_{DF} on which it is based, the accuracy of which is $\underline{B}_{GM}$. Two outcomes are thus possible: (1) $\underline{B}_{DF} \sim \underline{B}_{MF}$ and (2) $\underline{B}_{DF} > \underline{B}_{MF}$. The third possibility, $\underline{B}_{DF} < \underline{B}_{MF}$, has already been excluded. In either case, $\underline{B}_{MF}$ provides the valid, lower-bound estimate sought for $\underline{B}_{GM}$, a point to be demonstrated in the following.

Observe that the procedure just outlined is similar to the use of linear regression. For that particular problem, the goal is to obtain the best fit between available data and a straight line that serves as a model for the data. Uncertainty or errors in the data can be estimated by the mismatch that is found between that data and the best-fit straight line. In our case, the model is a rational function rather than a pole series, but the concept is the same. Both of these problems are examples of the more general one of finding the model that best represents the data of interest, a problem also known as system identification. A general discussion of modeling data is given in [4].

For illustrative purposes, consider the pole-spectrum example just presented. Based on the results of Figure 11.1(b), it appears reasonable to select a candidate FM with $D = 7$ and an N varying from 6 to 9 in order to achieve a relative maximum in $\underline{B}_{MF}$, using the requisite data set S_{DF} needed to quantify the FM parameters.

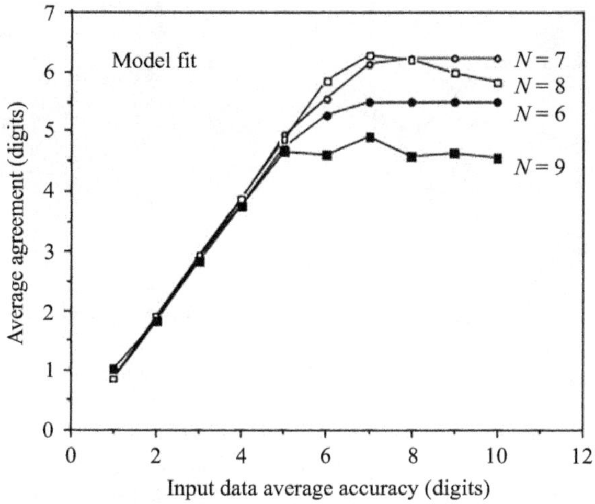

Figure 11.2 The average agreement achieved between a FM and additional samples of the GM for the example of Figure 11.1, as a function of the GM sample accuracy. The FM denominator-polynomial order is 7 and its numerator-polynomial order is a parameter. This result indicates the feasibility of estimating the accuracy of the GM samples from the agreement they exhibit with the fitting model.

Using the new data set S_{MF}, $\underline{B}_{MF}$ can then be computed over F_{FM}. In order to illustrate the effect of GM uncertainty, uniformly distributed random noise is added to all GM samples in order to determine the degree of correlation between the noise level and $\underline{B}_{MF}$. The results of this experiment are plotted in Figure 11.2.

The quantity $\underline{B}_{GM}$, controlled by the dynamic range of the additive noise, is plotted on the horizontal axis with $\underline{B}_{MF}$ plotted on the vertical axis. There is essentially a one-to-one correlation between $\underline{B}_{GM}$ and $\underline{B}_{MF}$ up to values between 5 and 6, beyond which variations are seen that can be deduced from Figure 11.1(b) as being due to differences in the FM numerator-polynomial order. Note that this computation involves no *a priori* knowledge of $\underline{B}_{GM}$, its value rather being inferred from $\underline{B}_{MF}$. Thus, when using simulated frequency-response data, it appears that the MBPE procedure does provide a quantitatively valid way for estimating the noise level, or equivalently the accuracy, of the data being modeled. In the next section, the technique is applied to an EM problem.

11.5 Application to an electromagnetic transfer function

The problem to be considered is the frequency-dependent current excited on the illuminated side of an infinite, circular cylinder for a normally incident, TE plane wave. This problem has an analytical solution, which is used here, but has also become a benchmark solution for various CEM models. Using the analytical

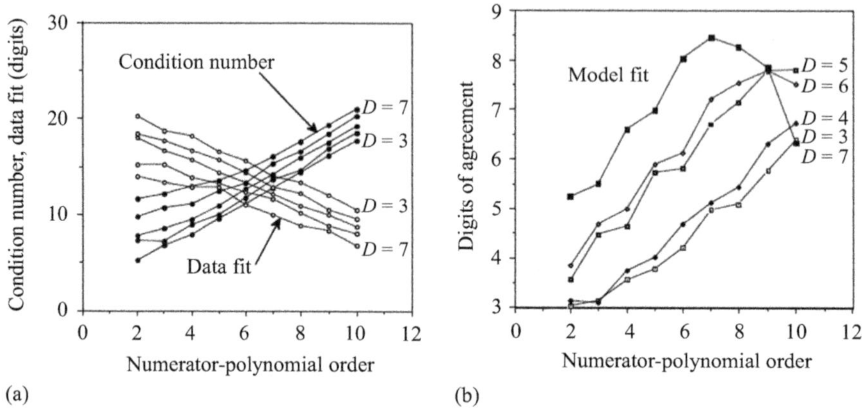

Figure 11.3 The results here are similar to those of Figure 11.1 except that the generating model is the analytical, series solution for the current induced on the illuminated side of an infinite, circular cylinder by a normally incident plane wave and sampled between ka = 3 and 6. The lower condition numbers for this problem compared with the pole-series example of Figures 11.1 and 11.2 result in more accurate agreement between the fitting and generating models in (b). A computer precision of 24 digits was used.

solution permits performing a computation whose accuracy can be modified in a controlled fashion by adding noise to, or otherwise affecting, the solution. First discussed is the convergence parameter that controls the number of terms summed in an infinite series to control its accuracy with all other calculations otherwise done in 24-digit CP. The frequency range covered by the results which follow is for $ka = 3$ to 6 where ka is the cylinder circumference in wavelengths.

Results for the CN and DF are shown in Figure 11.3(a), and results for the MF as a function of FM numerator-polynomial order with the FM denominator-polynomial order a parameter are shown in Figure 11.3(b). As found for the pole spectrum above, the CN increases and the DF decreases with N, with their sum approximately equal to the CP. Similarly, the MF is found to increase with N until, as is seen for $D = 7$, the effect of the increasing CN causes it to decrease after reaching a peak value of about 8.5 digits.

For the parameters used to model this particular response, it appears that using $D = 7$ and letting N vary from 3 to 7 will yield the best MF as the GM accuracy is varied, which leads to the result shown in Figure 11.4.

The dependence of the $\underline{B}_{MF}$ on the input-data accuracy is similar to that already seen in Figure 11.2. Of most significance is that there is again a nearly one-to-one correspondence between $\underline{B}_{MF}$ and $\underline{B}_{GM}$, indicating the potential utility of the procedure for estimating the accuracy of that data, and hence, of the GM that produced it in the case of CEM. The effect of decreasing the input-data accuracy to 4 digits for the infinite-cylinder GM, as shown in Figure 11.5, further demonstrates the

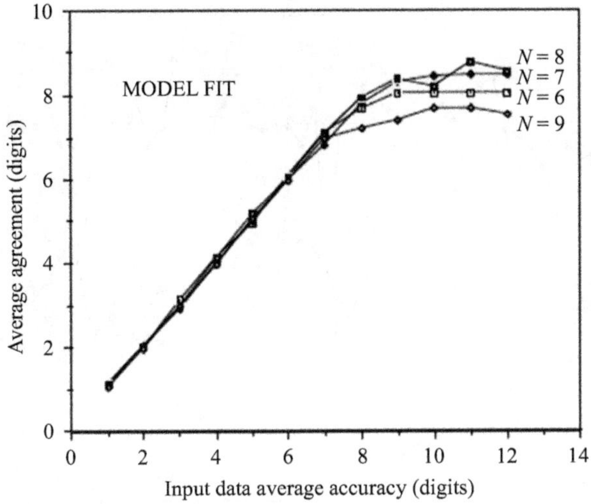

Figure 11.4 Average agreement achieved between a fitting model and additional samples of the EM generating model for the example of Figure 11.3, as a function of the generating-model sample accuracy. The fitting-model denominator-polynomial order is 7 and the numerator-polynomial order is a parameter. This result indicates the feasibility of estimating the accuracy of CEM samples from the agreement they exhibit with the fitting model.

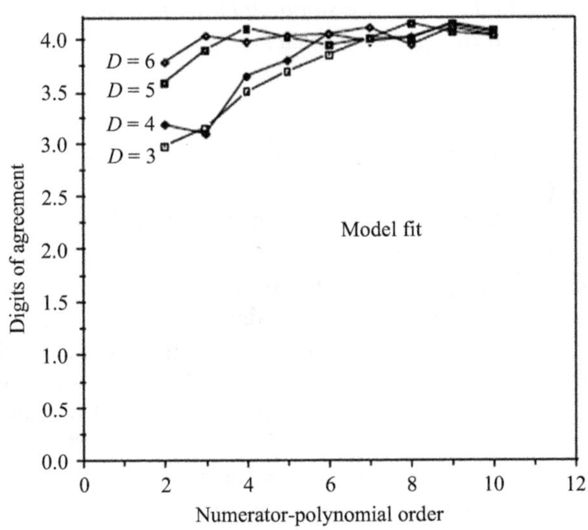

Figure 11.5 The results here repeat those of Figure 11.3 for the infinite-cylinder generating model, but the input-data accuracy was reduced to 4 digits. It can be seen that if the fitting-model, denominator-polynomial order is 5 or 6, the fitting-model agreement to the input data is within 3×10^{-4} of the input-data accuracy, for numerator-polynomial orders of 4 or greater.

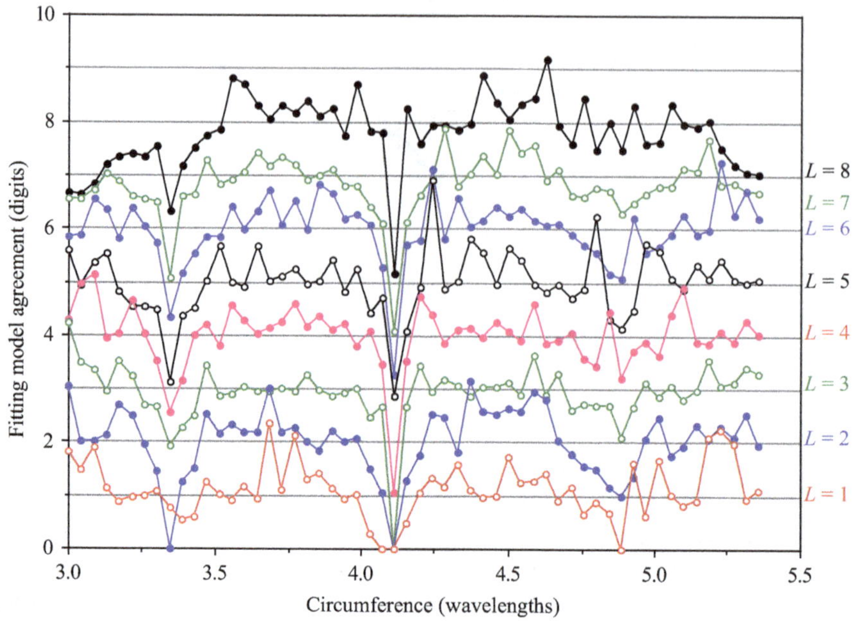

Figure 11.6 The fitting-model agreement for each additional data sample of the cylinder generating model not used for the fitting-model computation, with the input-data accuracy a parameter ranging from 10^{-1} *to* 10^{-8} *is presented here. These results show that "noise" in the input data (i.e., the stopping term in the summation) is accurately estimated by the fitting model.*

feasibility of determining the generating-model accuracy from its agreement with the FM.

A more detailed demonstration of this possibility is illustrated in Figure 11.6 where the frequency dependence of the MF is shown with the input-data accuracy a parameter. The parameter L is the digit specification for the convergence of the series used in computing the infinite-cylinder test current, that is, the series is computed until $|T_i/\mathrm{Sum}(i)| < 10^{-L}$, where T_i is the ith term in the series and Sum(i) is the series sum up through that term. The individual $B_{i\mathrm{MF}}$ for each L are seen to cluster randomly about that value of digit, thus yielding a $\underline{B}_{\mathrm{MF}}$ value being nearly equal to L as shown in Figure 11.4.

The results of Figure 11.6 are summarized in Figure 11.7 where the FM accuracy is plotted versus the convergence parameter L, again in digits. A best-fit straight line to the plot is included in the figure that shows the FM accuracy varies approximately as $0.93 \times L$. While this is not a 1-to-1 agreement, it does show that the FM provides a useful quantitative indication of the GM accuracy.

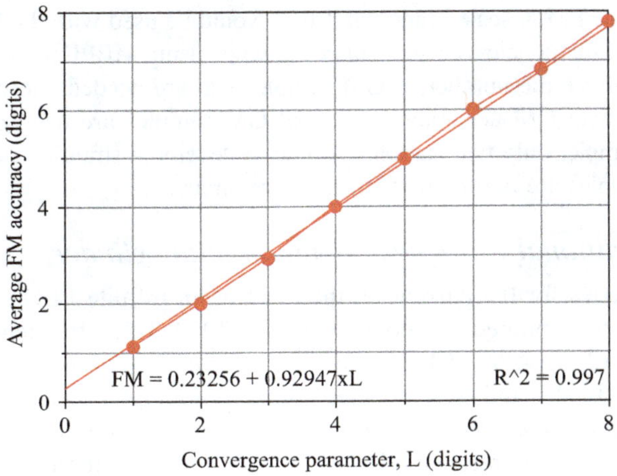

Figure 11.7 The average FM accuracy variation with the series convergence parameter over a range of 8 orders of magnitude

11.6 Further examination of MBPE accuracy tests

It was observed previously that using MBPE for assessing the accuracy, or uncertainty, of a GM using a rational-function FM is comparable to classical linear regression. In that case, the FM is a straight line whose slope and intercept on the y-axis are determined by minimizing the RMS difference between the straight-line FM and an overdetermined set of data, which constitutes the GM.

The approach introduced in this chapter for frequency-response EM data is basically equivalent, but somewhat more involved approach. A single FM, as for linear regression, might be used for a frequency range that brackets only 1 or 2 resonances. If several resonances are involved, two or more overlapping FMs may be used [1].

There are two reasons for this. One is to control ill-conditioning of the matrices whose solution yields the coefficients of the FM nth-order numerators and dth-order denominator polynomials. Note that maintaining n and d to be less than 6 to 8 is advised when using a CP of 24 digits as is done here. The other is to enable comparison of results from the overlapping FMs to indicate where a new frequency sample is needed to attain an agreement between overlapping FMs of a minimum specified amount. An agreement of 2 digits usually suffices. Note that adding a new sample to a FM involves respectively increasing n or d by 1. This is done alternately between the two. If the order of a FM comes to exceed a specified limit, it is then itself divided into two overlapping FMs. When the desired convergence is achieved, an average FM_{AVE} is developed and compared with the GM samples not used for the FM computation [1].

Because the frequency exponents in the FM polynomials can be as high as 6 or 7, it is advisable to scale the frequency. Thus, rather than using a frequency range of

1–2 MHz for the FM, a scaled range of 1 to 2 would be used with the FM then scaled back to the GM frequencies. Observe also that when using MBPE to estimate a TF, the goal is to minimize the number of GM samples that are needed. However, when the goal is to estimate GM accuracy, additional GM samples are needed. In the linear regression example, only two samples suffice to develop a straight-line FM but additional GM samples are needed to estimate its accuracy.

11.6.1 Additional examples for the infinite cylinder

The infinite series for the current on the front of an infinite PEC cylinder for TE polarization was computed to a convergence of 10^{-4}. Three frequency bands in *ka* were modeled (3–6.64, 6–9.64, 9–12.64) with the extra 0.64 included for comparison purposes. Each band was modeled using 18 evenly spaced GM samples and 4 FMs. The successive FMs used GM samples (the open circles) numbered 1–15, 1–16, 3–18 and 4–18, respectively. Averaging the overlapping FMs yields the normalized current results of Figure 11.8. The digit difference between the average FM_{AVE} and additional GM samples is a little better than 4 digits, as shown in Figure 11.8. This result is compatible with the specified GM accuracy.

A different kind of situation is illustrated in Figure 11.9 where the digits of FM–GM match for the frequency dependence of a dipole conductance is shown. The GM data in this case is perturbed by adding random noise of $\pm 10^{-1}$ to $\pm 10^{-3}$ to accurately computed conductance. The data varies more widely as a result; the average overall vale is shown by the horizontal lines. The averaged agreement between the FMs for the various noise levels and noise-contaminated GM values of Figure 11.9 is consistent with the aggregated results plotted in Figure 11.7. The

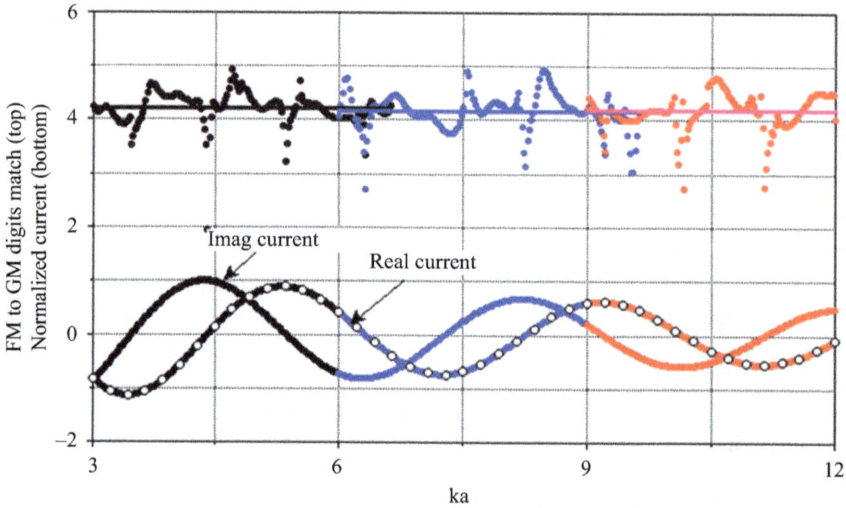

Figure 11.8 Normalized current on an infinite circular cylinder and the FM_{AVE}–GM match for a specified computation accuracy of 10^{-4}

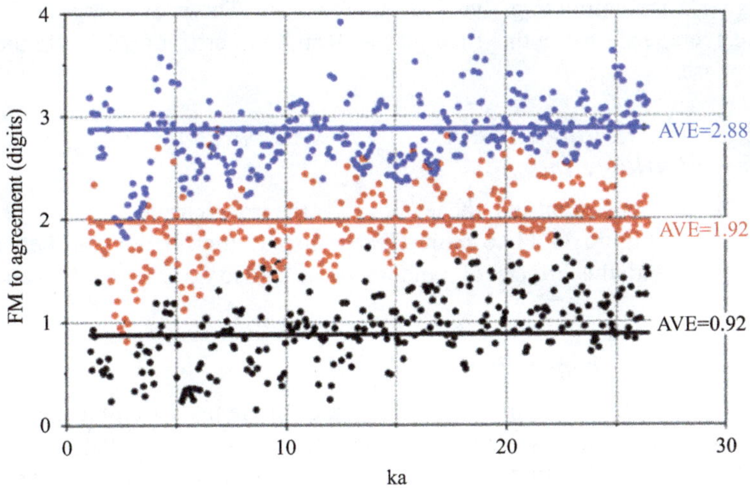

Figure 11.9 Digits agreement between FM_{AVE} and GM for dipole conductance with added noise

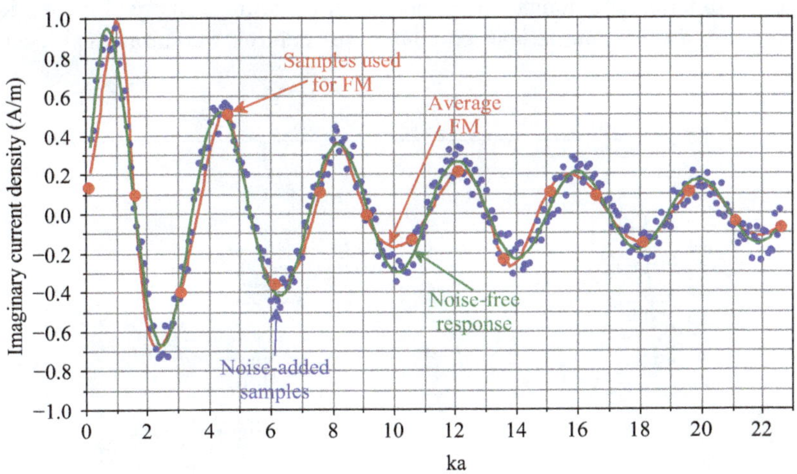

Figure 11.10 Fitting model results for the imaginary current on the infinite cylinder

horizontal best-fit straight lines to this data also show that these results are frequency independent.

Next shown in Figure 11.10 is the imaginary current on the front side of an infinite PEC cylinder illuminated by a TE, normally incident plane wave. The noisy data (the solid circles) comes from adding uniformly distributed random noise varying between ±0.1 to accurate current samples. The solid larger circles are the samples used for computing five overlapping FMs. These overlapping FMs produce the average shown in this final plot with an RMS error of 0.12 with the noise-free response.

11.6.2 More detailed results for the real current on the infinite cylinder

Results for the real current like those above for the imaginary current are presented sequentially in Figure 11.11 to more clearly display the relationship between the accurate GM and the effects of random noise added to it on the average FM estimate.

11.6.3 Results from NEC

The results presented thus far have involved the simple test case of a specified pole series and an analytical separation-of-variables solution for scattering from an infinite, circular PEC cylinder. The pole-series model served to demonstrate the relationship between the CN and MF and the cylinder provided a simple way to control the accuracy of the results by varying the convergence of the series solution as shown in Figure 11.12.

Examined next is an example of a moment-method model (NEC) for the input impedance of a dipole antenna where the spatial-sampling density of the model and the precision of the numerical computations control the numerical accuracy.

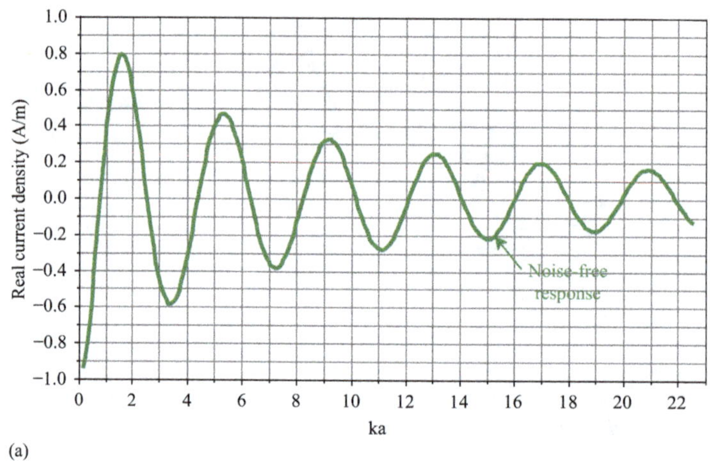

Figure 11.11 (a) The real current on the infinite cylinder computed using L = 8, (b) GM samples having uniformly distributed random noise of maximum amplitude ±0.1, (c) the noisy GM samples used for the FM computation, and (d) the average FM derived from the noisy GM samples

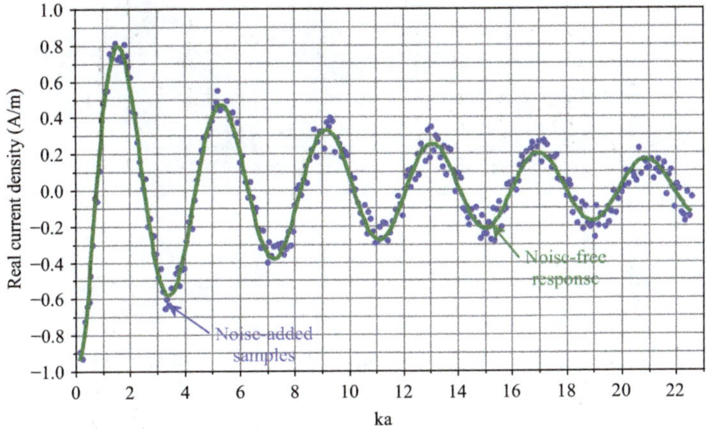

(b)

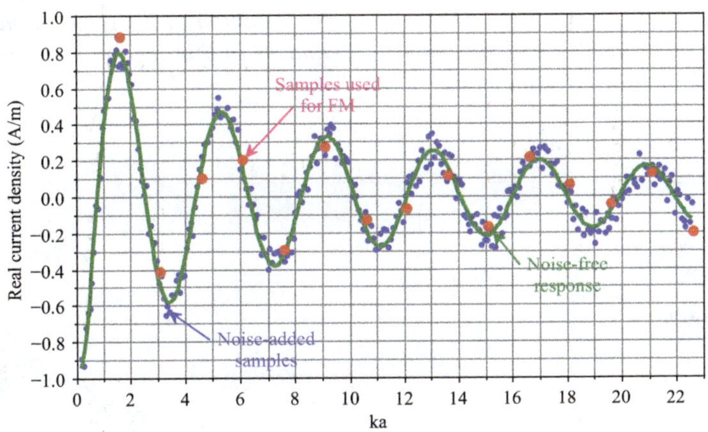

(c)

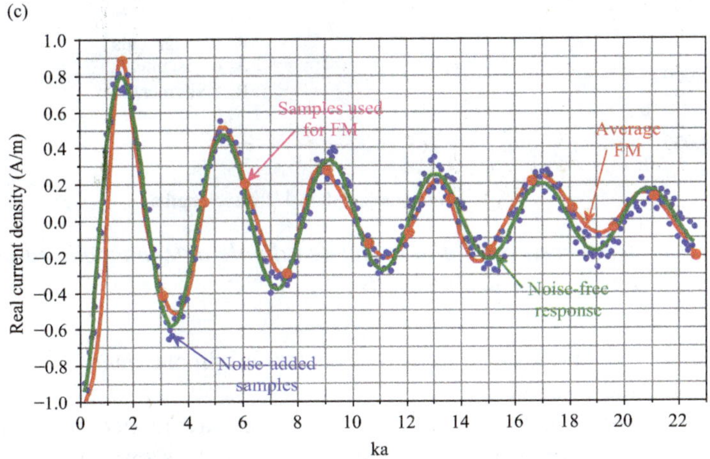

(d)

Figure 11.11 (Continued)

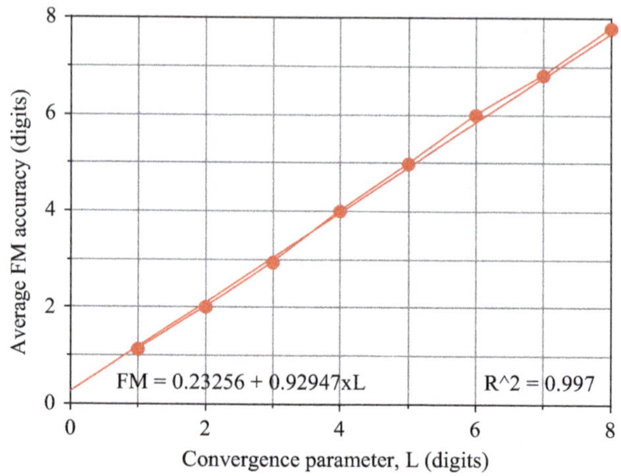

Figure 11.12 The average FM accuracy as a function of the convergence parameter L for the PEC cylinders in this section with the best-fit straight-line model

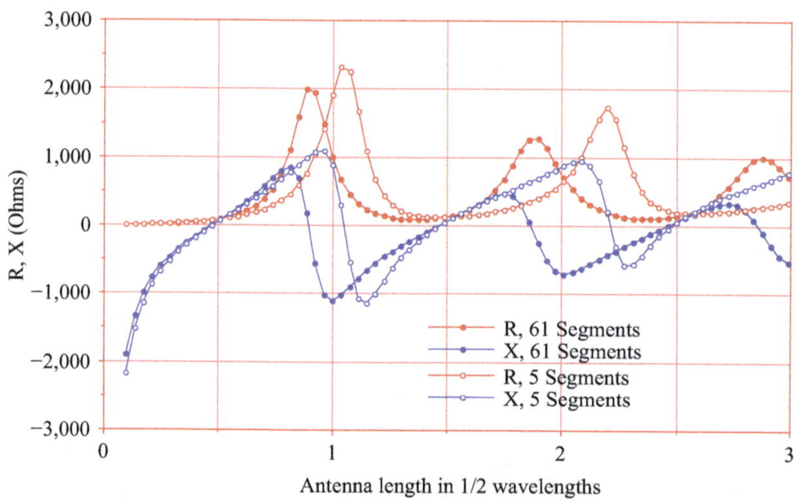

Figure 11.13 Input resistance of a center-fed dipole antenna modeled using 61 and 5 segments, respectively

Figure 11.13 shows impedance plots of a center-fed dipole antenna modeled with NEC using 61 and 5 segments.

The respective FM-to-GM accuracy estimates are plotted in Figure 11.14 where both models exhibit similar numerical values. This immediately raises the question about how is this possible as the 61- and 5-segment results in Figure 11.13

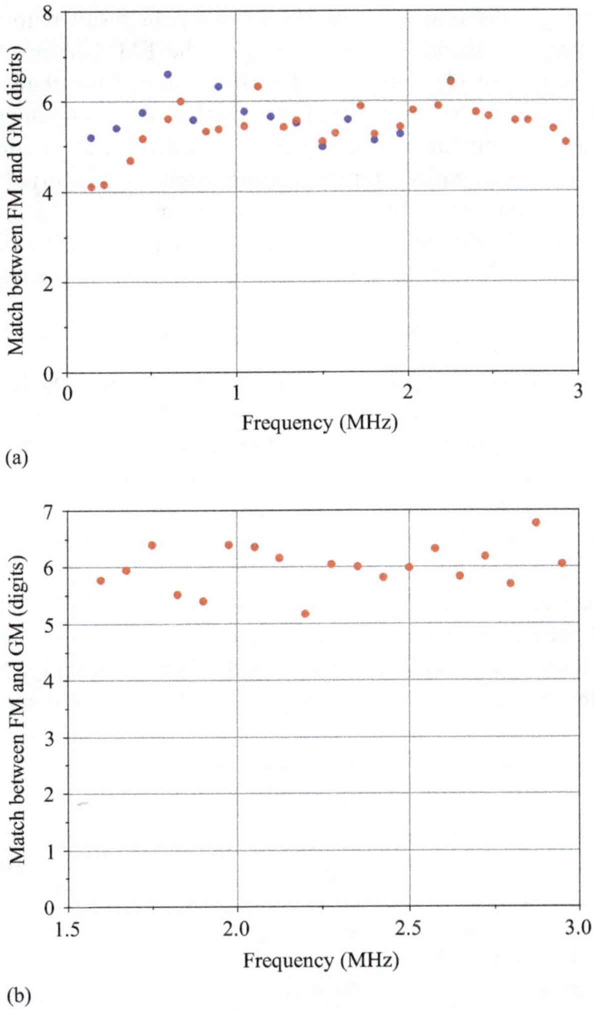

Figure 11.14 Match in digits between the FM and GM using 61 segments (a) and 5 segments (b)

are very different showing that the latter must be clearly the less accurate of the two. This is because wire-modeling experience over the years has long ago established that sub-domain models such as NEC required 6 or so segments per wavelength for acceptably accurate results. What might be an explanation for this result?

There are basically two issues involved here. One is the sampling rate per wavelength required for a numerical model to provide reasonable physical fidelity. The other is the precision required in the numerical computations. The 5-segment dipole model clearly does not satisfy the nominal 6 segments per wavelength sampling density over most of the length range covered in Figure 11.13, which explains its growing discrepancy with the 61-segment result with increasing frequency. However, the quantitative similarity of the FM–GM results for the two models shown in Figure 11.14 suggests that this comes from their common compute precision. It might also be inferred that the 5–6 digit agreement shown there represents an upper limit for a NEC result obtained from an adequately sampled numerical model. Justification for this conclusion is supported by the more straightforward demonstration based on the analytical results for the cylinder current presented above in Figure 11.11. In the following section, another approach is described based on adding random noise to NEC, or other numerical models.

11.6.4 Uncertainty testing using additive noise

A further test of the Kluskens data [5] discussed for scattering from a dielectric cylinder in Chapter 9 is illustrated in Figure 11.15 where noise of variable, known levels is added to the initial GM data samples. This may seem to be counter-intuitive and to raise the question of what does contaminating the data with noise accomplish? The basic motivation is that adding random noise of a known, but variable amount to the GM samples provides a way for MBPE to determine that noise level and thus the GM accuracy.

For this particular case, the noisy data is processed using four FMs having $n = 7$ and $d = 6$ for models 1 and 4 and $n = 7$ and $d = 7$ for models 2 and 3, to ensure overlap of at least two FMs everywhere, and where alternate GM samples are reserved for testing the DF. The average FM DF is shown as a function of

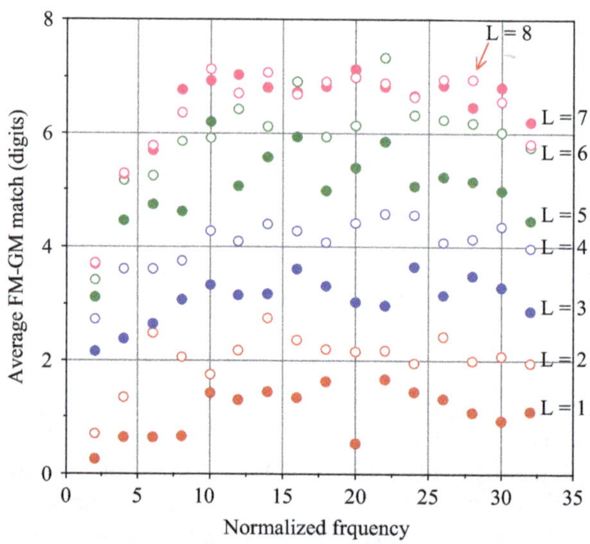

Figure 11.15 Data fit between average FM and GM samples as a function of normalized frequency for the Kluskens data with additive random noise of maximum value 10^{-L}. The data fit is seen to saturate at about 6.5 digits for $L > 6$.

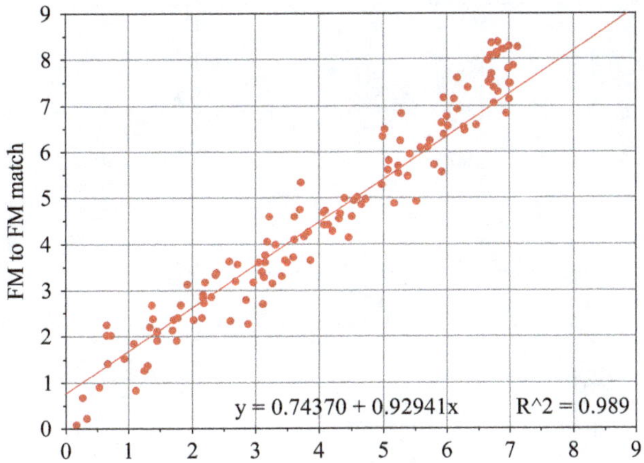

Figure 11.16 A scatter plot of the FM-to-FM agreement compared with the FM-to-GM agreement for the dielectric cylinder data of Figure 11.15 [5]

frequency with the parameter L denoting the level of the random noise of maximum value 10^{-L} added to the GM samples. For values of L up to 6, the individual data-fit values vary about an average of approximately L, but for $L = 7$ and 8 saturate at about 6.5, comparable to the results seen in Figure 11.4. Note that using various 10^{-L} noise levels of increasing values of L is similar to summing an series for increasing values of L as shown previously in Figures 11.6 and 11.7 even to the similar best-fit curves of Figures 11.7 and 11.16.

It should be mentioned that this test could begin with a maximum noise level such as 10% of the original GM data and successively decreasing the noise level in 10^{-1} steps as suggested above. An obvious alternative is to begin with a small noise level that is successively increased factors of 10. In the first case, the one used here, L would be increased until a maximum average digit count in the FM–GM match is reached. The result of this experiment is presented in Figures 11.15 and 11.16.

Presenting the data of Figure 11.15 in a scatter plot as in Figure 11.16 demonstrates that the proportionality between the overlapping FMs is quantitatively similar to that between the FMs and GM. This shows that FM-to-FM agreement is a good indicator of the actual GM accuracy as shown by the best-fit straight line. However, if the desired maximum uncertainty in the estimated FM spectrum is 0.01, it indicates that the target FM-to-FM agreement should be reduced to 0.929×0.01. Somewhat surprisingly, the best-fit straight line to the scatter plot of Figure 11.16 has a slope of 0.929, the same as for the cylinder current plot of Figure 11.12.

11.6.5 Uncertainty testing using an overdetermined system

The result in Figure 11.17, using scattering data for a dielectric cylinder [5] and a microstrip antenna [6], is for the case where 27 successive GM samples are used for the same four FMs as used for Figure 11.5, but with each FM data matrix now

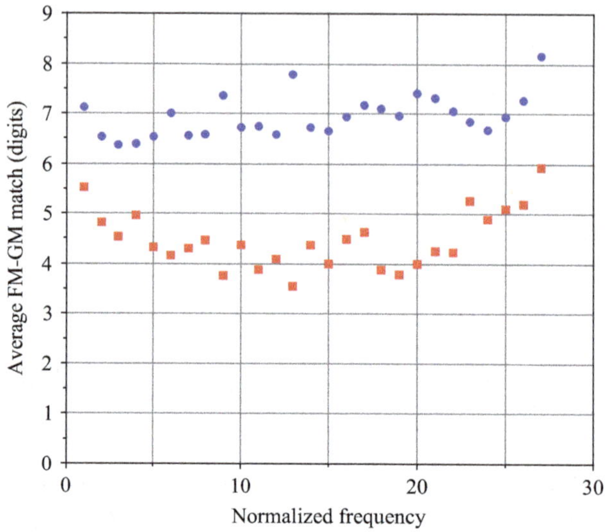

Figure 11.17 Data fit between the average FM and GM samples as a function of normalized frequency for the Kluskens dielectric cylinder data [5] (solid circles) and Cloete microstrip-antenna data [6] (solid squares). Each computation employed four FMs and 10 extra GM samples with the overdetermined data matrix solved using a pseudo-inverse [7].

overdetermined by 10 GM samples. It can be seen that the cylinder FM–GM match oscillates around an average of about 6.8, again consistent with previous results. This indicates that the cylinder data has a normalized uncertainty no worse than $2\text{–}3\times10^{-7}$. The corresponding value for the micro-strip antenna from this test indicates an uncertainty of somewhat less than 10^{-4}, somewhat higher than implied by Figure 11.4(b). Note that using an overdetermined data matrix in MBPE is essentially equivalent to linear regression where a straight line is fit to more than two data points while minimizing the mean-squared error between them.

11.6.6 Examples of estimating the uncertainty of experimental data

Examples of using MBPE to estimate the uncertainty of experimental data are presented in Figures 11.18–11.21 using results measured at the David Florida Laboratory [8] for the field backscattered from a PEC cube. The actual data comprises 801 samples from 2 to 18 GHz. Initially, complex data was employed, yielding the results shown in Figure 11.18, where rather poor agreement is obtained between the FM ($n = 7$, $d = 6$) and the measured data.

Anticipating that the data might be known more accurately in magnitude than in phase (which was used to obtain the real and imaginary components), the FM was recomputed using only the magnitude data instead, with the outcome shown in

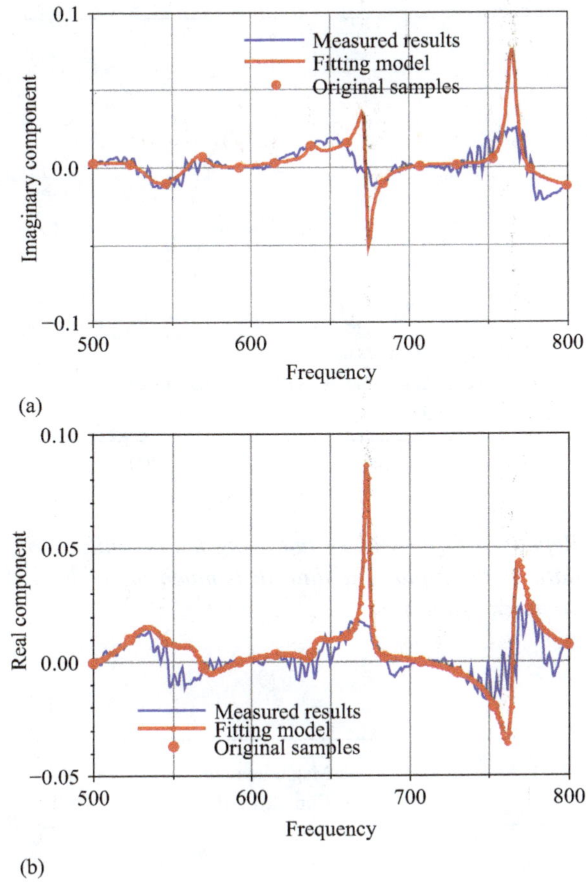

Figure 11.18 *Results obtained from a 14-coefficient complex FM with n = 7 and d = 6 (the dark solid line) using data samples from a PEC cube (shown by the solid circles), with the narrow line showing the original data (from Mishra). Although the FM seems to average the oscillations seen in the experimental data, the overall agreement is poor.*

Figure 11.19. For simplicity, the computation continues to use a complex FM but since the input data is now pure real, the imaginary part of the FM is automatically zero. Graphically, it is clear that the match obtained between the FM and magnitude data is much better than when using the complex data.

The result of using MBPE to estimate the accuracy of measured RCS-magnitude data is illustrated in Figure 11.19. The solid line is the average of five FMs as obtained by data samples indicated by the 14 large, solid circles. The “o’s” are the unused data whose deviation about the average FM is multiplied by 3 to

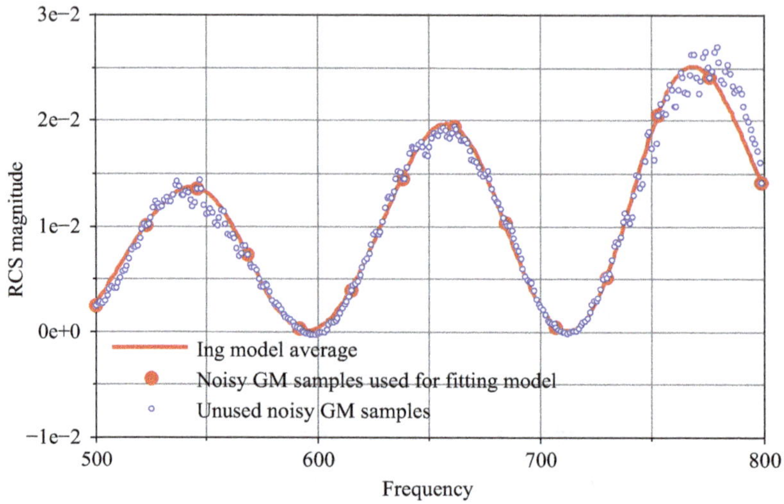

Figure 11.19 Repeat of Figure 11.18 but using the magnitude of the measured data as the input. The data fit is much better here than when using complex input data.

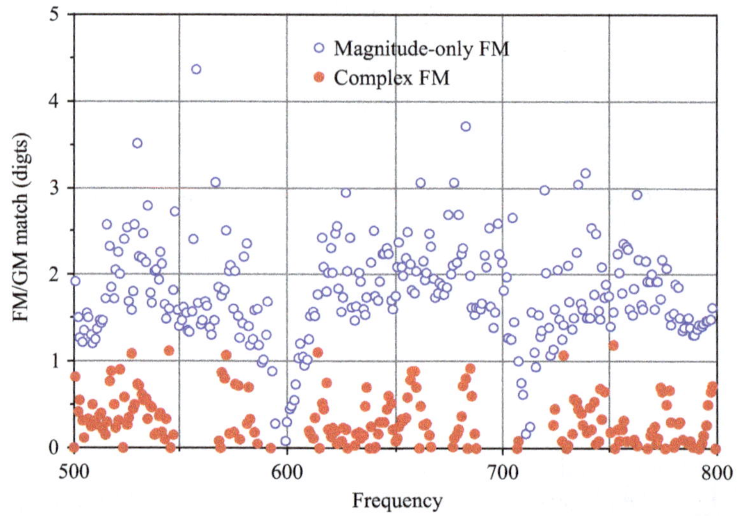

Figure 11.20 The match in digits obtained using the complex FM (solid symbols) and magnitude-only FM (open symbols) for the cube RCS [8] with the frequency in MHz. Average agreement for the magnitude-only FM in is about 2 digits, with minima occurring in the vicinity of minima in the RCS.

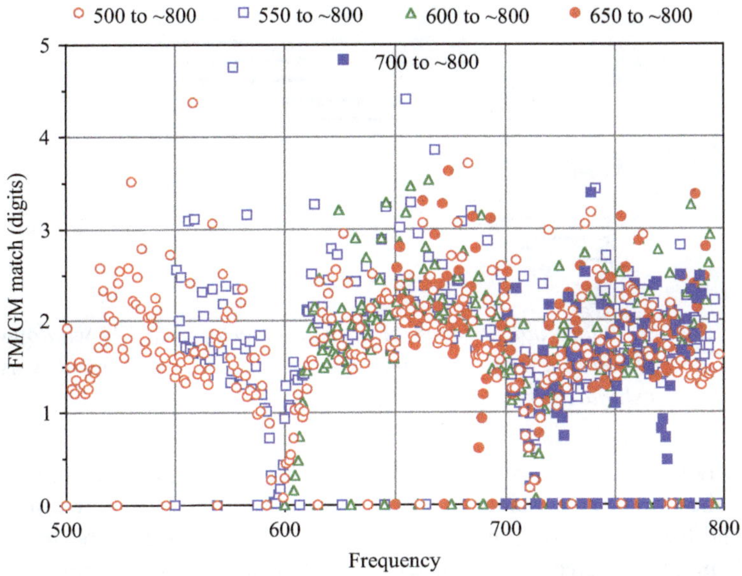

Figure 11.21 *The match in digits obtained using a magnitude-only FM of various window widths for the experimental field scattered from a PEC cube [8] in each case using n = 7 and d = 6. Window widths were used of 500–800 (open circles), 550–800 (open squares), 600–800 (open triangles), 650–800 (solid circles), and 700–800 (solid squares).*

show the difference more clearly. The RMS difference is $\sim 3\times10^{-4}$. Fitting the complex version of the data is much less accurate.

11.6.7 Implications for adaptive frequency sampling

Adaptive frequency sampling is considered in Chapters 6 and 9. This topic is examined further here from the perspective of how adaptive sampling also provides a way for estimating spectrum-sampling accuracy. This is because pairs of two FM samples are compared over the spectrum window being evaluated to determine if and where the next GM sample should be obtained. This comparison provides a quantitative measure of the maximum uncertainty of the average FM up to that point in the adaptation process.

Initial sampling of the GM begins with low-order FMs, possibly using as few samples and an overlap arrangement like that shown in Figure 11.22. After the first pass through the GM observation range, the FMs are sampled much more finely in angle or frequency, a procedure that is generally much faster than these initial GM samples. These finely sampled results from the overlapping FMs are then compared over the observation range to find the maximum difference, or maximum fitting-model error (FME) between them. If the FME were to be less than or equal to the error criterion that is specified, the estimation process would be complete.

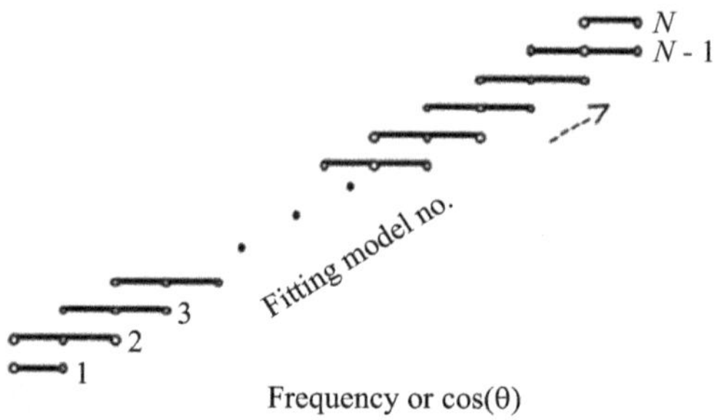

Figure 11.22 A conceptual illustration of using overlapping FM windows to estimate a frequency spectrum. A similar arrangement is used to estimate a radiation pattern.

Generally, this is not expected to be the case, and a new GM sample is computed at the observation point where the maximum difference between two FMs occurs and the FMs sharing this new data are then updated. For spectrum estimation, this involves increasing n or d with each GM update. The comparison step is then repeated using the updated FMs and also those with which they overlap. The scanning process is repeated to find a new maximum FME, and if not acceptable, the process of adding a new GM sample and updating FMs continues.

Because the matrix from which the FM parameters are obtained may eventually become ill-conditioned as n and d are successively increased to a total of 15 or so, the affected FM window is divided into two overlapping windows to form two new FMs of lower order. Thus it can often happen that the final number of FMs and windows can exceed, sometimes substantially, the number used originally. When the maximum difference between all FMs is less than the specified error level, the process is terminated and an average FM is computed as a final result, using whatever observation-variable spacing is desired. It can happen at times during this process that the maximum FME increases as the FM parameters are updated. While monotonic convergence to a final result may sometimes not occur, this does not invalidate the process so long as convergence to a specified eventually is achieved.

The largest acceptable FME specified between the paired FMs for spectrum estimation has typically been 0.01 normalized to the maximum magnitude of the quantity being modeled for an equivalent of a minimum 2-digit agreement. This shows that the GM samples must be accurate to at least 2 digits or else the FM results would not converge. Note that FM samples co-located with GM samples generally agree to 20–22 digits when using 24-digit compute precision.

Since the paired FMs can agree to significantly more than 2 digits as demonstrated in Figure 11.14 the process used for adding FM samples to develop the frequency-spectrum estimate includes a built-in accuracy test. Note that if the FME were to set smaller than the actual GM accuracy then the paired FM samples cannot be expected to satisfy the convergence test since their numerical cannot be expected to exceed it.

For pattern estimation, a different approach has been used because the dynamic range of patterns can be of interest over 10s of dBs from the pattern peak. In order to adapt the acceptable FME as the pattern side lobes decrease in value instead increased in proportion to the decrease in sidelobe level.

11.7 Concluding comments

A major deficiency in CEM is the almost-universal absence of any cost-effective way to estimate the accuracy of the solutions provided by numerical models. The rationale for the procedure described in this chapter is to exploit knowledge of the macroscopic behavior of EM observables as a means of measuring the uncertainty of results computed from CEM models and thereby to estimate their accuracy. The procedure is based on the known, resonance-like frequency dependence of EM observables, a behavior well approximated by a rational function that can thus serve as an FM for such observables. By quantifying the FM using a set of frequency samples from the first-principles or GM, the computed match, or fit, between this FM and additional GM samples provides an accuracy estimate of the data produced by the GM. This possibility has been demonstrated here for two problems, one a pole spectrum and the other the current excited on an infinite circular cylinder by an incident plane wave.

It should be appreciated that quantifying the FM requires solution of a "data" matrix whose size is determined by the orders of the numerator and denominator polynomials that comprise it, and which, even though generally involving fewer than 20 equations, can have a CN exceeding 10^{10}. Because the error estimate derived from the FM procedure can be no better than the fit achieved between the FM and the original GM data, the CP required will generally exceed 16 digits to overcome the effects of data-matrix ill-conditioning. The results presented here were obtained using 24-digit CP. It should also be appreciated that the FM is not unique and furthermore, that some experimentation is required concerning the bandwidth covered by GM samples and the parameters used for the FM. Finally, it is worth noting that the procedure outlined here can not only provide an error estimate for the data obtained from a GM, but it may be also possible to improve the accuracy of the GM samples themselves by averaging the parameters of overlapping, windowed FMs. Both of these procedures, of course, derive from the fact the oversampled GM data is available whose redundancy provides the additional information that makes such operations possible.

References

[1] E. K. Miller, and G. J. Burke, "Using Model-Based Parameter Estimation to Increase the Physical Interpretability and Numerical Efficiency of Computational Electromagnetics," *Computer Physics Communications*, Vol. 68, pp. 43–75, 1991.

[2] A. Ralston, *A First Course in Numerical Analysis*, McGraw-Hill Book Co., New York, 1966.

[3] E. K. Miller, G. J. Burke, and E. S. Selden, "Accuracy-Modeling Guidelines for Integral-Equation Evaluation of Thin-Wire Scattering Structures," *IEEE Transactions on Antennas and Propagation*, AP, Vol. 19, pp. 534–536, 1971.

[4] W. H. Press, S. A. Teukolsky, W. T. Vetterling, and B. P. Flannery, *Numerical Recipes*, Cambridge University Press, 1992.

[5] M. Kluskens, Naval Research Laboratory, Washington, DC, private communication, 1998.

[6] J. H. Cloete, Stellenbosch University, South Africa, private communication, 1998.

[7] E. K. Miller, "Developing Adaptive Models and Estimating the Uncertainty of Presampled Spectral Data," in *Proceedings of 15th Annual Review of Progress in Applied Computational Electromagnetics*, Naval Postgraduate School, Monterey, CA, 1999, pp. 278–285.

[8] D. Shantnu Mishra, Florida Laboratory, Ottawa, Canada, private communication, 1996.

Chapter 12

Using Prony's method to develop pole-based models of linear sources[1]

Abstract

Prony's method is applied to the far-field radiation patterns of straight cylindrical wires (antennas and scatterers) of varying lengths and radii to identify their equivalent source distributions. In antenna configurations, the poles with the largest residues consistently appear near the feed region and at the wire ends, while smaller-residue poles are found elsewhere along the wire. Physically, the dominant poles correspond to radiation due to charge acceleration at the feed (excitation) point and at the wire tips, along with a weakly attenuated guided current wave. In scattering configurations (plane-wave incidence on a wire), for near-axial (broad-side) incidence the largest-residue poles again localize at the wire ends, indicating end-diffraction dominance. Examined also is how the pole residues change with wire radius. Thicker wires tend to radiate more from the feed region (as in an effectively infinite wire), whereas very thin wires radiate predominantly from their ends (as in a semi-infinite case). These findings demonstrate the potential of far-field pattern fitting to "image" the source regions of simple objects.

12.1 Introduction

The inverse problem of electromagnetics is to determine the source distribution that produces a given field. If the sources lie on a perfectly conducting surface and the fields are measured arbitrarily close to that surface, the solution is straightforward. In practice, however, one often has access only to the far-zone fields, observed at large distances from the object. In that case, one might expect to resolve only those source components that radiate into the far field. (In principle, analytic continuation can recover near-field behavior if the boundary conditions on the object are known.) In this chapter, we explore the use of far-field data to locate and characterize the actual radiating sources on a linear conductor—a process we may call *imaging*.

Earlier work demonstrated that Prony's method, which fits a sum of exponentials to data, can image discrete linear source distributions from their far-field

[1]Excerpted and revised from UCRL-52822, "Imaging of linear source distributions," E. K. Miller, D. L. Lager, J. T. Okada, 1979.

patterns [1]. It can also synthesize a finite set of equivalent point sources to approximate a continuous source distribution. Here, we apply the same approach to the case of a straight cylindrical wire. We compute the wire's far-field pattern (for various lengths, radii, and excitations) and then use Prony's method to recover a set of source poles and residues that reproduce that pattern. Our study shows that the effective source resolution is on the order of two sources per wavelength. In all antenna cases examined, the recovered poles indicate strong radiation at the antenna feed (excitation point) and at the wire ends, consistent with the physics of accelerated charges. Additional weaker sources are also recovered; these can be interpreted as arising from the traveling-wave current on the antenna or from the continuously driven portions of a scatterer. We discuss the implications of these results for imaging simple objects at the end of this chapter.

12.2 Background

Microwave holography experiments by Izuka and Gregoris [2] provide a motivating example. They scanned the near field of a long monopole antenna and then optically reconstructed the "image" of the antenna in free space (Figure 12.1). The reconstruction revealed two distinct bright points, which corresponds to radiating sources at the antenna feed (the center) and at the end of the monopole. This observation is consistent with classical electrodynamics: according to Panofsky and Phillips [3], the far-zone radiation (falling off as $1/R$ with distance) from a point charge moving with (nonrelativistic) velocity v and acceleration a is proportional to the charge's acceleration (12.1). Thus, when charge is accelerated by the feed or reflected at the antenna end, significant radiation is produced. In other words, the

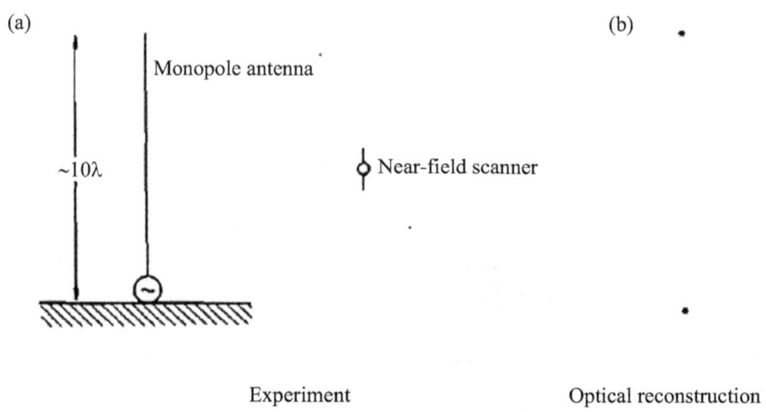

Figure 12.1 Schematic of the Isuka and Gregoris experiment [2]. Hologram generated by scanning field of long monopole antenna (a) is optically constructed to give two points (b). These point are associated with radiation from the feedpoint and end of the antenna.

feedpoint and the wire tip naturally act as dominant radiators in an antenna.

$$\mathbf{H}(\mathbf{r},t) = \frac{q\mu_o}{4\pi}\left[\frac{\mathbf{v}(\tau)\times\mathbf{R}}{R^3} + \frac{\mathbf{RxR}[\mathbf{Rxa}(\tau)]}{cR^4}\right] \tag{12.1a}$$

and

$$\mathbf{E}(\mathbf{r},t) = \frac{q}{4\pi\varepsilon_o}\left[\frac{\mathbf{R}}{R^2} + \frac{\mathbf{RxRa}(\tau)}{c^2R^3}\right] \tag{12.1b}$$

where

H(r,*t***)** is the magnetic field vector at time t and vector coordinate **r**,

E(r,*t***)** is the electric field vector,

ε_o = the medium permittivity,

μ_o = the medium permeability, c = the velocity of light.

R = **r** – **r'**(τ), the vector difference between the observation and source point coordinate vectors

$R = |\mathbf{R}|$,

$\tau = t - |\mathbf{r} - \mathbf{r}'|/c$, the retarded tune,

In an alternative discrete-source approach, one can fit the far-field data by an equivalent set of point sources using Prony's method. Consider S isotropic point sources located along a straight line at positions d_a, each of strength R_a. The resulting far-zone pattern as a function of angle θ measured from the wire axis is

$$f(x) = \sum_{\alpha=1}^{S} R_\alpha e^{(s_\alpha x)}, \text{ for} - 1 < x < 1 \tag{12.2}$$

With the substitution $x = \cos(\theta)$, (12.2) becomes

$$f(\theta) = \sum_{\alpha=1}^{S} R_\alpha e^{(ikd_\alpha \cos\theta)}, \text{ for } 0 \le \theta \le \pi, \tag{12.3}$$

which can be processed by applying Prony's method to samples of $f(x)$ to determine the set of S values of R_α and s_α. In practice, the samples of $f(x)$ [equal to samples of $f(\theta)$ weighted by $\sin(\theta)$] must be taken at uniform intervals in x. Applying Prony's method to these samples provides estimates of the complex poles s_α and residues R_α. Physically, a purely imaginary pole $s_\alpha = ikd_a$ corresponds to an isotropic point source at position d_α, while a nonzero real part $\text{Re}(s_a)$ indicates source directivity (an anisotropic source). For an actual point source, then $\text{Re}(s_a) \sim 0$ (except for numerical noise), whereas spurious curve-fitting poles might have larger real parts or small residues. Prony's method applied to a far field can yield two typical results.

In the first case, if the true source distribution is reasonably sparse (relative to L), the recovered pole set in Figure 12.2 (shown on an isometric plot in which the residue amplitudes are given on a three-decade log scale) will reproduce the actual source set. In that case, the output poles (and their residues) match the input

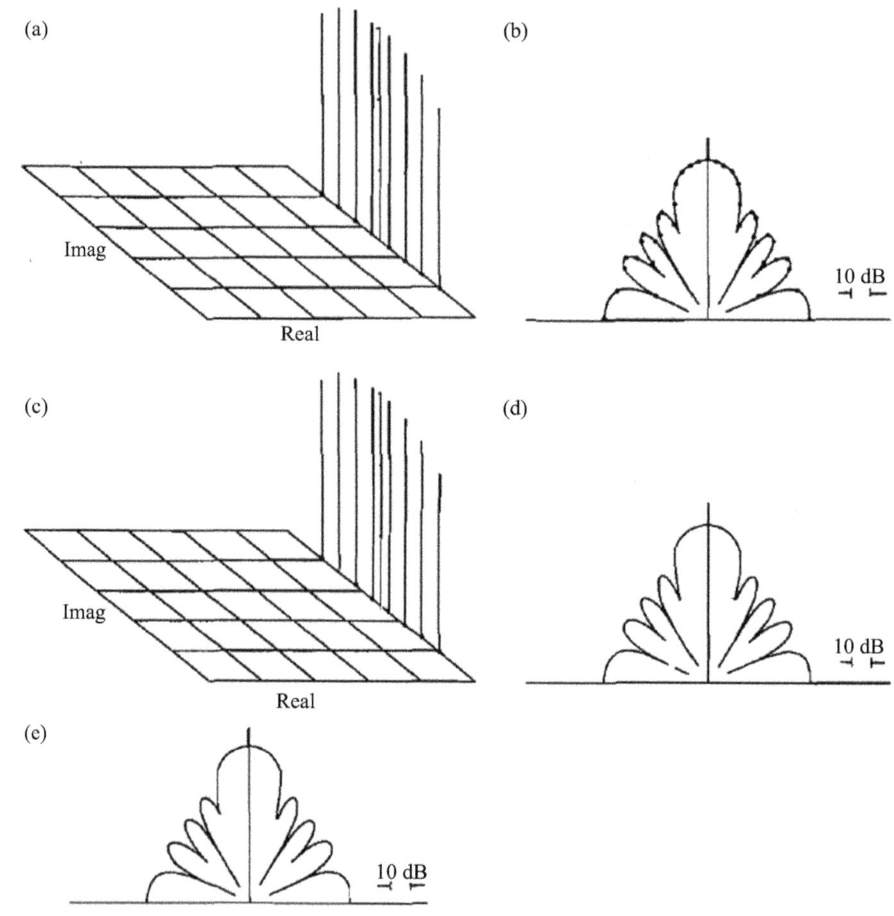

Figure 12.2 A specified source (or input pole) distribution (a) produces the radiation pattern (b) whose Prony analysis yields the output pole distribution (c) that produces the reconstructed pattern (d) and the constrained pattern (e), in which s_R are set to zero

distribution, and the reconstructed far field agrees with the original. Figure 12.2 illustrates this imaging case: an input distribution of point sources (Figure 12.2(a)) produces a known far-field pattern (Figure 12.2(b)), and Prony's method recovers an output pole set (Figure 12.2(c)) whose constrained pattern (Figure 12.2(d)) overlays the input pattern. This means the method successfully locates the true radiating points from the far-field data.

The pattern poles produced by this exercise can have the general form

$$s_a = s_{a,R} + js_{a,I}$$

From (12.3), it can be seen that

$$s_{a,I} \sim kd_a$$

and that

$$s_{a,R} = 0 \text{ for isotropic sources}$$

and

$$s_{a,R} \neq 0 \Rightarrow f_{a,R}(\theta) = \exp(s_{a,R}\, cos\theta).$$

Thus, a complex pole implies source directivity. For isotropic sources, a non-zero value for $s_{a,R}$ can occur because of numerical inaccuracy. The pole of actual sources will usually be that $s_{a,R} << 1$, while for "curve-fitting" poles, it will generally be true that $|R_a|$ is much less than the maximum residue value.

In the second case, if the source distribution is very dense or effectively continuous (beyond the far-field resolution limit of about two points per wavelength), the algorithm produces an equivalent discrete pole set with *fewer* sources. The radiation patterns of the input and output sets are essentially the same, but the output set contains only about $2L/l$ poles. Figure 12.3 demonstrates such a scenario: a dense

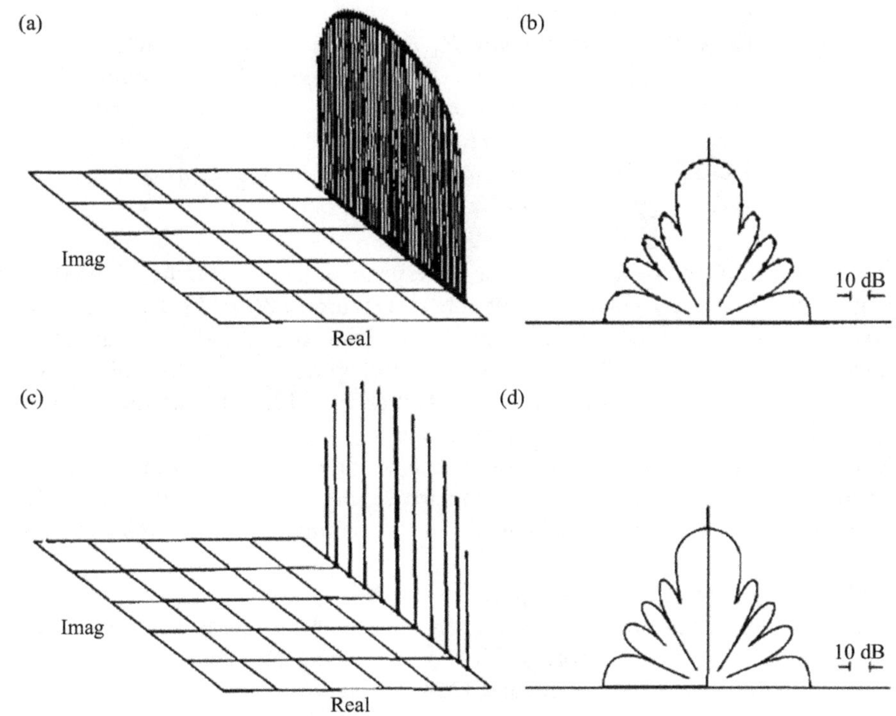

Figure 12.3 Development of a discrete-source equivalent of a "dense" or continuous aperture distribution where the input set (a) produces the radiation pattern (b) and the output pole set (c), whose reconstructed pattern is given in (d)

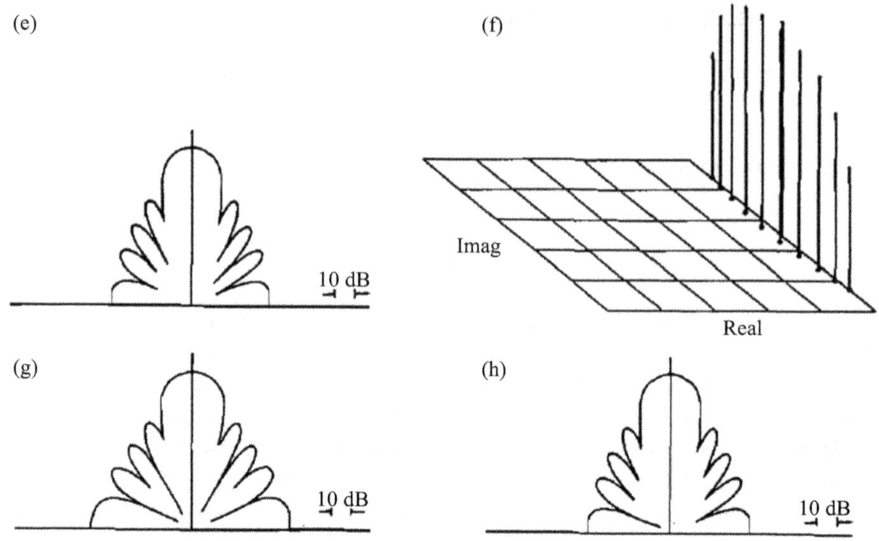

Figure 12.3 (Continued)
The constrained pattern is given in (e). As the number of poles, P, used in implementing Prony's method is varied, the number of non-curve-fitting output poles, P_o, can also vary, as in (f), which should be compared with (c). Normally, P_o is approximately equal to 2L/l, where L is the aperture width (the linear extent of the input pole set), and will not vary by more than one as P is changed, so long as $P > 2L/l$. The reconstructed and constrained patterns of the output pole set (f) are given in (g) and (h)

input distribution (Figure 12.3(a)) has a corresponding pattern (Figure 12.3(b)), and Prony's method yields a smaller output pole set (Figure 12.3(c)). As the number P of Prony test poles is varied, the number of significant output poles P_o can change slightly, but typically $P_o \sim 2L/l$ remains constant (Figure 12.3(f)) [1]. In all cases, the reconstructed (Figure 12.3(g)) and constrained (Figure 12.3(h)) patterns match the input. Thus, the output set is equivalent to the true distribution in terms of the far field, but it is not unique: different valid pole configurations can produce the same pattern (the example in Figure 12.3(e) is another solution for the same input). This situation is, in this regard, similar to that of a multimode cavity, which "jumps" between dominant modes.

In practice, multiple output pole sets of nearly equal validity can emerge (depending on the chosen fitting parameter P). The "best" set is often selected by criteria such as symmetry, pole stability with respect to P, or minimal residue values. The results in Figures 12.2 and 12.3 show representative clean cases where the solution is clear. In general, however, especially for realistic antenna currents, the output pole sets may be more complicated with examples of such complexity presented in the results below.

12.3 Application to current distributions

Building on the discrete-source results, Prony's method is applied to continuous current distributions on a straight current. We treat the current both as a center-fed antenna and as a passive scatterer. In each case, the current is obtained numerically (via a moment-method solution of the electric-field integral equation [4]) from which the far field is then computed. Following the sinusoidal example and for consistent comparison with its results, each antenna pattern is multiplied by sin(θ) before analysis (to factor out the 1/sin(θ) dependence). The cases considered are outlined in Figure 12.6: several lengths (3, 3.5, 4, 5, 7, 9-*l* with radius either r = 0.001 l or 0.01 l and a 9-λ current loaded at its center; and scattering cases for a 9-λ current under plane-wave incidence at various angles (with and without a center load).

12.3.1 Sinusoidal current

As a reference case, a sinusoidal current provides a convenient analytical test case to study the applicability of Prony's method to continuous distributions. Furthermore, such a current closely approximates that on a long wire. For a current of the form

$$I(x) = I_o sin[k(L/2) - |x|], -L/2 \leq x \leq L/2, \tag{12.4a}$$

where L is the length of the current filament and I_o is the maximum current, the far field can be written as

$$E_\theta(\theta) = Af(R)\left[e^{ikL}\, cos(\theta/2) - e^{-ikLcos(\theta/2)} - 2cos(kL/2)\right] /sin(\theta), \tag{12.4b}$$

Equation (12.4b) consists of three exponential terms (corresponding to poles at $z = -L/2$, $z = 0$, and $z = +L/2$). Hence, we expect Prony's method, when applied to $E(\theta)x\sin(\theta)$, to recover three significant poles located at the wire center and ends. This expectation is confirmed in Figure 12.4 for a 3λ wire. Figure 12.4(a) shows the analytically computed far-field pattern (multiplied by sin(θ). Applying Prony's method yields the output pole distribution in Figure 12.4(b), which indeed contains three poles: one at the center of the wire and one at each end $z = \pm 1.5l$. The reconstructed (constrained) pattern from these poles (Figure 12.4(c)) overlaps the original pattern almost exactly. Not surprisingly, the recovered pole positions are graphically indistinguishable from the wire's effective feedpoint and ends (the dashed lines in the figure).

The effect of numerical sampling on these results is examined next. This is done by numerically integrating the 31 sinusoidal current to obtain $E(\theta)$ while varying the number of integration points per wavelength. Figure 12.5 shows the results for 20, 100, and 500 samples per wavelength. All far-field patterns are nearly identical (after sin(q) weighting), but the output pole sets differ noticeably. With only 20 samples (Figure 12.5(a)), the end poles are seen to "split" into pairs of nearby poles. With 100 and 500 samples (Figures 12.5(b) and 12.5(c)), the poles

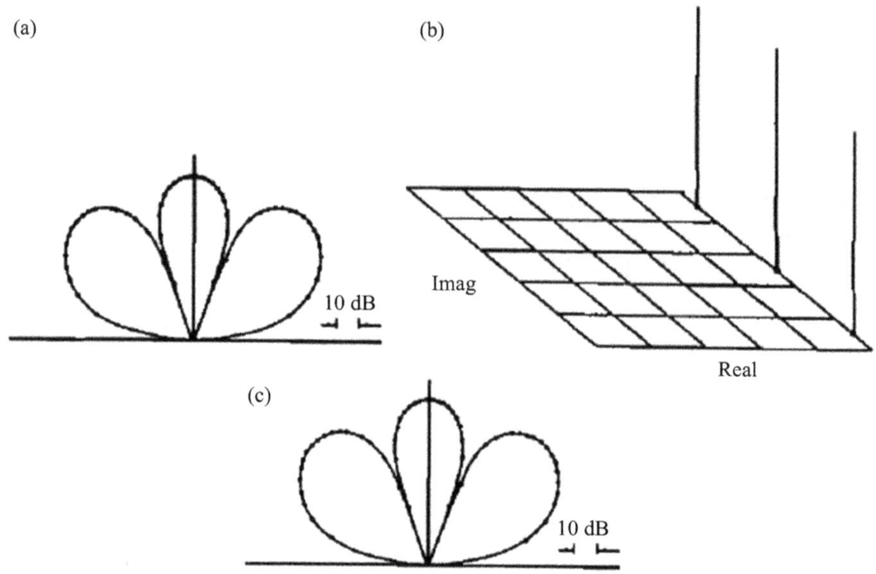

Figure 12.4 Results for analytically evaluated field of a 3λ sinusoidal current. The far field multiplied by sin(θ) (a) produces the output pole set as (b) and the constrained pattern in (c)

converge to the correct three locations. This behavior shows that limited sampling or measurement error can introduce spurious poles, especially at the wire ends. The lesson is that one must use sufficient resolution in the data to avoid artifacts in the pole reconstruction.

12.3.2 Currents on straight wires

The problem next considered is the current distribution on a straight wire, excited as either an antenna or a scatterer. The cases that were considered and for which representative results are to be presented are illustrated in Figure 12.6. In all cases to follow, the wire current is computed using a moment-method model based on the electric-field integral equation [4]. In line with the sinusoidal current just considered, all patterns are multiplied by $\sin(\theta)$ before using Prony's method.

12.3.2.1 Antenna case

To follow through on the sinusoidal current already examined, the next problem examined is a 3λ center-fed wire antenna, for which results are presented in Figure 12.7. Comparison of the pole distribution obtained in this case with those of the sinusoidal current in Figures 12.4 and 12.5 reveals both similarities and differences. They are similar with respect to the "end" poles, but quite different in the feed region. In particular, the feed region of the wire current comprises four poles, which are located symmetrically about the center of the antenna. Because of the

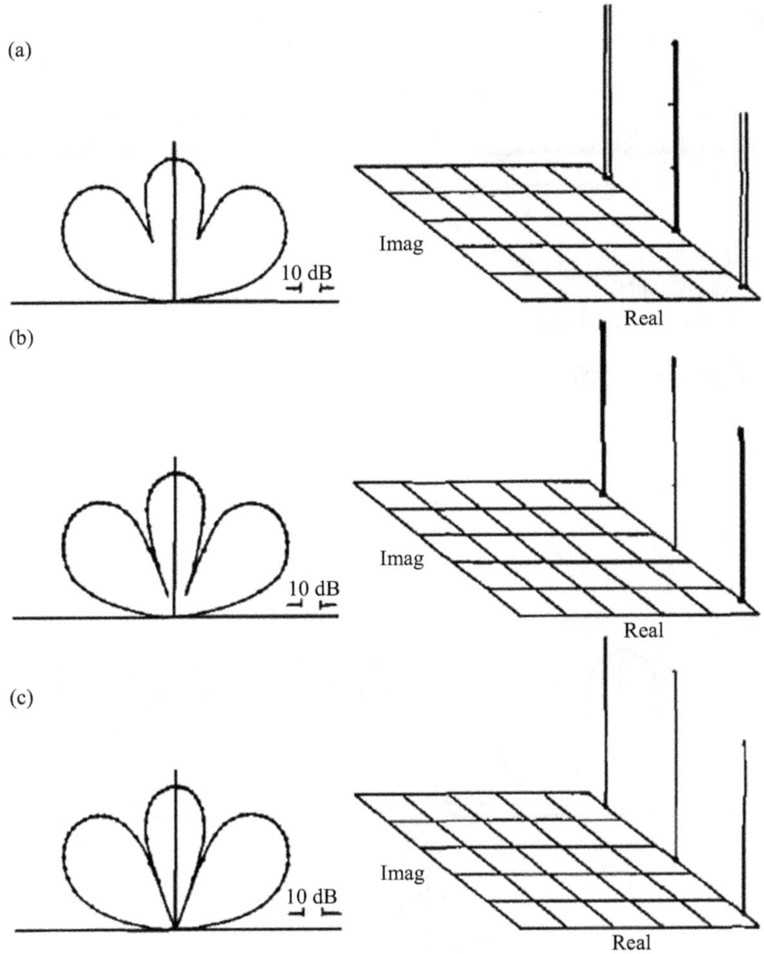

Figure 12.5 Results of numerically integrating a 3λ sinusoidal current to obtain its far field, using different numbers of integrand samples: (a) 20 samples, (b) 100 samples, (c) 500 samples per wavelength. Output source poles split as the accuracy of far-field samples decreases because of fewer integrand values

nonzero s_R components, the constrained pattern is quite different from the original input pattern, especially in the axial directions.

To gauge the importance of the pole directivity implied by the nonzero s_R values, presented in Figure 12.8 are the patterns $f(\theta) = \exp[(s_R k\cos(\theta)]$, for several values of s_R. The pattern anisotropy turns out to be quite small, less than 3 dB, for $s_R <= \lambda/20$. For the case illustrated in Figure 12.7, the two center poles have s_R values of about 0.04λ, while those next closest to the center have s_R values of about

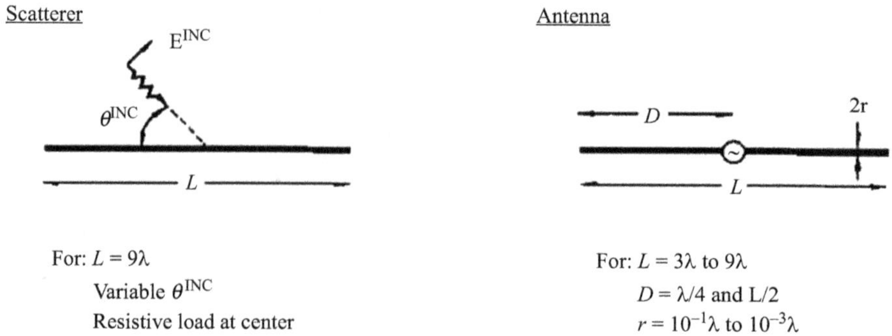

Figure 12.6 Summary of antenna and scatterer cases considered

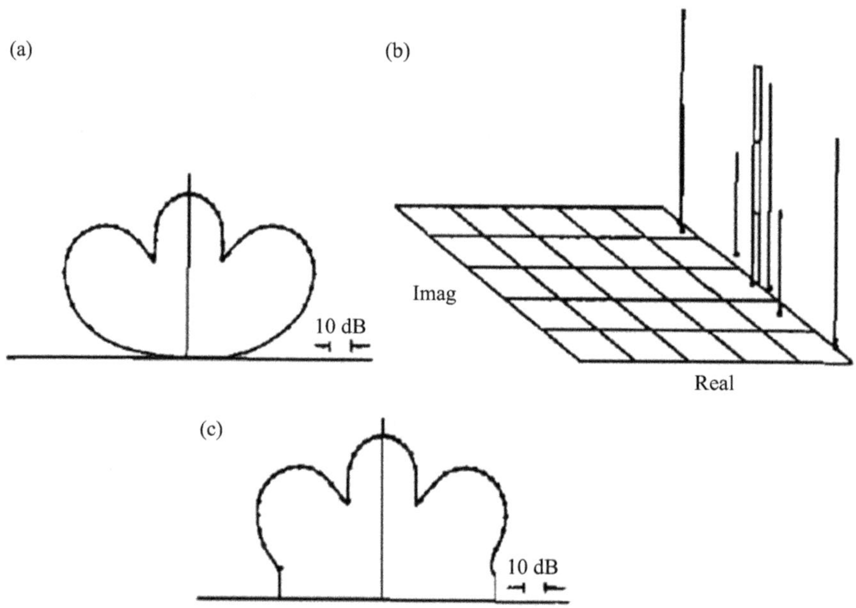

Figure 12.7 Three-λ antenna case for center-feedpoint and wire diameter $4 = 0.001\lambda$ (a) Far-field pattern (multiplied by $\sin(\theta)$, (b) output pole set, and (c) constrained pattern. Compare these with the results for the 3λ sinusoidal current (Figure 12.5).

0.1λ. Since the residue values of the latter are much smaller than those of the former, the influence of the latter's anisotropy is much reduced.

The influence of antenna length on the pole sets is illustrated in Figures 12.9 and 12.10. Lengths of 3, 3.5, and 4 wavelengths are included in Figure 12.9, and lengths of 5, 7, and 9 wavelengths in Figure 12.10. As measured by the numbers of poles and their residue magnitudes, the pole sets simplify somewhat as the antenna length increases.

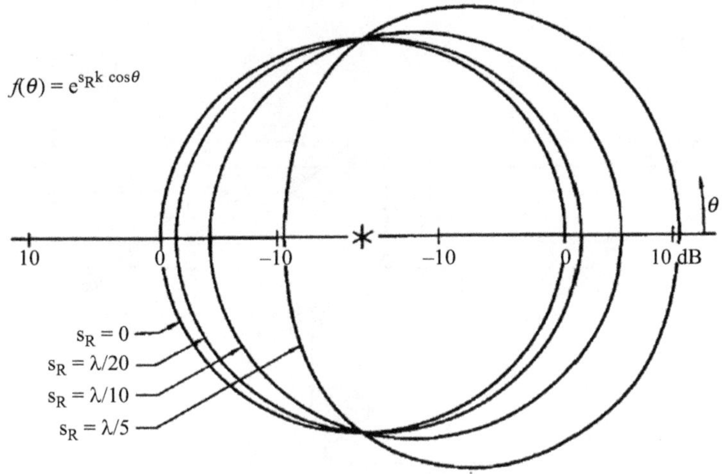

Figure 12.8 *Polar patterns showing the effect of a nonzero real pole component. Pattern is isotropic for $s_R = 0$ and has a directivity of about 15 dB for $s_R = \lambda/5$.*

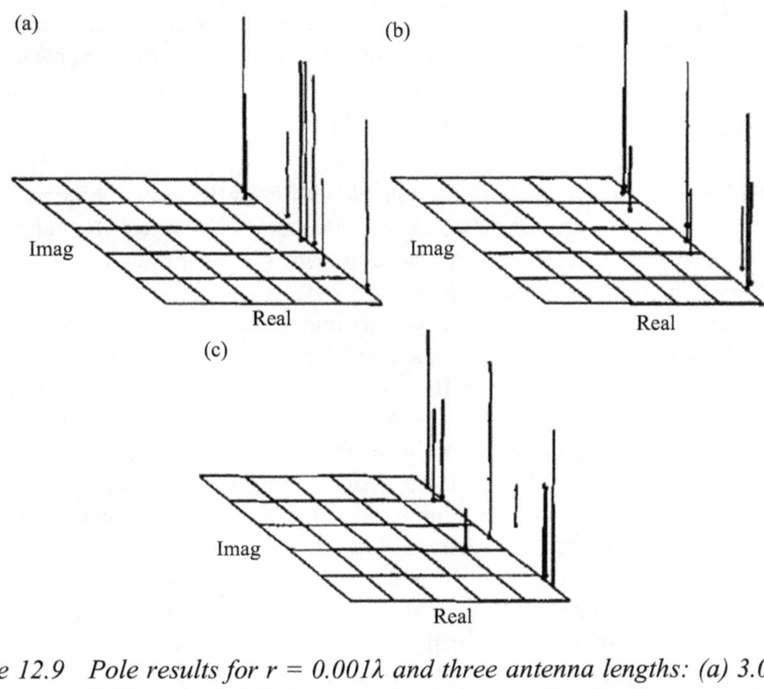

Figure 12.9 *Pole results for $r = 0.001\lambda$ and three antenna lengths: (a) 3.0λ, (b) 3.5λ and (c) 4.0λ Decreased relative amplitude of center pole for 3.5λ case is a result of an even-mode current and consequent deceased charge at feedpoint. These pole sets differ slightly, but their general character is similar, with dominant poles at the antenna feedpoint and ends.*

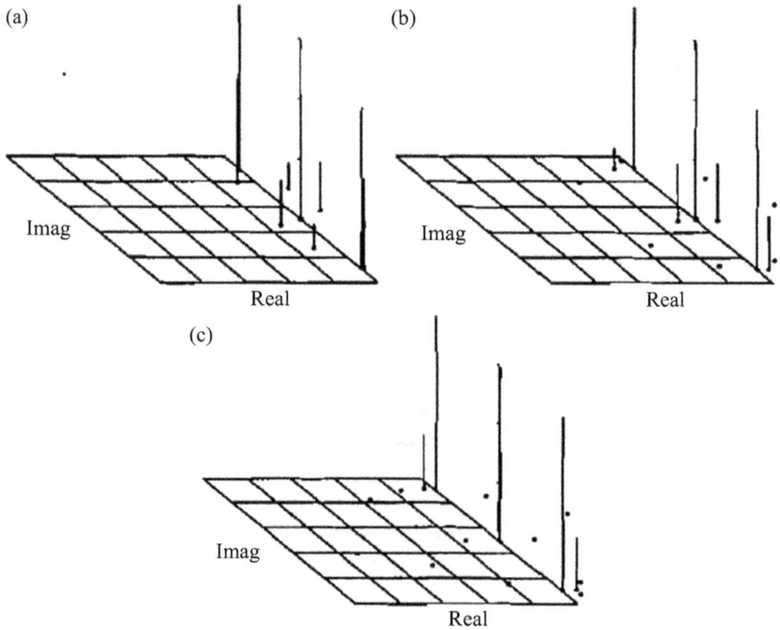

Figure 12.10 *Pole results for $r = 0.001\lambda$ and three antenna lengths: (a) 5.0λ, (b) 7.0λ, and (c) 9.0λ. The pole sets seem to simplify somewhat as antenna length increases. As in Figure 12.9, they are characterized by dominant poles at the feed point and wire ends.*

Note that for calibration purposes these pole plots have a vertical line at the wire center with 3 horizontal ticks denoting its 3-decade amplitude. Consistent throughout these results, however, is a correlation of the dominant (largest-residue) poles with the antenna ends and center (feedpoint location).

Further evidence that the center pole is due to the feedpoint radiation is presented in Figure 12.11. Here, the 9λ antenna case is rerun with the feedpoint moved to a point one-quarter wavelength from one end of the antenna. As expected, the center pole has disappeared and has been replaced by a new pole near one end of the antenna. Otherwise, the poles are similar to those of Figure 12.10 for the 9λ antenna. This new pole is clearly due to radiation from the relocated feedpoint.

This demonstrates that the dominant poles are those associated with the antenna ends and feedpoint. The origin of the other poles is not so easy to establish. However, closer examination of the pole sets in Figures 12.9 and 12.10 reveals some interesting patterns in their locations. For example, the nondominant poles near the center have real and imaginary components of the same sign. The converse behavior is true of the nondominant poles near the antenna ends. This leads to radiation patterns for such poles located the ends that have maxima facing toward the antenna center. Conversely, the each of the poles near the center has a maximum that faces toward the nearer antenna end. It is interesting to speculate that

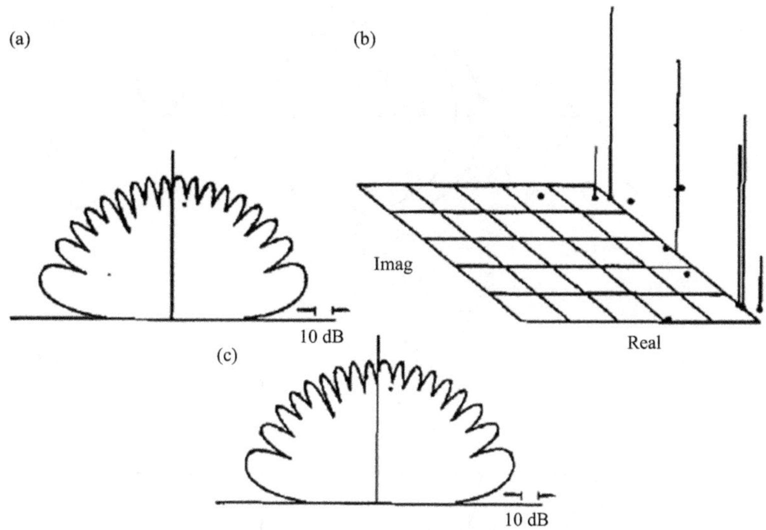

Figure 12.11 Results for a 9λ, a symmetrically excited antenna: (a) far-field pattern multiplied by sin(θ), (b) output pole set, and (c) constrained pattern. The feedpoint pole is now located λ/4 from one end of the antenna, where in fact the actual feedpoint is located, but otherwise the pole set is similar to that of Figure 12.10.

these poles may be associated with traveling wave radiation, but at this point nothing more definite can be said concerning that possibility.

To demonstrate the individual patterns of an actual pole set, some results are presented in Figure 12.12(a) for the 3λ antenna (from Figure 12.7). Both the spatial locations and relative amplitudes of the various pole patterns are shown, thereby providing a picture of this antenna's radiation characteristics. Upon combining the patterns of the four center poles and the pair of poles at each end of the antenna, the patterns of Figure 12.12(b) are obtained. All three of the latter patterns are essentially circular, demonstrating that the combined radiation from each region is effectively isotropic. Actually, because the nondominant poles contribute so little to the overall radiation, this observation applies well to the patterns of the dominant poles, if the two at the center are considered in combination.

One clue to the origin of the various poles may be provided by studying their dependence upon the wire radius. This is because the efficiency of traveling wave radiation is known to increase with increasing wire radius [5–8] and the dependency of feedpoint [8,9] and end radiation [8,10] on wire size is also known. Some results of such a study on 3.0λ, 3.5λ, and 5.0λ antennas are summarized in Figure 12.13. Plotted there as a function of wire radius are values of the various pole-residue magnitudes, relative to their values at a radius of 0.001λ. Also shown are corresponding results for the square root of the radiated power (a 1-V source strength was used throughout) and the feedpoint current.

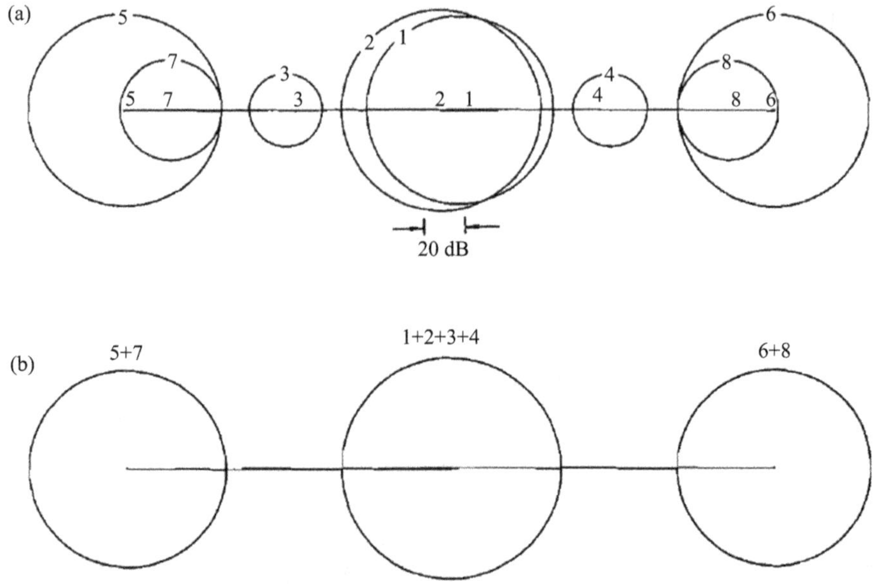

Figure 12.12 *Polar radiation patterns for the poles of the 3λ antenna in Figure 12.7(b), plotted on a plan view of the antenna. (a) Patterns of the individual poles, (b) combined patterns for the two sets of end-region poles and the feed-region poles.*

A comparison of the results for the three antenna lengths studied reveals some interesting trends. First, note that all plots for the 3λ and 5λ cases are very similar. Second, of the curves for the 3.5λ case, only that for the feed-region pole resembles its counterparts for the other two lengths. Perhaps the striking difference between the 3.5λ case and the others is the radius-independent nature of the end-region pole ratio, power ratio, and current ratio for the former. However, further examination of these curves reveals additional similarities, as shown below.

The ratios of the nondominant pole residues, normalized by the ratios of the square roots of the power, are plotted in Figure 12.14(a) for all three antenna lengths. Corresponding results are plotted in Figure 12.14(b) for the dominant end-region poles. In both cases, these normalized results for all three antenna lengths are much alike, indicating an underlying length-independent behavior in the dominant feed-region poles. Furthermore, the radius dependencies exhibited in these figures are compatible with the $\log(ka)$ dependence of traveling wave radiation [5–8] and with the dependence of end radiation on $A[1 + b(ka)^2]$, where A and B are constants [8,10].

Several tentative conclusions can be drawn from these results:

- The poles seem to represent an inherently length-independent manifestation of the radiation process.
- The radiated power is proportional to the amplitudes of the dominant poles.

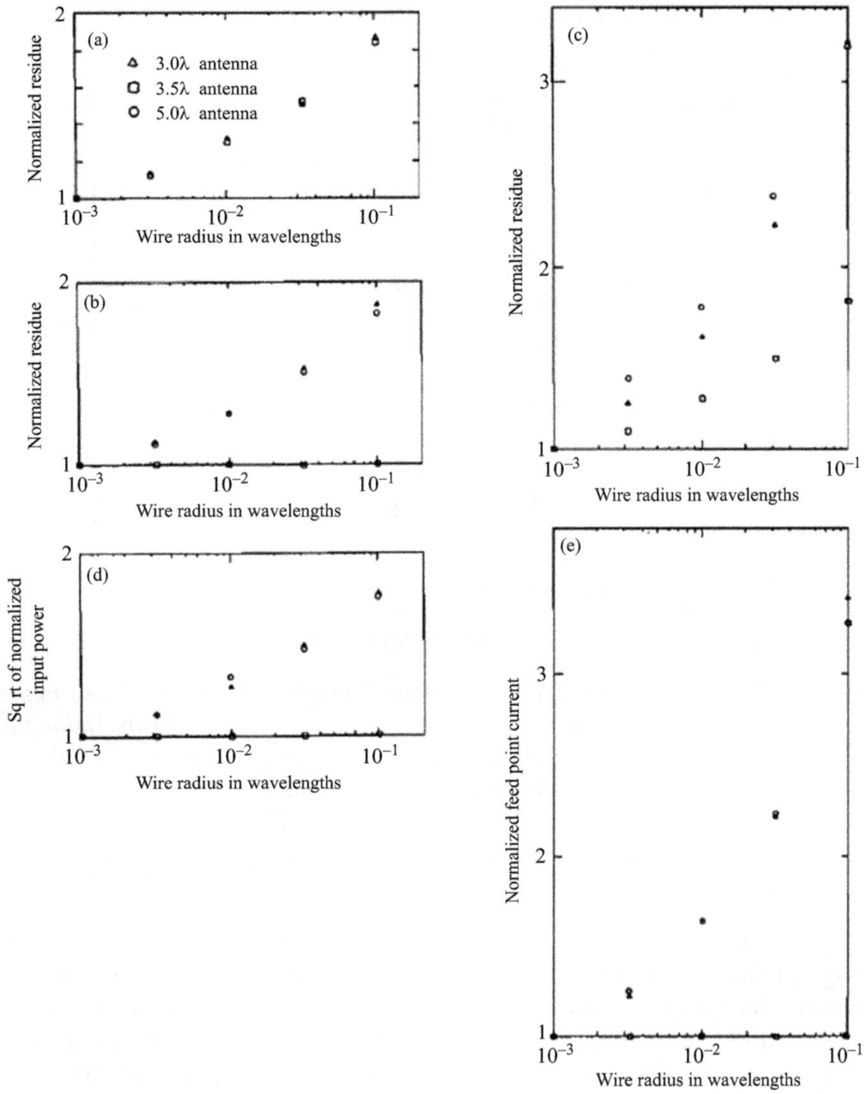

Figure 12.13 *Plots of various quantities associated with 3.0λ, 3.5λ, and 5.0λ antennas, relative to their values for a wire radius of 0.001λ, as functions of wire radius: (a) residues for dominant feed-region poles; (b) residues for dominant end-region poles; (c) residues for all nondominant (feed-region and end-region poles; (d) square root of antenna input power for a I–V source; and (e) feedpoint current magnitude*

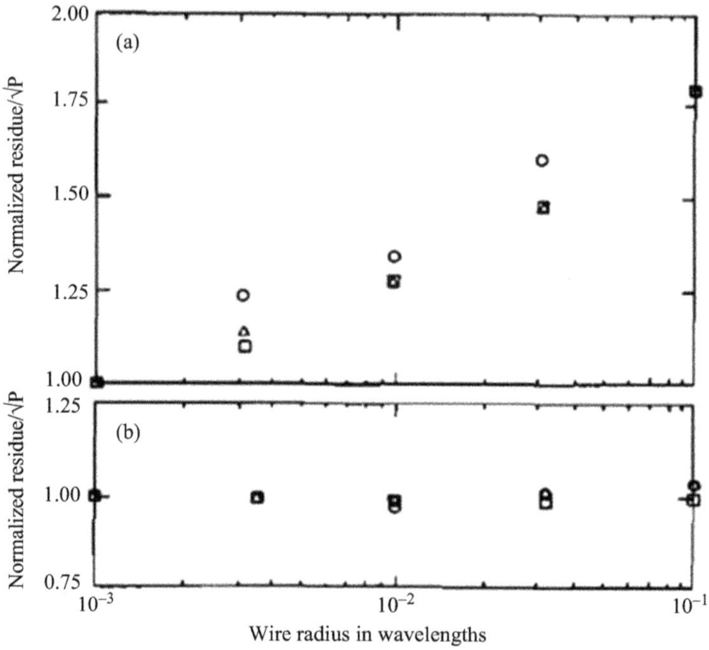

Figure 12.14 Residues divided by the square root of input power, relative to their values for a wire radius of 0.001λ, as functions of wire radius; (a) nondominant poles and (b) dominant end-region poles. Symbols have the same meaning as in Figure 12.13

– End radiation seem to be essentially independent of wire radius for the range of values considered here.

These finding suggest a simple, pole-based model for the behavior of a straight, thin-wire antenna. The wire may be viewed as a transmission line that connects the feedpoint and ends of the antenna, which is of course a well-established picture in any case. A standing current distribution is set up whose shape and amplitude depend on the antenna length. Both are related to the interference between the incident and reflected current waves at the wire ends and their relative phases at the feedpoint. Radiation from the ends is determined by the charge amplitude there. The corresponding pole magnitude is related to this amplitude. Radiation from the source region occurs more as though from an infinite antenna, but with the total charge there established by superposition of the incident and reflected waves. Finally, traveling wave radiation takes place as the waves propagate outward from the feedpoint and back again. Of the three radiation mechanisms, that due to the ends seems not important for the cases considered.

Evidence that source-region radiation is related to the infinite antenna is provided in Figure 12.15, which shows the feed-region pole ratios of Figure 12.13(a),

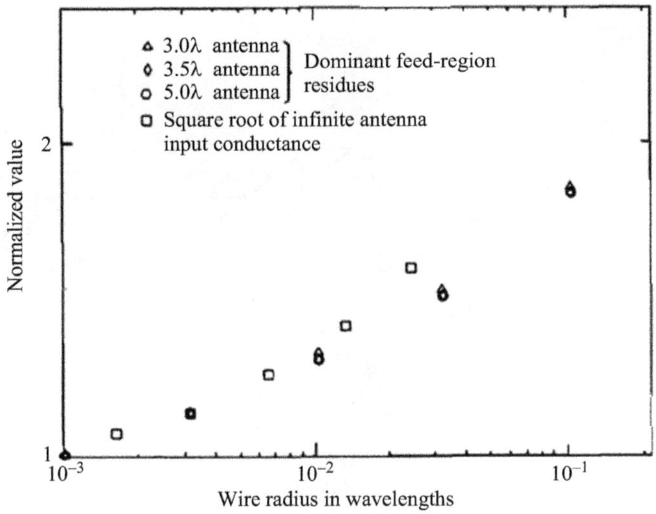

Figure 12.15 Plot of dominant feed-region pole residues and the square root of infinite-antenna input conductance, relative to their values for a wire radius of 0.001λ, as functions of wire radius

together with corresponding ratios of the square root of the radiated power for a 1 V source on an infinite antenna [9]. The curves are very similar, supporting the hypothesis that feed-region radiation from a finite antenna is similar to that from the infinite antenna.

12.3.2.2 Scattering case

A longer 9λ wire having a radius of 0.001λ is next considered as a passive scatterer when illuminated by a plane wave. We first look at the broadside incidence where the illuminating wave is perpendicular to the wire axis. The output pole set for a normally incident plane wave is shown in Figure 12.16. In contrast to the antenna, the poles here are spread all along the wire (16 of them in total). This is expected because the incident field drives the entire wire uniformly. The situation is analogous to the continuous-source case of Figure 12.3: a large number of sources (about $2L/\lambda$) contribute, and the output pattern matches the input.

Varying the incidence angle of the exciting plane wave leads to results like those shown in Figure 12.17. Several interesting characteristics of these pole sets can be identified. First, for all cases except broadside incidence, the poles associated with the wire ends have the largest residues (i.e., are "dominant" in our nomenclature). Second, the interior poles (those away from the ends) exhibit both increasingly larger real components and smaller residues as the incidence angle becomes smaller. Finally, when the excitation approaches axial incidence, the end poles remain significant. This latter result is of particular interest since it indicates that for near-axial incidence, only the ends of the wire are important in the scattering process, a conclusion compatible with geometrical diffraction models for this

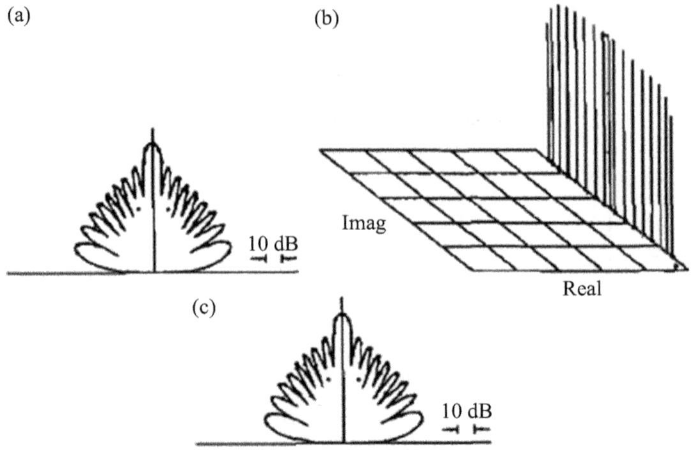

Figure 12.16 *Scattering results for a broadside plane wave incident on a 0.001λ radius wire 9λ long: (a) bistatic scattering pattern, (b) output pole set, and (c) constrained pattern*

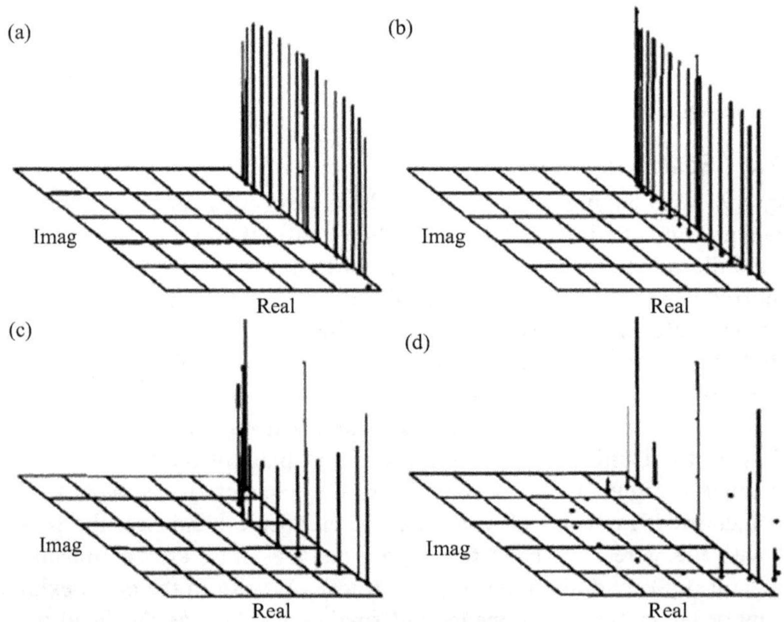

Figure 12.17 *Output pole sets for varying angle of incidence on a 9λ wire: (a) broadside incidence, (b) 5° from broadside, (c) 30° from broadside, and (d) 60° from broadside*

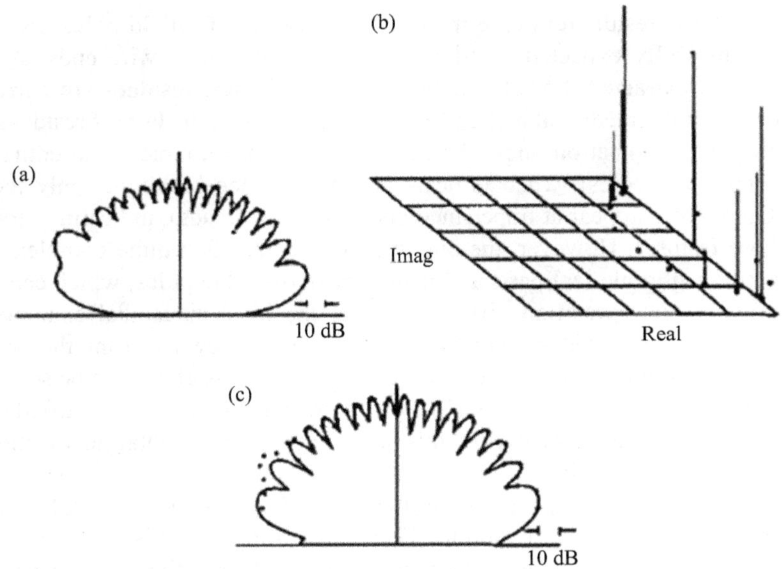

Figure 12.18 Scattering results for a plane wave incident 60° from broadside on a 9λ wire loaded by 300 Ω at its center: (a) bistatic scattering pattern, (b) output pole set, and (c) constrained pattern. Note the new pole at the center due to the load (compared with unloaded case of Figure 12.17(d))

situation. Similarly, for near-broadside incidence, there is the so-called specular flash, or reflected field, which is associated with scattering from the wire's entire length.

One other source of radiation from a straight wire may also be of interest—an impedance discontinuity. This is demonstrated in Figure 12.18 for oblique incidence ($\theta = 30°$) on a 9λ wire loaded at the center by a 300-Ω resistance. The additional pole is clearly due to the resistance load, showing that the current reflection that it causes produces radiation, presumably because of the charge acceleration associated with that reflection.

12.4 Discussion and conclusion

The original motivation for this study was to assess whether and how much of an object's source distribution can be identified from its far-field pattern. The holographic experiment of Izuka and Gregoris [2] provided one piece of evidence that source locations (feed and ends) can be imaged. The Prony-method results presented here on linear arrays of point sources [1] supported the feasibility of this approach. These findings extended that exploration to wire antennas and scatterers.

In general, these results reinforce the basic premise: the far-field poles associated with the physically expected locations (antenna feed region, wire ends, and any loads) can be extracted. For all antenna cases, the largest residues are correlated with sources at the feedpoint and ends. In scattering cases, the largest residues shift depending on illumination angle: broadside incidence illuminates the entire wire (producing many poles), whereas near-end-on incidence highlights only the end poles. Other poles appear at impedance discontinuities (loads), indicating radiation from those features. However, the output pole sets are often quite complex. They contain non-negligible real parts and many smaller-residue poles, which can make straightforward interpretation difficult. Unlike the ideal sinusoidal example, the reconstructed constrained pattern typically differs somewhat from the original when significant poles have nonzero real components. This effect can be somewhat mitigated by factoring out the $1/\sin(\theta)$ term for the antenna case (as justified by the analytic solution), but for scattering, this factor has no fundamental justification yet seems to work empirically.

In any case, this analysis presumes that beyond basic exponentials, the pattern analysis can be enhanced by including this factor. Ultimately, this procedure can be seen as constructing a Huygens-equivalent source distribution for the wire. As illustrated conceptually in Figure 12.19, we characterize the far field in terms of a set of exponential sources along the wire plus a known pattern factor. In effect we are "imaging" the wire by fitting its radiation with point sources. Such an imaging approach is valuable not only for solving the inverse source problem but also for providing insight into the radiation mechanism itself. The results shown here are not fully conclusive, but they demonstrate that Prony-based far-field modeling can indeed reveal where and how a simple object radiates. This suggests potential for applying similar techniques to more general inverse problems in electromagnetics.

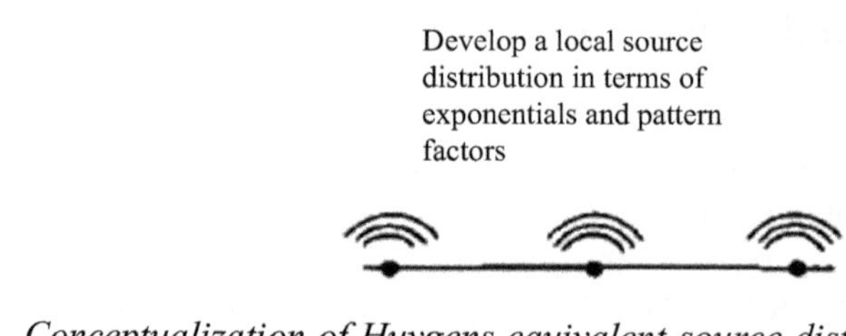

Figure 12.19 Conceptualization of Huygens equivalent-source distribution as established from characterizing far field in terms of exponentials

References

[1] E. K. Miller, and D. L. Lager, Radiation Field Analysis and Synthesis Using Prony's Method," *Electronics Letters*, Vol. 14, p. 180, 1978.

[2] K. Izuka, and L. G. Gregoris, "Application of Microwave Holography in the Study of the Field from a Radiating Source," *Applied Physics Letters*, Vol. 17, No. 12, p. 509, 1970.

[3] W. K. H. Panofsky, and M. Phillips, *Classical Electricity and Magnetism*, Addison- Wesley, Inc. Reading, MA, 1956.

[4] E. K. Miller, *Computer Models for Antennas*, Lawrence Livermore Laboratory, Liver- more, CA, UCRL-51515, 1974.

[5] Y. M. Chen, and J. B. Keller, "Current on and Input Impedance of a Cylindrical Anten- na," *journal of Research*, Vol. 1, No. 66D, p. 14, 1961.

[6] K. S. Kunz, "Asymptotic Behavior of the Current on an Infinite Cylindrical Antenna," *Journal of Research*, Vol. 4, No. 67D, p. 4, 417, 1963.

[7] L. C. Shen, T. T. Wu, and R. W. P. King, "A Simple Formula of Current in Dipole An- tennas," *IEEE Transactions on Antennas and Propagation*, **AP-16**, Vol. 5, p. 542, 1968.

[8] J. Bach Anderson, "*Admittance of Infinite and Finite Cylindrical Metallic Antenna," Radio Science*, Vol. 3, No. 6, p. 607, 1968.

[9] E. K. Miller, "Admittance Dependence of the Infinite Cylindrical Antenna Upon Exciting Gap Thickness," *Radio Science*, Vol. 3, No. 12, p. 1431, 1967.

[10] S. W. Lee, and R. Mittra, "Admittance of a Solid Cylindrical Anttenna," *Canadian Journal of Physics*, Vol. 47, p. 1959, 1969.

Chapter 13

Inversion of one-dimensional scattering data using Prony's method[1]

Abstract

A one-dimensional configuration is the simplest geometry to infer from scattered fields; yet, it has practical applications to such problems as scattering from inhomogeneous half-spaces and propagation on nonuniform transmission lines. Whether the medium parameters vary continuously or discretely with position, the problem's numerical description can usually be developed using a finite-difference approximation. As such, the scattered and transmitted fields can be represented as exponential series, whose exponents are related to the electrical thicknesses of the layers that make up the model. If the exponents or poles are derivable from field data, then the inverse problem is formally solvable. This discussion considers application of Prony's method, a procedure for obtaining the poles of exponential signals, to such one-dimensional problems. Analysis of both the time-domain and frequency-domain data is studied. The effect of medium characteristics, number of layers, and other factors is examined. It is shown that Prony's method has merit for certain classes of one-dimensional inverse problems.

13.1 Introduction

Since the advent of the singularity expansion method (SEM) [1], there has developed a growing interest in the concept of poles in electromagnetics. The poles in the SEM arise as complex resonance frequencies, associated with the damped, sinusoidal ringing of impulsively excited objects—much as in the case of an acoustical analog, a tuning fork struck by a hammer. But poles occur in many other forms as well, owing to the fact that so many physical phenomena are describable as solutions of second-order, linear, differential equations. For example, the far field of a spatial source distribution is an integral (or sum) of exponential terms. The response of a receiving array to incident plane-wave fields can be similarly described.

Considered in this discussion is another problem for which poles provide a natural and useful characterization: the reflection of plane waves from a layered

[1]Adapted from Radio Science, E. K. Miller and D. L. Lager, Vol. 17 (1), Inversion of one-dimensional scattering data using Prony's method, pp. 211–217, 1982.

medium. The description is equally valid for a nonuniform transmission line. But for these kinds of problems, two distinct but related pole sets are found, depending on whether transient or spectral reflected fields are used. Of particular interest is the possibility that pole sets derived from such data may be useful for the inverse problem—that is, given the pole sets, to determine the medium parameters. Some numerical results for the direct problem are presented, and their implications for the inverse problem are discussed here.

13.2 Analytical background

It is useful for what follow to introduce the exponential expression given by

$$f(x) = \sum_{\alpha} R_{\alpha}^{(x)} e^{s_{\alpha}^{(x)} x}, \tag{13.1}$$

where $s_{\alpha}^{(x)}$ and $R_{\alpha}^{(x)}$ are the poles and residues of $f(x)$, whose Laplace transform is

$$F(X) = \sum_{\alpha} \frac{R_{\alpha}^{(x)}}{X - s_{\alpha}^{(x)}}. \tag{13.2}$$

The quantities x and X are the transform variables, and $f(x)$ and $F(X)$ form a transform pair. For clarity, the superscript (x) on s_{α} and R_{α} denotes the domain in which the description is exponential as in (13.1).

The exponential form of (13.1) is relevant because previous work in [2] has shown that the reflection coefficient for impulse, plane-wave illumination from a planar layer is given by (see Figure 13.1):

$$2\pi r(t) = R_{12}\delta(t) + T_{12}T_{21}R_{23}(R_{21}R_{23})^{-2} \sum_{\alpha=-\infty}^{\infty} e^{j\omega_{\alpha} t}. \tag{13.3a}$$

The exponential terms ω_{α} are the roots of

$$1 - R_{21}R_{23} \exp\left(-2j\omega\sqrt{\mu_2\varepsilon_2}D_2 \cos\theta_2\right) = 0, \tag{13.3b}$$

and it is assumed for now that the medium is nondispersive (the response of lossy, dispersive media is considered as "Numerical results," below).

The corresponding reflection coefficient in the spectral domain can also be written in exponential form as

$$R(\omega) = R_{12} + R_{23}T_{12}T_{21} \sum_{\alpha=1}^{\infty} (R_{21}R_{23})^{\alpha-1} \exp(-j2\alpha\omega\mu_2\varepsilon_2 D_2 \cos\theta_2). \tag{13.4}$$

It is intriguing to see that descriptions of the problem expressed in two different domains that form a transform pair can, as in this case, both be exponential. The same situation holds for a transmission line, as reported in [3]. Of interest to us in a pole context is the fact that two different sets of poles result. From (13.1),

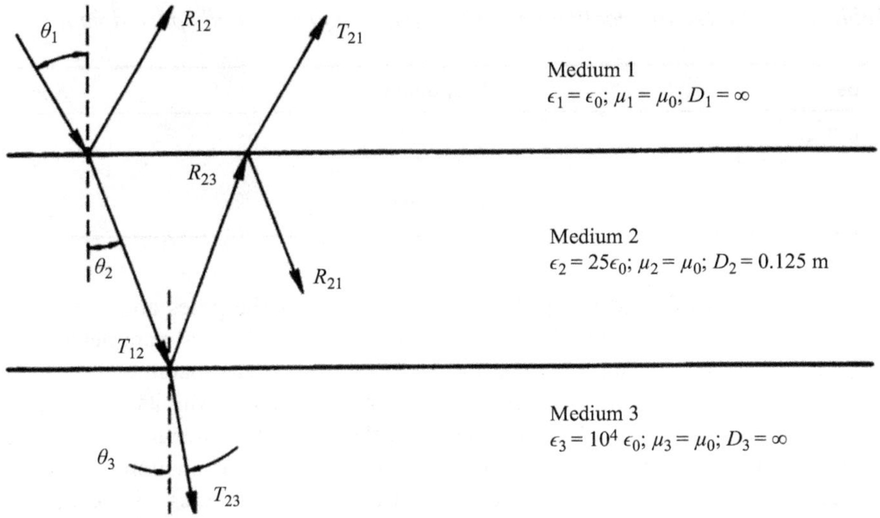

Figure 13.1 The geometry for the one-layer problem and the notation used. The first subscript on the reflection ® and transmission (T) coefficients denotes the medium from which the wave originates, and the second denotes the medium with which it interacts. The electrical permittivity is ε, μ is the magnetic permeability, and D is its physical thickness.

(13.3), and (13.4), they are seen to be

$$s_\alpha^{(t)} \equiv \omega_\alpha = \frac{-j\ \ln|R_{21}R_{23}| \pm \alpha 2\pi}{2\sqrt{\mu_2\varepsilon_2}D_2\ \cos\theta_2} \tag{13.5a}$$

for the time-domain poles, and

$$s_\alpha^{(\omega)} = -j2\alpha\sqrt{\mu_2\varepsilon_2}D_2\ \cos\theta_2 \tag{13.6a}$$

for the frequency-domain poles. For simplicity in the following, it is assumed that $\mu_2 = \mu_0$ and that $\theta_2 = 0$.

The corresponding residues are given by

$$R_\alpha^{(t)} = T_{12}T_{21}R_{23}(R_{21}R_{23})^{-2}, \alpha = 1, 2, \ldots \tag{13.5b}$$

and

$$R_\alpha^{(\omega)} = T_{12}T_{21}R_{23}(R_{21}R_{23})^{\alpha-1}. \tag{13.6b}$$

Note that from (13.1 and 13.2) and (13.5 and 13.6), the reflected fields can be put in the form summarized in Table 13.1.

Table 13.1 Reflection coefficients in the time, frequency, and space domains

Time	Frequency	Space
$\sum R_\alpha^{(t)} \exp\left[s_\alpha^{(t)} t\right]$	$\sum \frac{R_\alpha^{(t)}}{\omega - s_\alpha^{(t)}}$	–
–	$\sum S_\alpha^{(\omega)} \exp\left[s_\alpha^{(\omega)} \omega\right]$	$\sum \frac{R_\alpha^{(\omega)}}{D - s_\alpha^{(\omega)}}$

Comparing (9.5) and (9.6), it can be deduced that the poles and residues of both contain information about the layer's electrical and physical parameters. This information, however, appears in different ways in the two-parameter sets. For example, $R_\alpha^{(t)}$ is a constant independent of α, whereas $R_\alpha^{(\omega)}$ exhibits a geometric-series α dependence. And $s_\alpha^{(t)}$ is a function of the interface reflection coefficients, electrical thickness, and angle of incidence, whereas $s_\alpha^{(\omega)}$ is a function of the latter two quantities only. Further, for a lossless (real μ and ε) layer, $s_\alpha^{(\omega)}$ is purely imaginary, while $s_\alpha^{(t)}$ can be complex in general. Perhaps most interesting is the differing dependence of both pole types on $\sqrt{\mu\varepsilon}D$, since $s_\alpha^{(t)}$ has a real part which is inversely proportional to the layer's electrical thickness, while $s_\alpha^{(\omega)}$ has an imaginary part which is directly proportional to the electrical thickness.

Therefore, since they contain this information, both sets of poles and residues must lend themselves to solving the inverse problem, that is, given $r(t)$ or $R(\omega)$, the layer's parameters can be determined, in principle at least. However, the step of going from $s_\alpha^{(t)}$, $R_\alpha^{(t)}$ or $s_\alpha^{(\omega)}$, $R_\alpha^{(\omega)}$ to μ_2, ε_2, and D_2 may not be a trivial one and is not the primary subject of this discussion, although it will be touched on briefly below. Instead, the primary interest is devoted to the extraction of poles from reflected field data, for the viability of this approach to the inverse problem requires that such be available.

13.3 Pole extraction using Prony's method

The exponential series of (9.3) and (9.4) are candidates for processing via Prony's Method [4], a technique for obtaining the poles and residues from sampled values of such functions, as outlined elsewhere in this book. For completeness, a key step in the development of PM is included here.

Derivation of the poles from (13.1) by PM involves sampling $f(x)$ at $M+1$ points $x_i = i\delta$, $I = 0,1, \ldots, M$. On rewriting (13.1) as (omitting superscripts)

$$f_i \equiv f(x_i) = \sum_{\alpha=1}^{N} R_\alpha x_\alpha^i, \tag{13.7}$$

where

$$x_\alpha = e^{s_\alpha \delta}.$$

Equation (13.7) then consists of

$$
\begin{aligned}
f_0 &= R_1 + R_2 + \ldots + R_N \\
f_1 &= R_1X_1 + R_2X_2 + \ldots + R_NX_N \\
f_2 &= R_1X_1^2 + R_2X_2^2 + \ldots + R_NX_N^2 \\
&\vdots \\
f_{M-1} &= R_1X_1^{M-1} + R_2X_2^2 + \ldots + R_NX_N^{M-1} \\
f_M &= R_1X_1^M + R_2X_2^M + \ldots + R_NX_N^M.
\end{aligned}
\tag{13.8a}
$$

Now, suppose that X_α represents the roots of an Nth-order polynomial, that is,

$$\sum_{\alpha=0}^{N} a_\alpha x^\alpha = (x - x_1)(x - x_2)\cdots(x - x_N) = \prod_{\alpha=}^{N}(x - x_\alpha);\, a_N = 1, \tag{13.8b}$$

Upon multiplying the first equation in (13.8a) by a_0, the second equation by a_1, etc., and forming the sum of the first N equations, a new equation can be derived, which, upon using (13.8b), can be reduced to

$$a_0f_0 + a_1f_1 + \cdots + f_N = 0.$$

If this operation is repeated $M - N$ times by successively beginning with the second equation of (13.8a), &, until the $(M - N)$th equations is used, the following $M - N + 1$ linear equations will result:

$$
\begin{aligned}
a_0f_0 + a_1f_1 + \cdots + a_{N-1}f_{N-1} &= -f_N \\
a_0f_1 + a_1f_2 + \cdots + a_{N-1}f_N &= -f_{N+1} \\
&\vdots \\
a_0f_{M-N} + a_1f_{M-N+1} + \cdots + a_{N-1}f_{M-1} &= -f_M
\end{aligned}
\tag{13.8c}
$$

The linear system in (13.8c) provides the basis for obtaining the N coefficients in the polynomial in (13.8b), from which the N pole values are found as

$$s_\alpha = \frac{1}{\delta}\ln X_\alpha. \tag{13.8d}$$

If $M + 1 - 2N$, the system is square of order N and can be solved directly for the N real a's. For $M + 1 > 2N$, the method of least squares can be used. Finally, the $R_\alpha's$ can be found by various means, such as fitting the first N equations of (13.8a) or by applying the least-squares technique to the entire set.

It is important to note that Prony's method requires for its implementation that all the exponents change in uniform steps of the independent variable. This is in order that the successive samples of $f(x)$ can be written as sums of constants raised to successively higher integer powers. From (13.5a), it may be surmised that if ε_2 or

μ_2 were to be frequency dependent due to dispersion or loss, time-independent solutions for ω_a would still exist. Therefore, Prony's method remains applicable to $r(t)$, (13.3a), for finding the ω_a poles.

However, for the case of dispersive and/or lossy media, the poles $s_a^{(\omega)}$ in (13.61) can become frequency dependent. Consequently, unless that dependence is explicitly known so that frequency steps can be used which produce uniform changes in the exponents of (13.4) for $R(\omega)$, Prony's method is no longer applicable. Thus, although the $s_a^{(\omega)}$ have somewhat more attractive features than the $s_a^{(t)}$ for establishing the medium parameters, the $s_a^{(t)}$ may provide the more workable approach for handling realistic (i.e., lossy) media. Nevertheless, our attention here is directed primarily to the frequency-domain pole set, as the time-domain case had been considered initially by as the time-domain case has been considered initially by Lytle and Lager [3].

It is also worthwhile commenting on the effect on the reflected field when the incident field is nonimpulsive. If $\overline{r}(\overline{t})$of (13.3a) is the impulse response, then the reflected

$$\overline{r}(\overline{t}) = \int_{-\infty}^{\infty} r(t-\tau)d\tau. \tag{13.9}$$

Substituting (13.3a) into (13.9) reveals that the poles are unchanged from their impulse-response values, while the residues are now given by

$$\overline{R_a}(t) = T_{12}T_{21}R_{23}(R_{21}R_{23})^{-2}\int_{-\infty}^{\infty} \exp(-j\omega_a\tau)i(\tau)d\tau. \tag{13.10}$$

The corresponding frequency-domain response for a nonimpulsive incident field, or equivalently, one whose spectral content is frequency dependent with amplitude $I(\omega)$, is

$$\overline{R_a}(\omega) = R(\omega)I(\omega). \tag{13.11}$$

Thus, if $I(\omega)$ is known, $R(\omega)$ can be found from $\overline{R_a}(\omega)$, and the poles and residues can be estimated as described above. If, however, $i(I(\omega)$ is not known, the results is that the residues of (13.4) become frequency dependent, so that, as in the case of the dispersive medium, Prony's method is no longer applicable. The, however, optimization-based identification techniques may be applicable.

13.4 Numerical results

Figures 13.2 and 13.3 show the pole sets that arise from a dielectric layer of relative permittivity $\varepsilon_r = 25$ and thickness $D_2 = 0.125$ m sandwiched between half-spaces of $\varepsilon_r = 1$ (free space) and $\varepsilon_r = 10^4$. The reflection coefficients $r(t)$ and $|R(\omega)|$ are also shown, where the incident field arriving from the $\varepsilon_r = 1$ half-space is normal to the planar surfaces. The isometric plots show the pole locations in the complex plane, with the residue magnitudes depicted on a three-decade logarithmic vertical scale.

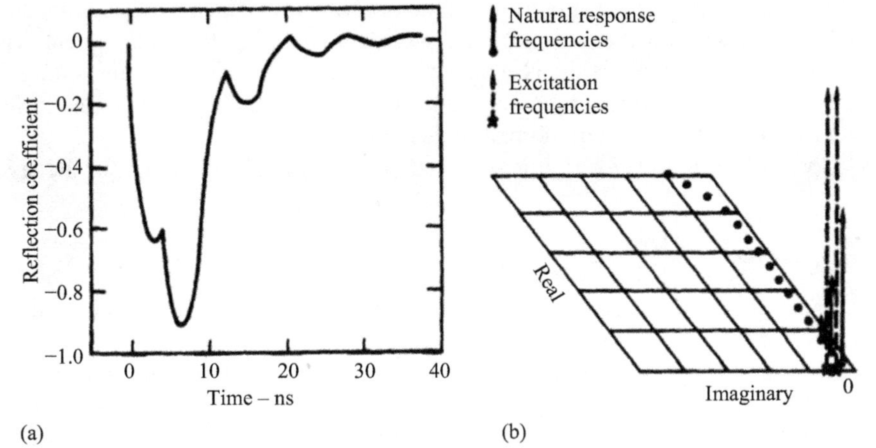

Figure 13.2 Time-domain results for the geometry of Figure 13.1. (a) The transient reflection coefficient for an incident plane wave for double exponential time variation, (b) the resulting pole-residue plot. (Note that the pole sets for the time-domain case are harmonic along the real-frequency axis, whereas for the frequency-domain case (Figures 13.3–13.7), they are harmonic along the imaginary axis.) (from [2]).

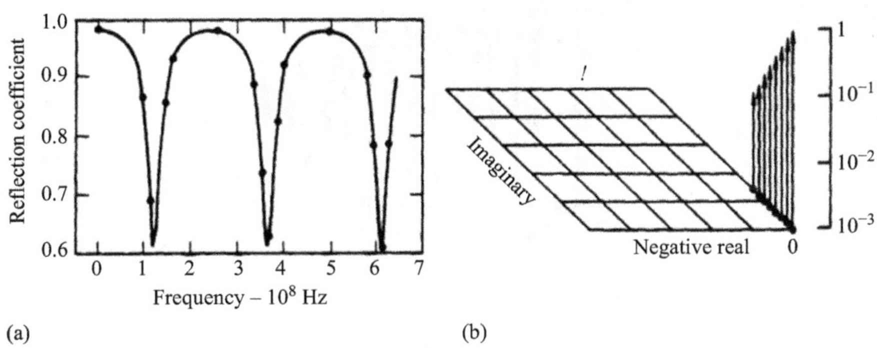

Figure 13.3 Frequency-domain results for the geometry of Figure 13.1: (a) the amplitude of the frequency-dependent reflection coefficient, (b) the resulting pole-residue plot (these plots in Figures 13.3–13.7 exclude the curve-fitting poles, which are not in the $R(\omega)$ data)

Note that in obtaining the time-domain poles, the data are values of a real function sampled at time intervals, whereas to find the frequency-domain pole, the data are values of a complex function sampled at equal-frequency intervals.

In the time-domain case (Figure 13.2), the poles associated with the incident field are so large as to mask the residue magnitudes of the layer-caused poles. For the frequency-domain case (Figure 13.3), however, the residue magnitudes are seen to fall off linearly on the vertical log scale, demonstrating their agreement with (13.6b). Also observable is the absence of loss in the $s_\alpha^{(\omega)}$ poles, whereas the $s_\alpha^{(t)}$ locations exhibit loss through having nonzero real components. The results to follow will be for the frequency-domain poles only.

Upon referring to (13.6a), it is seen that $\sqrt{\mu_0 \varepsilon_2} D_2$ can be obtained for the layer from

$$\sqrt{\mu_0 \varepsilon_2} D_2 = j s_\alpha^{(\omega)} / 2\alpha.$$

Thus, each of the nonzero poles extracted from $R(\omega)$ can be used to provide an estimate for the layer's electrical thickness.

In Figure 13.4, the reflection coefficient magnitude and pole set which result for normal incidence on two layers having relative permittivities of 9 and 16 and thicknesses of 0.208 and 1 m, respectively, are shown, and which are sandwiched between the same two half-spaces as for Figures 13.2 and 13.3. It may be seen that the pole set is composed of several groups. The first group is due to multiply reflected waves in the upper ($\varepsilon_r = 9$) layer (the pole at the origin comes from the field reflected at the first interface). The second group arises from wave which first make on round trip through the low (and electrically thicker) layer and then are

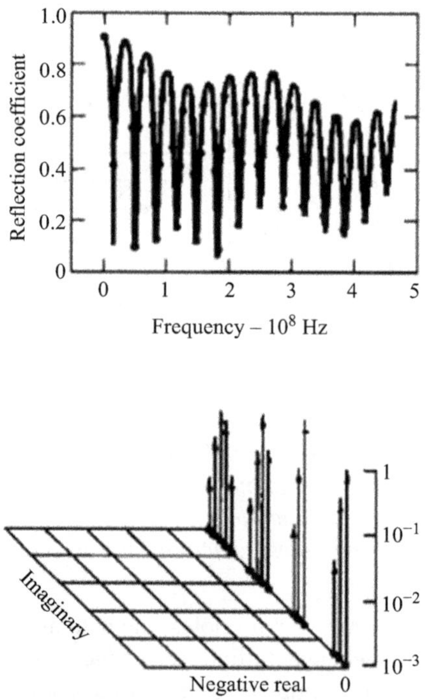

Figure 13.4 Reflection-coefficient and pole-residue results for a two-layer case, $\varepsilon_2 = 9\varepsilon_0$; $\varepsilon_3 = 16\varepsilon_0$; $D_2 = 0.208$ m; $D_3 = 1.00$ m

multiply reflected in the upper layer. Subsequent groups are caused by successively higher-order reflections in the lower layer. Note that the residue amplitudes for these groups do not monotonically decrease as in Figure 13.3, because of the increasing number of reflection combinations that yield the same total electrical phase change as the number of "bounces" increases.

Although lossy media will be dispersive in general and thus will exclude use of Prony's method to find the poles of $R(\omega)$, it is possible to handle data having frequency-independent complex exponents. This behavior would correspond to a medium having a fixed loss tangent. Consequently, the geometry used in connection with Figure 13.4 was rerun with loss tangents of 0.05 and 0.02 for the upper and lower layers, respectively. The reflection-coefficient amplitude and extracted pole set that result are in Figure 13.5. Comparison of Figures 13.4 and 13.5 reveals the same general pole patterns in both cases, except that in Figure 13.5 the poles now have nonzero real components, showing the effect of loss. Upon increasing the loss tangent of the upper layer to 0.1 and 0.5, the results shown in Figures 13.6 and 13.7 are obtained. The most noticeable difference in Figures 13.5–13.7 is the increasing obliqueness of each pole-group set relative to the imaginary axis, due to the increasing loss in the upper layer.

Upon referring to Figure 13.8, in which is sketched a pole set characteristic of those in Figures 13.5–13.7, a simple formula for evaluating the loss tangent can be

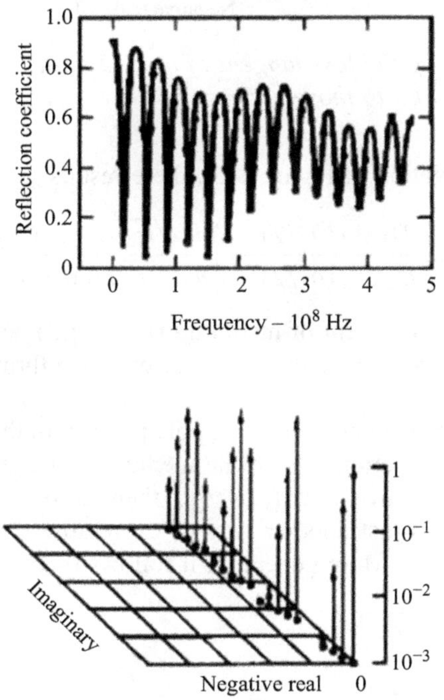

Figure 13.5 Introduction of a frequency-independent loss tangent into the problem of Figure 13.4 [$\tan(\delta_2) = 0.05$, and $\tan(\delta_3) = 0.02$] leads to the results presented here

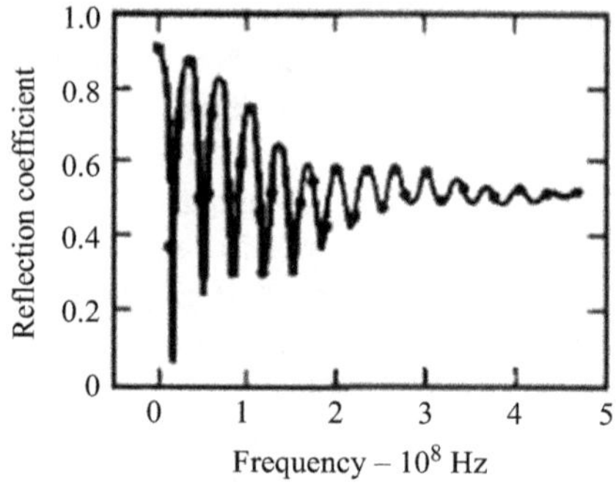

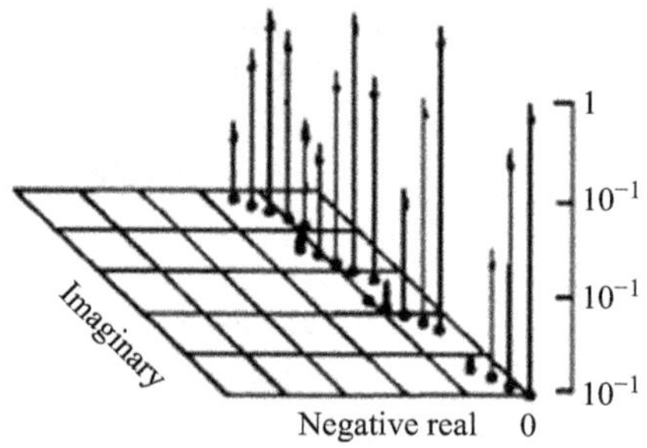

Figure 13.6 Increasing the loss tangent of layer 2 to 0.1 changes the results of Figure 13.5 to those given here

derived. Using the notation of Figure 13.8, there results

$$\tan\theta = \frac{\Delta n_2\sqrt{|\varepsilon_2|}D_2 \sin(\delta_2/2) + \Delta n_3\sqrt{|\varepsilon_3|}D_3 \sin(\delta_3/2)}{\Delta n_2\sqrt{|\varepsilon_2|}D_2 \cos(\delta_2/2) + \Delta n_3\sqrt{|\varepsilon_3|}D_3 \cos(\delta_3/2)}, \tag{13.12}$$

where $\tan(\delta_i)$ is the loss tangent of medium i ($i = 2$ or 3), and Δn_i is the change in the number of round trips made by the reflected wave through medium i for two different poles.

By selecting various combinations of pole pairs from the derived set, it is thus possible to evaluate the parameters of the medium. For example, if $\Delta n_3 = 0$, then $\tan(\theta) = \tan(\delta_2/2)$, and conversely if $\Delta n_2 = 0$, then $\tan(\theta) = \tan(\delta_3/2)$.

Equation (13.12) is just another way of expressing (13.6a) for the case of a lossy, two-layer medium. More generally, it follows that

$$s_\alpha^{(\omega)} = -2j\sum_i \sqrt{\mu_0\varepsilon_i}D_i n_i, \tag{13.13a}$$

where the constraint

$$\alpha = \sum_i n_i \tag{13.13b}$$

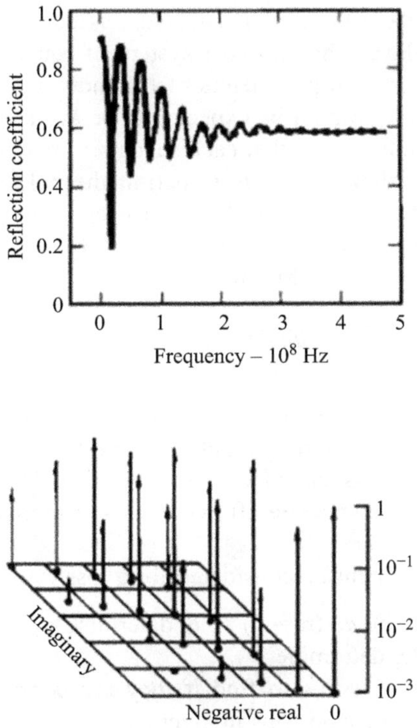

Figure 13.7 Further increasing the loss tangent of layer 2 to 0.5 has the effect shown here

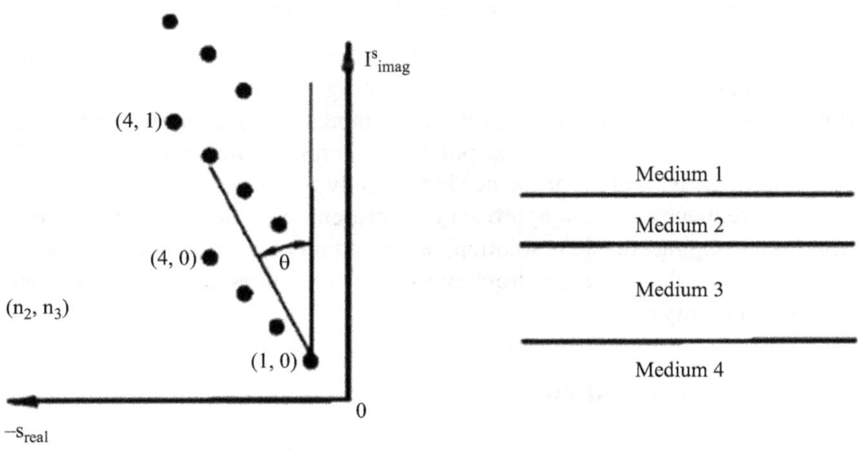

Figure 13.8 Explanation of notation for derivation of loss tangent (13.12)

applies with n_i the number of round trips in layer i associated with $s_\alpha^{(\omega)}$. Clearly, if n_i were known for each layer, then a linear system of equations could be generated from (13.13a) for finding the parameters of the individual layers. Unfortunately, this information generally cannot be expected to be available.

It can be surmised, however, that each successive extractable pole (one whose residue amplitude is sufficiently large to permit the pole to be accurately determined from the reflection coefficient data) must represent an integer increment in round trips through one or more of the medium's layers. Thus, by subtracting pole pairs as given in (13.13a), there follows

$$\left[s_k^{(\omega)} - s_l^{(\omega)}\right] = -2j \sum_i \sqrt{\mu_i \varepsilon_i} D_i \Delta n_i. \tag{13.14}$$

If the Δn_i can be established, then estimates of the electrical thicknesses can be made. Estimates of these quantities could also be made from the reflection and transmission coefficients associated with each residue. The problem is essentially one of disentangling the composite effects of two or more layers on a given pole and its residue.

One approach that might be considered for a lossless medium is as follows:

- From the residue of the *dc* ($\alpha = 0$) pole, determine ε_2.
- From the $\alpha = 1$ pole, determine D_2.
- Check the $\alpha = 2, \ldots$ poles to find if they are consistent with successively increasing numbers of round trips in layer n_2.
- The first pole which fails step 3 must represent a single round trip through D_2 and D_3; from this pole's residue ε_3 can be found and its location yields D_3.
- Continue in this fashion, using the remaining poles to identify additional layers.
- Use the remaining residues to obtain additional confirmation that the interpretations are consistent with the model being developed.

This exercise would appear to become less reliable as α increases, leading to ambiguity regarding the number of passes through each of the layers which the various poles represent. A key to developing a model for the medium from the pole data would seem to be locating those poles that represent a single pass through 1, 2, 3, & layers. Clearly, more work is needed to study this aspect the problem. Such a pole-based frequency-domain approach may present more difficulty in this respect than a time-stepping transient solution, which admits the possibility of obtaining the parameters of deeper layers from late-time (r (t) values, albeit with a possibly growing uncertainty).

13.5 Discussion and conclusion

An approach to the one-dimensional inverse problem has been discussed here. It is based on the term-wise exponential character of either the transient or the spectral reflected field, whose exponents and amplitudes contain information about the electrical and physical parameters of a layered medium.

Results from the simple cases considered, which consisted of at most two contrasting layers between two different half-spaces, demonstrate that poles indeed can be extracted from the reflected field. These results also show that in spite of the geometrical (and hence electrical) simplicity of the problems treated, the pole sets obtained are not simple to interpret. While the poles and residues do provide a link to an inverse solution—that is, determining the medium parameters from the reflected field—there is no obvious procedure for carrying out the inversion, although one sequential procedure for doing it was outlined as above.

The fact is that the poles and residues, and thus the reflected field as well, can contain much redundant information, as in the examples presented; this is demonstrated by the existence of many more pole and residue values than the minimum number needed to establish the medium parameters. This situation might be described as one having overdetermined data—that is, where the number of data parameters (poles and residues), or data rank, exceeds the number of medium parameters, or medium rank. An inverse problem is not solvable uniquely unless the data rank equals or exceeds the medium rank. When it does, as is the case above, better parameter estimates might be expected to be possible, but at the expense of incurring increased effort to sort out the confusion created by the redundancy—or worse, developing an incorrect inverse model. Thus although the pole-residue description may provide one approach to the inverse problem for one dimension, much additional work will be needed to make it practicable.

References

[1] C. E. Baum, "The Singularity Expansion Method," in *Transient Electromagnetic Fields*," edited by L. B. Felsen, Chapter 3, Springer, New York, 1976.

[2] F. Hildrebrand, *Introduction to Numerical Analysis*, McGraw-Hill, New York, 1956.

[3] R. J. Lytle and D. L. Lager, "Using the Natural-Frequency Concept in Remote Probing of the Earth," *Radio Science*, Vol. 11, pp. 199–209, 1976.

[4] C. Tai, "Transient on Lossless Terminated Transmission Lines," *IEEE Transactions on Antennas and Propagation*, SPS-26. Vol. 26, P 556, 1978.

Chapter 14

Derivative sampling of computational data[1]

Abstract

A need to determine the electromagnetic response of an object over a specified frequency band is frequently encountered in numerical modeling. This is typically done using many successive evaluations of a moment-method (MM) model over the frequency range of interest, leading to possibly excessive computer time if many frequency samples are required. One way of reducing the number of frequencies at which the MM calculation must be done is by introducing *a priori* knowledge of the expected frequency behavior in the form of a model, a pole series for example. The model could also be based on frequency samples or on frequency derivatives of the response, exploiting the fact that these derivatives can be obtained recursively using significantly less time than that needed to develop the initial impedance and admittance matrices. Both of these approaches are discussed here.

14.1 Introduction

The electromagnetic response of an antenna, propagation path, or a scatterer is most often needed over a spectrum of frequencies rather than at just one or a few isolated frequency points. This is increasingly the case as the operating bandwidths of systems increase or their transient responses are required.

For various reasons, including the fact that most electromagnetic textbooks and courses continue to emphasize time-harmonic analysis, development of wideband spectral responses or even a short-pulse transient response is most often based on using a set of discrete frequency samples. Unfortunately, the samples used and the spacing thereof in frequency seem to be determined more often by the esthetic appearance of straight-line or other low-order interpolation rather than by exploiting the underlying physics of the phenomenon being modeled. The result can be that many more frequency samples are used than should be necessary from

[1]Excerpted and edited from G. J. Burke and E. K. Miller (1988), "Use of Frequency-Derivative Information to Reconstruct an Electromagnetic Transfer Function," *The Fourth Annual Meeting of the Applied Computational Electromagnetics Society*, Monterey, CA; Magnetic article.

sampling requirements based on information theory, with a consequent but avoidable increase in the cost of the associated computation or even a measurement.

As an alternative to this interpolator approach, perhaps best described as curve-fitting, two sampling techniques are developed here that incorporate a model of the expected behavior from electromagnetic physics, using model-based parameter estimation (MBPE). An example of this idea, applied in a quite different context is that of replacing the Sommerfeld integrals that occur in the interface problem by simple analytical approximations whose amplitudes are determined numerically by matching them to accurately computed Sommerfeld integrals [1]. A more thorough discussion of MBPE applications in electromagnetics is given in [2].

One MBPE approach uses ordinary discrete frequency samples to develop a functional representation of the sought-for frequency response. The second takes advantage of an evidently previously unused property of the moment-method impedance and admittance matrices to obtain derivatives of the frequency response for much less computational cost than that required to obtain the original response at that same frequency. Each approach utilizes the mathematical property that the frequency response is well approximated by a rational function or ratio of two polynomials (the model) in which the variable is the complex frequency and in which the parameters are the polynomial coefficients. The problem then becomes one of determining numerical values for the unknown coefficients that appear in the rational function by using whatever appropriate data might be available. As will be shown, the possibility then also arises of developing a generalized moment-method admittance matrix whose coefficients are themselves functions of complex frequency, leading to further interesting ramifications for obtaining and representing both spectral and temporal responses. The use of both frequency sampling and frequency-derivative sampling is summarized below, followed by representative numerical examples, which illustrate application of the latter.

It is relevant to observe that Shannon [3] published the following remarks concerning derivative sampling of time-series data: "If the function is limited to the time interval *T* and the samples are spaced *1/2W* s apart, there will be a total of *2TW* samples in the interval. … The *2TW* numbers used to *specify* the function need not be the equally spaced samples used above. For example, the samples *can* be unevenly spaced, although, if there is considerable bunching, the samples must be known very accurately to give a good reconstruction of the function. The reconstruction process is also more involved with unequal spacing. One can further show that the value of the function and its derivative at every other sample point are sufficient. The value and first and second derivatives at every third sample point give a still different set of parameters which uniquely determine the function. Generally *speaking*, any set of *2TW* independent numbers associated with the function *can* be used to describe it."

One limitation of time-series or any other derivative sampling is that the underlying structure of the data needs to be known to compute analytical derivatives. Numerical derivative samples might be obtained using finite differences of successive close-enough data samples. This might be worth considering for measured or computed data where use could be made of what might otherwise appear to be over-sampled results. Our attention in the following is focused on spectral or frequency-dependent electromagnetic data whose underlying structure is well represented by a rational function. For completeness, brief summaries of waveform-domain and spectral-domain function sampling and derivative sampling are included.

14.2 Model-based parameter estimation

14.2.1 The basic waveform (exponential-series) and spectral (pole-series) models

Many kinds of mathematical transformations are used in solving physical problems. One of the more useful of these, and one which is ideally suited to the exponential forms of interest here, is the Laplace transform. For the generic waveform-domain series having W exponential terms as a function of waveform-variable x is given by

$$f(x) = f_p(x) + f_{np}(x) = \sum_{\alpha=1}^{W} R_\alpha e^{(s_\alpha x)} + f_{np}(x) \tag{14.1a}$$

The single-sided Laplace transform leads to the generic spectral-domain pole series as a function of transform-variable X given by

$$F(X) = F_p(X) + F_{np}(X) = \sum_{\alpha=1}^{W} \frac{R_\alpha}{(X - s_\alpha)} + F_{np}(X) \tag{14.1b}$$

where the terms waveform and spectral domains are used as the most commonly encountered physical examples given by the forms above, respectively. The subscripts p and np refer to the pole and nonpole parts of $f(x)$ and $F(X)$ respectively about which more will be discussed below. Note that knowledge of the W $s_a's$ (the poles) and the W $R_a's$ (the residues) provides complete information both for $f(x)$ and $F(X)$ in the absence of the nonpole term. This means that rather than needing many sampled values to reconstruct a waveform or spectrum, all that needs to be stored are the two sets of numbers (generally complex) s_a and R_a. The result can be a greatly reduced number of bits required for storing such data while also providing a function for extrapolating or interpolating $f(x)$ or $F(X)$, continuous in the independent variable. Also note that upon expanding (14.1b) into least-denominator form, it can be written as

$$F(X) = \frac{N(X)}{D(X)} + F_{np}(X) = \frac{N_0 + N_1 X + X_2^2 \cdots + N_W^{W-1} +}{D_0 + D_1 X + D_2^2 \cdots + D_W^W} + F_{np}(X) \tag{14.1c}$$

It has been found that adding a constant or a constant plus a linear term in X when using overlapping, windowed FMs improves the MBPE model. This might be due to approximating the influence of nonpole terms and/or by including the influence of contributions from outside the FM window. Observe that PM requires the values of x in (14.1a) must increase in uniform steps.

14.2.2 Parameter estimation

Practical implementation of (14.1) or (14.2) requires that s_a and R_a be derivable from whatever analytical, computed, or measured data or information that is available. A procedure for doing this from waveform data was developed in 1795 by Prony [4], [5]. Prony's method in various modified and extended forms may be recognized in many signal-processing methods in present use ([6], for example). As outlined below, it can be used as originally developed, for processing waveform data, and in a slightly different, but closely analogous version, for processing spectral data using in both cases what is termed function sampling. Further modifications permit processing of waveform and spectral data where the data samples are derivatives with respect to the independent variable, referred to as derivative sampling. The waveform-domain version of Prony's method is discussed in Chapter 4, and here its spectral-domain counterparts for both function sampling and derivative sampling are summarized.

Note that throughout the discussion below, W denotes the number of effective pole terms present in the model from which the data is *generated*, and that P denotes the number of pole terms employed in the model whose parameters are to be *estimated*. For valid results to be obtained, it is necessary that $P \geq W$, but since W is generally unknown *a priori*, it should also be regarded as a parameter to be determined. Also, the parameters of the generation model are indicated by ordinary letters and those of the estimation model by overscript carats, respectively. Finally, D is used to denote the number of data samples employed in the estimation process.

14.3 Processing in the waveform domain

14.3.1 Introduction

Obtaining the pole-based parameters of an exponential-series, or waveform-domain, model is the original problem for which Prony's method was developed. Because other variants of Prony's method share its fundamental characteristics, it is outlined in detail below. Then presented is an alternate procedure that uses waveform-derivative rather than waveform-function sampling, for estimating the same parameters.

14.3.2 Using waveform-function sampling

Determination from waveform-function sampling of the poles in (14.1) using Prony's method involves sampling $f(x)$ at D equally spaced data points $x_i = id$, $i = 0$, 1, &, $D-1$ over an interval of width $D = d(D-1)$ from which a linear system of equations is derived in terms of an intermediate set of "linear-predictor" coefficients. Using uniform steps in the waveform variable x is necessary because the data model must be recast in polynomial form where the powers of the individual terms in the polynomial are determined by the change in observation variable between successive terms.

It is important to recognize that for the method to work, the data must be available in uniform steps of the exponent, so that (14.1) can be written in polynomial form as in (14.5), which limits application of this form of Prony's method. Uniform sampling is *not* a requirement of the spectral-domain version of Prony's method, as will be shown below. Note also that the technique just outlined can be used for data whose exponential terms are products of two or more exponents depending on different variables, for example, where

$$f(x,y) = \sum_{\alpha=1}^{W_x}\sum_{\beta=1}^{W_y} R_{\alpha\beta}\exp\left(s_\alpha^{(x)}\right)\exp\left(s_\beta^{(y)}\right) \tag{14.2}$$

a form that occurs in connection with rectangular-mesh sampling [2].

14.3.3 Using waveform-derivative sampling

As discussed in Chapter 3, waveform-derivative sampling is another option. If the waveform from which the parameters are to be estimated is sampled in terms of its derivatives rather than as a sequence of equally-space observations, the basic procedure just outlined can also be employed. While the use of waveform-derivative sampling might seem unusual, it is included for completeness and because spectral-derivative sampling offers some attractive advantages [7], which might also eventually be found for the former as well. Begin by writing the *i*th derivative of (14.1) without the nonpole term as

$$f^{'}(x) = \frac{d^i f(x)}{dx} = \sum_{\alpha=1}^{W} s_\alpha^i R_\alpha e^{s_\alpha x} \tag{14.3}$$

Proceeding in the same way as for the case of waveform-function sampling, but using the polynomial

$$\sum_{\alpha=0}^{P} \hat{b}_\alpha s^\alpha = 0; s = s_i, i = 1,2,\ldots,P \tag{14.4}$$

the following linear system for the predictor coefficients can be derived

$$\begin{bmatrix} f^{(0)} & f^{(1)} & \cdots & f^{(P-1)} \\ f^{(1)} & f^{(2)} & \cdots & f^{(P)} \\ \vdots & \vdots & \vdots & \vdots \\ f^{(D-P-1)} & f^{(D-P)} & \cdots & f^{(D-2)} \end{bmatrix} \begin{bmatrix} \hat{b}_0 \\ \hat{b}_1 \\ \vdots \\ \hat{b}_{P-1} \end{bmatrix} = -\begin{bmatrix} f^{(P)} \\ f^{(P+1)} \\ \vdots \\ f^{(2P-1)} \end{bmatrix} \tag{14.5}$$

which is the counterpart of (14.5) of Chapter 4 with function samples replaced by derivative samples and where the *P*th coefficient in the characteristic equation has been set to unity. But where the poles are found from logarithmic transforms of the roots of the characteristic equation for waveform-function sampling, they are themselves found to be the roots of the characteristic equation for waveform-derivative sampling.

14.4 Processing in the spectral domain

14.4.1 Introduction

Waveform-domain sampling starts with an exponential-series generating function, whereas spectral-domain sampling starts with its pole-series counterpart. While this might seem to be a minor difference, it is significant in that the need for uniformly spaced samples that applies in waveform sampling is not a requirement in the spectral domain. This is because the spectral-domain model is already of polynomial form as will be seen below, a potential advantage. However, the spectral-domain linear system whose solution yields the model parameters has its data samples multiplied by the spectral variable raised to integer powers from 0 to P, producing a possibly very large dynamic range in the matrix coefficients. As one potential compensation for the latter possibility, it can be feasible to reduce the model order, that is, the number P of poles in the estimator, below the actual number of poles W in the generator by using a reduced spectral range for the estimator model. Spectral-function sampling and spectral-derivative sampling are discussed in turn below.

14.4.2 Using spectral-function sampling

Determination from spectral-function sampling of the poles in (14.2) using the pole-series version of Prony's method involves sampling $F(X)$ at D data points $X = X_i$, $i = 0, 1, \&, D-1$ over a spectral interval of width $D = \mathrm{Im}(X_{D-1} - X_{0)}$, from which a linear system of equations is also derived. But whereas in the waveform (time) domain it is possible to model the free-response portion of the waveform by using only function samples, which occur after the excitation is no longer affecting the response, this is not strictly feasible in the spectral domain. Although reasonable results can be found using only the pole-based part of (14.2) [8], more robust performance will be obtained if the nonpole portion is modeled as well, which is analogous to including the driven part of the response in the waveform domain. Thus, with $X = s + j\omega$ a general complex frequency (but which for sampling on the $j\omega$ axis would use $s = 0$) there results.

14.4.3 Using spectral-derivative sampling

The same rational-function model can be used when frequency-derivative samples are employed rather than frequency samples. Starting with (14.1c) in product form while dropping the non-pole term and upon differentiating t times with respect to X and dropping the explicit X dependence, there result the following equations:

$$
\begin{aligned}
&FD = N \\
&F'D + FD' = N' \\
&F''D + 2F'D' + FD'' = N'' \\
&F'''D + 3F''D' + 3F'D'' + FD''' = N''' \\
&\vdots
\end{aligned}
\tag{14.6}
$$

so that an expression for the tth derivative of the spectral response can be written as

$$F^{(t)} = N^{(t)} - \frac{\left[F^{(t-1)} - \left(D + \cdots C_{t,t-m}F^{(m)}D^{(t-m)} + \cdots + FD^{(t)}\right)\right]}{D} \tag{14.7}$$

where $C_{r,s} = r!/[s!(r-s)!]$ is the binomial coefficient. The system of $t+1$ equations in (14.7) (there are t frequency-derivative samples plus the original frequency sample) provides the information from which the model coefficients can be found if $t \geq n+d+1$, assuming as before that one of the coefficients is specified. With no loss of generality, (14.6) can be simplified by setting $s = 0$ since X can represent the variation in spectral variable about the point where the spectral-derivative samples are obtained. Having the derivative-sample sequence from (14.6), the following matrix representation for the unknown numerator- and denominator-polynomial coefficients has the form

$$\begin{bmatrix} 1 & 0 & \cdots & 0 & 0 & 0 & \cdots & 0 \\ 0 & 1 & \cdots & & -F_0 & 0 & \cdots & 0 \\ 0 & 0 & 1 & 0 & -F_1 & -F_0 & \cdots & 0 \\ 0 & 0 & \cdots & 1 & -F_2 & -F_1 & & \\ & & & & & -F_2 & & \\ & & & & & & & 0 \\ 0 & 0 & \cdots & 0 & -F_{D-1} & -F_{D-2} & \cdots & -F_{D-d} \end{bmatrix} \begin{bmatrix} N_0 \\ N_1 \\ \vdots \\ N_n \\ D_1 \\ \vdots \\ D_d \end{bmatrix} = \begin{bmatrix} F_0 \\ F_1 \\ F_2 \\ \vdots \\ F_{D-1} \end{bmatrix} \tag{14.8}$$

where the samples are now $F_m = m!F^{(m)}$. As noted above for frequency sampling, at least one coefficient of the two polynomials must be specified to render the system inhomogeneous. The choice here is to use $D_0 = 1$. As was the case for spectral-function sampling, spectral-derivative sampling also produces a linear system that contains both the numerator and denominator coefficients. An example of spectral estimation using derivative samples is presented in Figure 14.1. There two plots are

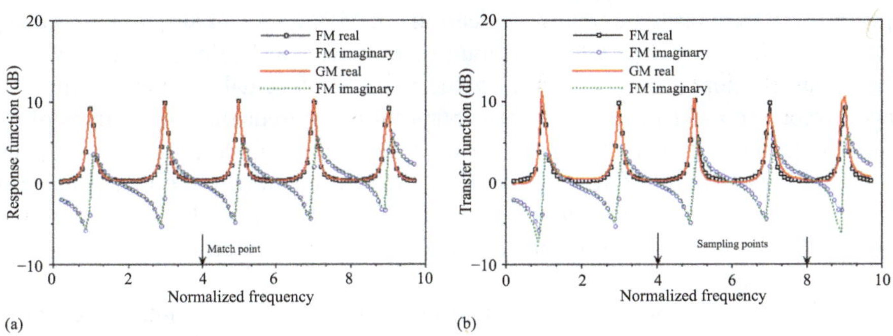

Figure 14.1 Result of derivative sampling of a 5-pole spectrum using 1 frequency sample and 9 derivatives (1a) and 2 frequencies with 4 derivatives (1b) and for each $n = 5$ and $d = 4$

shown for a 5-pole specified spectrum, one using 1 frequency sample and the first 9 derivatives in (a) and the other using 2 frequency samples and the first 4 derivatives in (b).

This linear system, solution of which yields the coefficients of the rational-function model for a transfer function based on frequency-derivative samples, is somewhat similar to that presented above for the case of frequency sampling. The approach summarized here is known as Pade approximation [9], although it might be regarded as a further generalization of the various extensions of Prony's method as discussed above and elsewhere [2]. All of these variations of the basic Prony's method have the goal of using computed or measured data to develop a numerical model for some phenomenological response which can be approximated by either a series of exponentials or a series of poles, although as noted above in (14.1), some nonpole (or nonexponential) terms can also be included. The maximum number of poles that can be employed in the spectral estimator model might be limited by the dynamic range of the coefficients, which appear in the model matrix for both function and derivative sampling. In either case, however, a compensating factor is the possibility that a lower-order (i.e., reduced number of model poles) model can be used because the data samples are most affected by the nearest poles of the spectrum being sampled. In the next section, how the frequency derivatives of the response can be found as a generalization of a moment-method model is discussed.

14.5 Computing frequency derivatives in a moment-method model

14.5.1 The basic idea

As noted above, the parameters needed to numerically quantify the waveform-domain or spectral-domain models can in principle be obtained from whatever information is available. Because most moment-method modeling based on integral equations is done in the frequency domain, subsequent discussion is limited to spectral- (or frequency-) domain applications of MBPE. As an alternative to using only frequency sampling, which although straightforward offers no further computational efficiency beyond reducing the number of actual samples required to reconstruct a transfer function, the advantage of using frequency derivatives of the response is demonstrated. The basis for the technique is due to earlier work by one of the authors (Burke) as reported in [10]. As will be shown, the additional programming complexity that arises in computing frequency derivatives from a moment-method model can be compensated by a more than proportional reduction in overall computer time.

Writing the moment-method equations that arise from an integral-equation formulation in matrix form, the impedance equation is obtained

$$\sum_{i=1}^{N} Z_{i,j}(\omega)I_j(\omega) = V_i(\omega) \tag{14.9a}$$

where $Z(\omega)$, $I(\omega)$, and $V(\omega)$ are the usual frequency-dependent impedance matrix and the current and voltage vectors, respectively. A solution for the current can then be formally written as an admittance relation

$$I_i(\omega) = \sum_{j=1}^{N} Y_{i,j}(\omega)V_j(\omega) \tag{14.9b}$$

where the matrix $Y(\omega)$ is the inverse (or equivalent) of $Z(\omega)$. Note that the MBPE approach discussed here, whether using frequency-function or frequency-derivative sampling, could be implemented using LU factorization, iteration, or any other method to obtain a numerical solution of the relevant linear system.

Because a solution for $I_i(w)$ or the fields it produces may be needed over a frequency range which spans many poles or resonances of $Y(w)$, it is usual to repeat these steps at enough frequency samples that a smooth representation of the transfer function can be derived. This requires that each computation essentially repeats the entire solution process, resulting in a computer time generally proportional to $AN^2 + BN^3$, the terms of which account for filling the $Z(\omega)$ matrix and solving the $Y(\omega)$, respectively. The overall computer time is therefore approximated by $F(AN^2 + BN^3)$ if F frequency samples are needed. It is therefore advantageous not only to minimize the number of unknowns N consistent with accuracy requirements but also to minimize F.

Two possibilities for working toward the goal of reducing, if not absolutely minimizing, F are discussed below. As might be expected from the preceding presentation, one approach is to employ the spectral-function model and sampling, and the other is to use a spectral-derivative model and sampling. Again, it should be emphasized that the analytical model employed for estimating the transfer function can be derived from whatever kinds of samples, function or derivative, are available.

Differentiating the impedance equation with respect to frequency, the following equation is obtained:

$$\sum_{j=1}^{N} \left[Z_{i,j}(\omega)I_j'(\omega)\right]Z_{i,j}'(\omega) = V_i'(\omega) \tag{14.10a}$$

where the prime denotes a frequency derivative. A solution of the differentiated impedance equation for the differentiated current can then be written as

$$I_i'(\omega) = \sum_{j=1}^{N} Y_{i,j}(\omega)\left(V_j'(\omega) - \sum_{k=1}^{N} Z_{j,k}'(\omega)I_k(\omega)\right) \tag{14.10b}$$

where it can be seen that while the differentiated impedance matrix appears as part of a modified right-hand side of the differentiated admittance equation, $I'(\omega)$ is given in terms of an undifferentiated admittance matrix. Computing the differentiated current thus requires an additional number of computations beyond those needed for solution of the undifferentiated current proportional to N^2 rather than the N^3 that would apply to obtain another frequency sample (assuming that LU decomposition is used rather than iteration).

Continuing this process, and omitting the explicit frequency dependences for clarity, the *n*th frequency derivative of the current is obtained as

$$\mathrm{I}_{\mathrm{i}}^{(\mathrm{n})} = \sum_{\mathrm{j}=1}^{\mathrm{N}} \mathrm{Y}_{\mathrm{i,j}} \left[\mathrm{V}_{\mathrm{j}}^{(\mathrm{n})} - \sum_{\mathrm{m}=1}^{\mathrm{n}} \mathrm{C}_{\mathrm{n,m}} \left(\sum_{\mathrm{k}=1}^{\mathrm{N}} \mathrm{Z}_{\mathrm{j,k}}^{(\mathrm{m})} \mathrm{I}_{\mathrm{k}}^{(\mathrm{n-m})} \right) \right] \tag{14.11}$$

where again $C_{n,k}$ is the binomial coefficient and the superscript in parenthesis indicates differentiation with respect to frequency of the order indicated. This equation is essentially identical in form to (14.9b), and emphasizes the similarity between the differentiated rational function and the moment-method models.

It is especially important to observe that information about the *n*th frequency derivative of the current continues to require a computer time proportional to N^2. Expressed in another way, each additional frequency derivative of the solution vector for the current can be computed in a number of operations proportional to $A(n,N_{\mathrm{rhs}})/N$ where A is a function which depends on the order of the derivative and the number of right-hand sides for which the solution is sought. If the frequency derivatives provide information comparable to that available from the frequency samples themselves, it can be appreciated that there could be a substantial computational advantage to using the solution derivatives in estimating the transfer functions. In the next section, it is shown how the National Electric Code (NEC) can be modified to do these computations.

14.5.2 An implementation using NEC

To implement frequency-derivative interpolation (FDI) as summarized above, the derivatives of the impedance matrix must be evaluated. In a code such as NEC, the fields which comprise the impedance matrix can have two sources of frequency dependence. One of these is due to the frequency dependence of the field expressions themselves, and the other can arise, as in NEC, from the basis and testing functions. NEC employs point sampling, that is., delta-function testing, which is not frequency dependent but does use the three-term sinusoidal current expansion function

$$\mathrm{I}_{\mathrm{i}}(\mathrm{s}) = \mathrm{A}_{\mathrm{i}} + \mathrm{B}_{\mathrm{i}} \sin\left[\mathrm{k}_{\mathrm{s}}(\mathrm{s} - \mathrm{s}_{\mathrm{i}})\right] + \mathrm{C}_{\mathrm{i}} \cos\left[\mathrm{k}_{\mathrm{s}}(\mathrm{s} - \mathrm{s}_{\mathrm{i}})\right] \tag{14.12}$$

where k_s is normally set equal to the medium wavenumber k. This is done because analytical expressions are then available for five of the six field components needed for general problems. But the frequency dependence of the basis function which thereby occurs can greatly complicate implementation of FDI.

In implementing FDI in NEC, it was decided to hold k_s constant in the basis functions to simplify the evaluation of derivatives. This step is justified by noting that the basis functions, although constructed from the functions of (14.12) which depend on k_s, have the form of B-splines which are insensitive to k_s for small $k_s d|$ [12]. In addition, the solution is relatively insensitive to small changes in the shape of the basis functions. The derivation begins with equations for the field of the current components in (14.12) for $k_s \neq k$, derived the equations for the derivatives, and then set $k_s = k$. A further simplification was effected by using the VLF version of NEC (NEC3VLF) in which the numerically cancelling fields due to point charges at segment ends are analytically removed from the field equations. With the point-charge fields removed and $k_s \neq k$, the field components for a current of the form

$$I(z') = I_0 \begin{pmatrix} \sin(k_s z') \\ \cos(k_s z') \end{pmatrix} \tag{14.13}$$

due to a segment extending from $-d$ to d on the z-axis of a cylindrical coordinate system are

$$E_\rho(\rho, z) = \frac{j\eta I_0}{4\pi\rho}\left\{\left[k_s(z-z')\begin{pmatrix} \cos k_s z' \\ -\sin k_s z' \end{pmatrix}\frac{e^{-jkR}}{kR} - j\begin{pmatrix} \sin k_s z' \\ \cos k_s z' \end{pmatrix} e^{-jkR}\right]_{-\delta}^{\delta} - \frac{(k^2-k_s^2)}{k}\int_{-\delta}^{\delta} dz'(z-z')\begin{pmatrix} \sin k_s z' \\ \cos k_s z' \end{pmatrix}\frac{e^{-jkR}}{R}\right\} \tag{14.14a}$$

and

$$E_z(\rho, z) = \frac{j\eta I_0}{4\pi}\left\{\left[-k_s\begin{pmatrix} \cos k_s z' \\ -\sin k_s z' \end{pmatrix}\frac{e^{-jkR}}{kR}\right]_{-\delta}^{\delta} + \frac{(k^2-k_s^2)}{k}\int_{-\delta}^{\delta} dz'\begin{pmatrix} \sin k_s z' \\ \cos k_s z' \end{pmatrix}\frac{e^{-jkR}}{R}\right\} \tag{14.14b}$$

The needed frequency derivatives of the fields are thus obtained by differentiating the above expressions with respect to k and then setting $k_s = k$. Consequently, derivatives of the integrals up to an order one less than the order of the derivative of the field remain when k_s is set equal to k. Considering in addition that E_z due to the constant term in the NEC basis is needed, there are five integrals to evaluate for each order of differentiation.

The integrals and their derivatives were evaluated by expanding the exponentials and integrating the terms analytically. In this process, each derivative shares some terms with the expansion for the previous derivatives. If numerical integration were used, the derivatives would share the $\exp(-jkR)$ term which accounts for much of the computation time. Hence, the additional effort is less than proportional to the number of derivatives. The remaining steps in evaluating the derivatives of equations (14.14a) and (14.14b) with respect to k are straightforward.

14.6 Some initial tests

The modified version of NEC described above was applied to three antenna problems as a preliminary test of the frequency-derivative approach. Selected as one of the simpler examples that might be attempted is a center-excited, straight-wire antenna. A more challenging case is presented by a straight horizontal wire located very near a conducting ground plane, which produces a much sharper resonance. The most difficult problem attempted is that of a forked monopole antenna having arms that are slightly different in length. Results from each case are summarized below.

14.6.1 Dipole antenna in free space

Results are presented in Figure 14.2 for a center-excited dipole antenna modeled with NEC using 21 segments. A Taylor series and rational model were fit to the current and four derivatives ($t = 4$, $n = d = 2$) at a point L/λ_o which is moved across the first resonance. Directly computed values were obtained at 0.01 intervals in L/A. The Taylor series result is seen to follow the correct result over a relatively narrow region around the expansion point, while the rational-function model provides a good match over a much wider interval. Of course, the radius of convergence of the Taylor series is limited by the distance to the nearest pole in the complex plane as is evident in the decreased convergence interval as the expansion point moves through the resonance.

Both first and second resonances are shown in Figure 14.3, with the Taylor series and rational-function models fit at $L/\lambda_o = 1.0$. Both resonances are recognized in the rational model, but not with high accuracy. Presumably a better result would be obtained with a higher-order rational function fit to more derivatives.

14.6.2 Horizontal dipole antenna close to a ground plane

Results for a horizontal antenna near a perfectly conducting ground plane are presented in Figure 14.4. This case was selected to test the frequency-derivative model for a response function having a higher-Q resonance. The solid curve was computed with a sampling interval of 0.001 in L/λ. It can be seen that for a relative bandwidth of about 10^{-3}, the rational function model provides a good match to the actual results even when sampled as far as 250 resonance widths away. The Taylor-series results are not shown here because they are so far from the correct curve.

14.6.3 Unequal-arm vee-monopole antenna

A still more challenging case is presented by a vee or forked monopole antenna having arms of unequal lengths. As shown in Figure 14.5, this antenna has, in

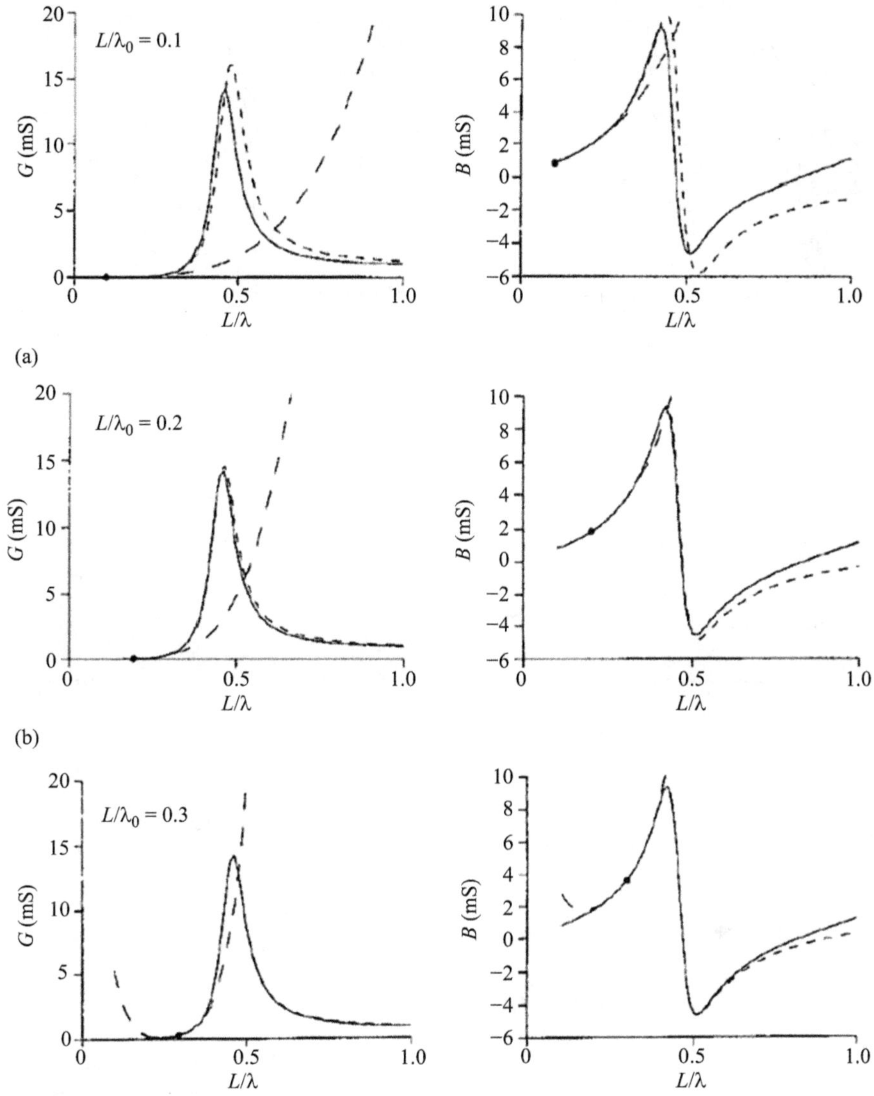

Figure 14.2 *Input admittance of a dipole antenna showing the first and second resonances of the dipole with L and thickness parameter Ω = 10 modeled with 21 segments. Results from a rational function model (short dashes) and Taylor series (long dashes) derived from 4 derivatives at $L//\lambda_o$ are compared with directly computed values (the solid line)*

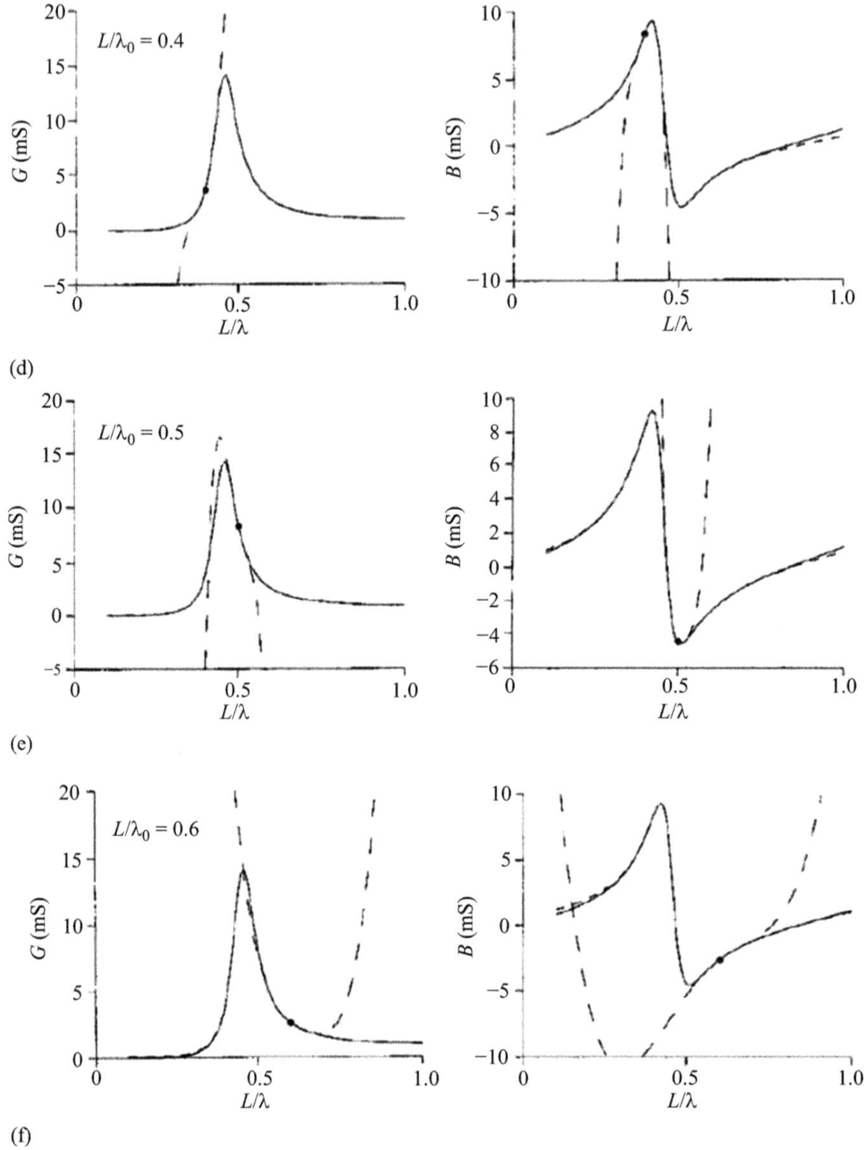

Figure 14.2 (Continued)

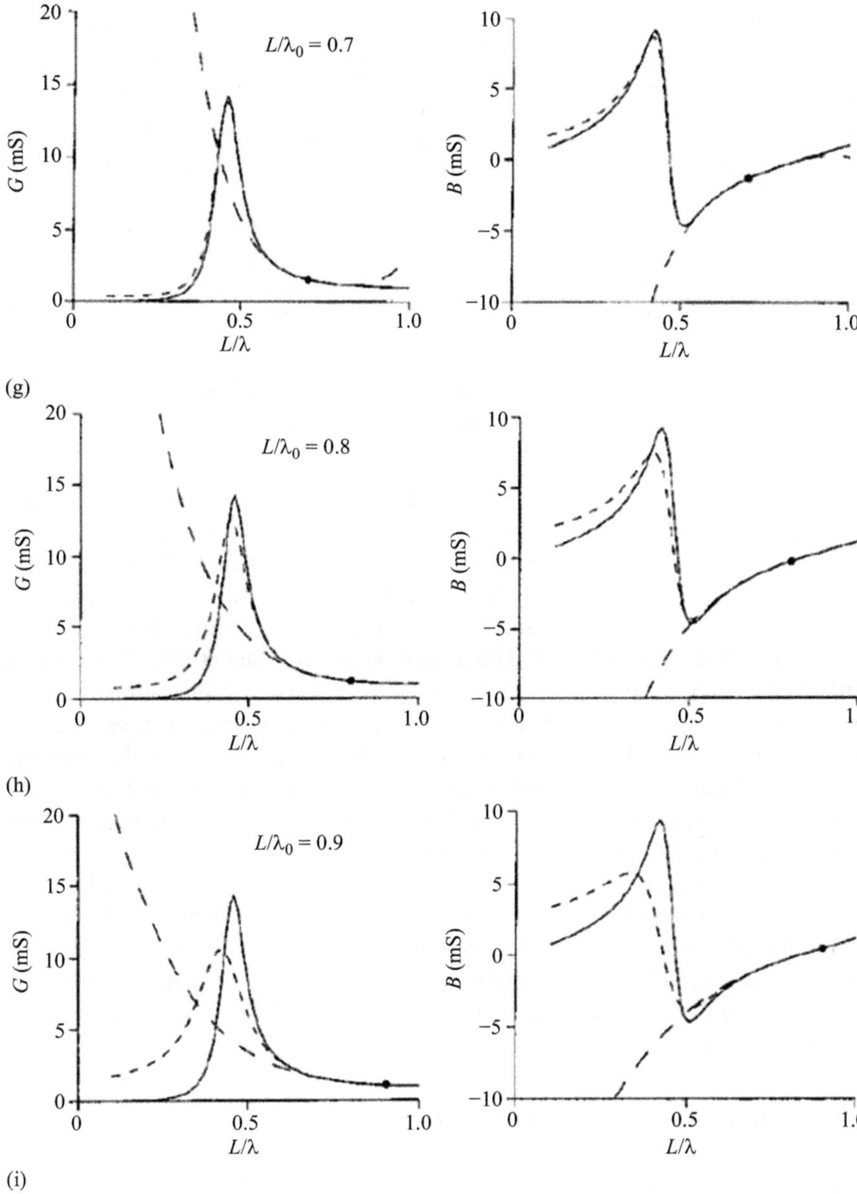

Figure 14.2 (Continued)

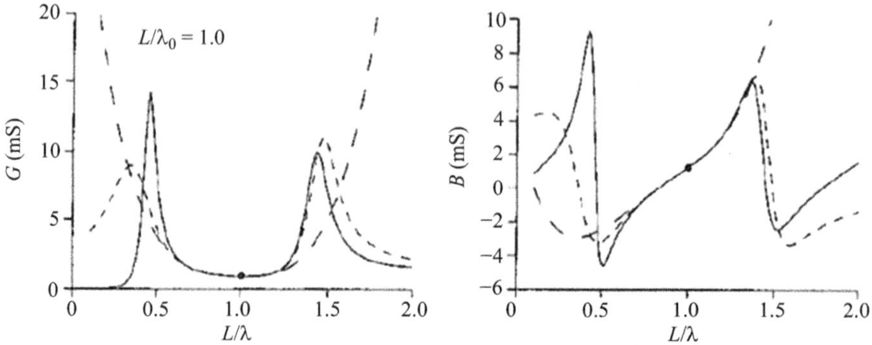

Figure 14.3 Input admittance of a dipole antennas as in Figure 14.2 showing the first and second resonances of the dipole

addition to the monopole resonance, a very sharp parasitic resonance due to a differential (transmission-line) current mode in the arms. The parasitic resonance corresponds to a pole very close to the real axis with a small residue, while the monopole resonance has a pole more distant from the real axis with a larger residue. This situation presents a severe challenge to the rational-function model derived from the current and four derivatives. When matched at $L/\lambda_o = 0.6$ [L is the length of the longer arm with the shorter of length $(14/15)L$], the rational-function model represents the monopole resonance with good accuracy. The second resonance is also detected, albeit with very poor accuracy. At $L/\lambda_o = 0.65$, the frequency of the second resonance is determined with reasonable accuracy, but the conductance is still negative. At $L/\lambda_o = 0.7$, the parasitic resonance is modeled with excellent accuracy, as seen in the detail plot of Figure 14.5. Accuracy is lost in the monopole resonance, however. The sampling intervals in L/λ_o for the solid curves were 0.003 for Figure 14.5 and 0.0005 for Figure 14.6. Performance of the rational-function model is not as good in this case as in the previous examples, although a very substantial reduction in sampling density can be achieved. Also observe that the negative conductance and changes in the result with L/λ_o provide an indication of where an additional sample point is needed.

14.7 Other representative results

The FDI version of NEC described above was applied to some other antenna problems as further tests of the frequency-derivative approach. A center-excited, straight-wire antenna was chosen as one of the simpler examples that might be attempted. A more demanding case is that of a straight horizontal wire located very near a conducting ground plane, which produces a much sharper resonance. Also included are more results for the forked monopole antenna having arms which are slightly different in length. Results from each case are summarized below. Also included is a scattering application from [11].

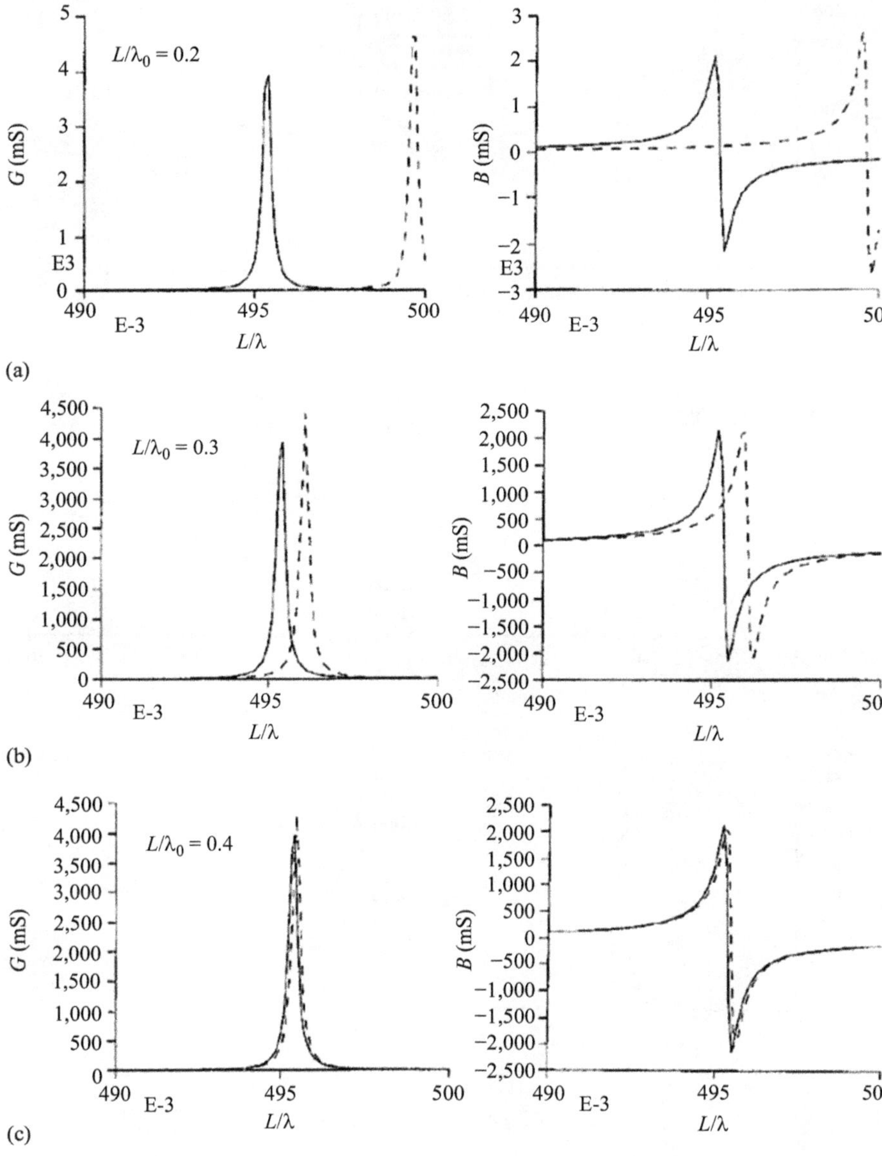

Figure 14.4 *Input admittance of a dipole antenna with length L and thickness parameter Ω = 15 at a height L/50 above a perfectly conducting ground. Results from a rational-function model (dashed line) derived from 4 derivatives at L/λ_o are compared with directly computed values (solid line)*

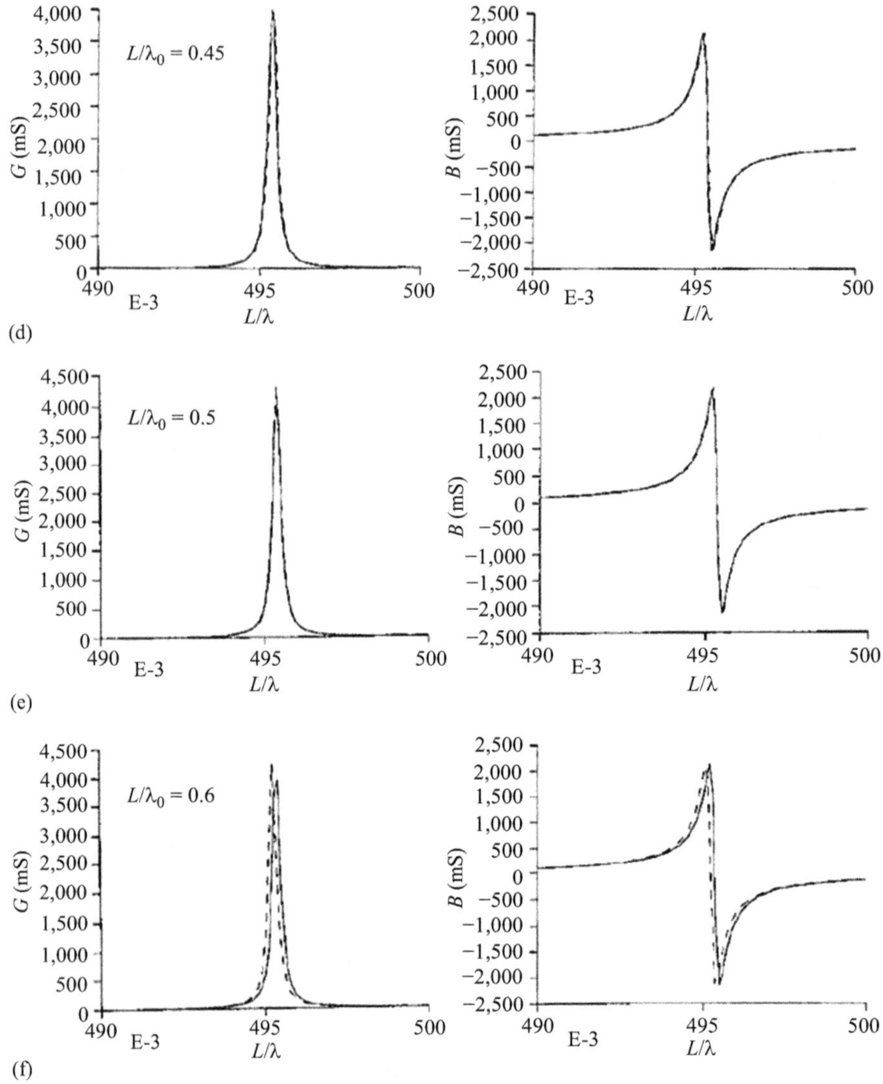

Figure 14.4 (Continued)

14.7.1 Dipole antenna in free space

Results are presented in Figure 14.7 for a center-excited dipole antenna modeled using FDI NEC. A rational model was fit to the current and four derivatives over a frequency range that includes the antenna's first two resonances. A frequency sample and 4 derivatives were computed at L/λ values of 0.1 and 2.0 for a total of 10 data samples to enable use of a FM having $n + d + 1 = 10$. Results are shown for two different FMs with FM#! using $n = 4$ and $d = 5$ and FM#2 using $n = 5$ and $d = 4$.

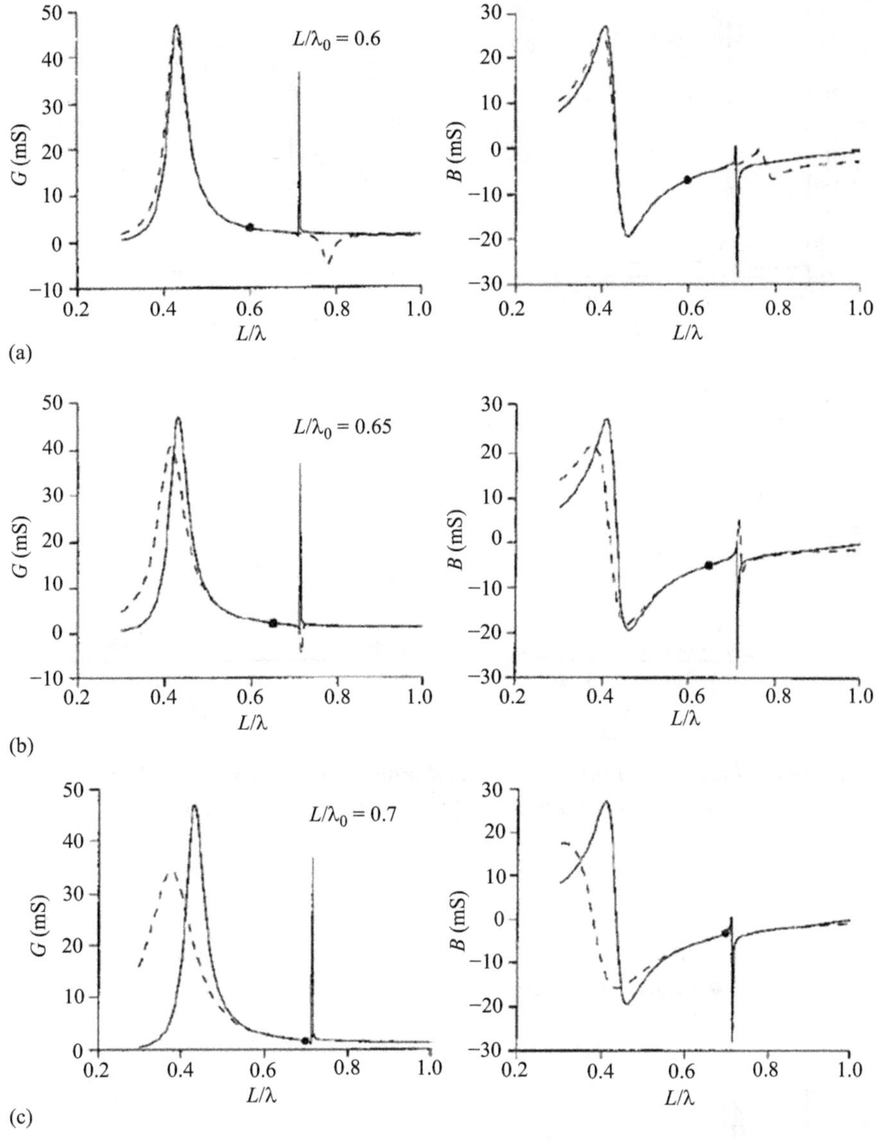

Figure 14.5 Input admittance of a forked monopole antenna with legs of length 140 and 150 m separated by 2.6°. Results from a rational function model (dashed line) derived from 4 derivatives at L/λ_o are compared with directly computed values (solid line).

While the FM#1 FDI (estimated) results might be considered acceptable, the FM#2 results are graphically indistinguishable from the finely sampled NEC results.

An explanation for this difference in the results of FM#1 and FM#2 is provided by the nonpole term in the general form of the spectral response. A FM having $n = d$

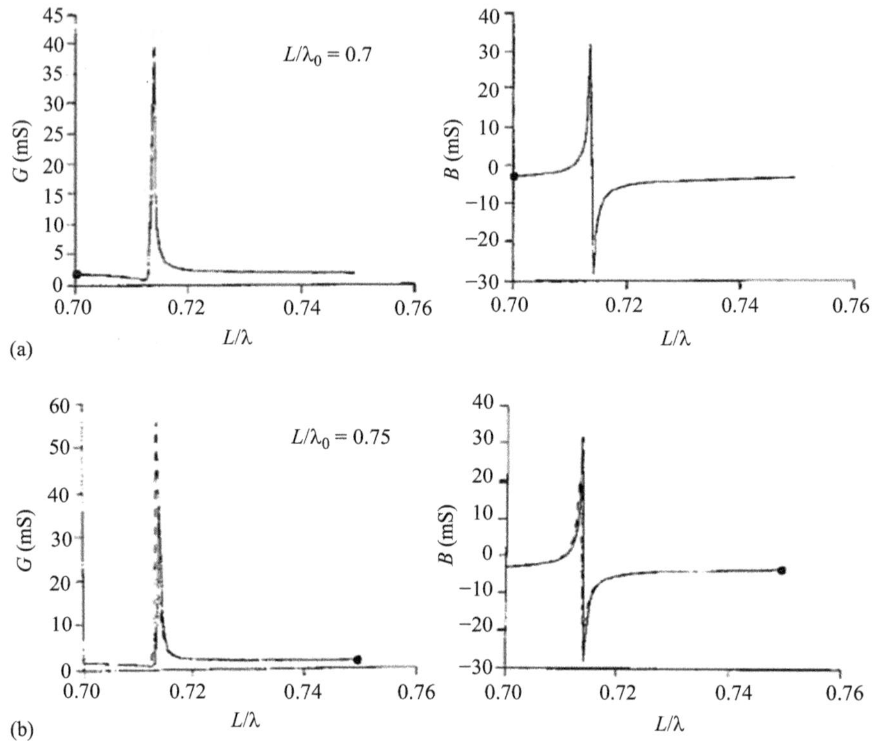

Figure 14.6 Input admittance of a forked monopole antenna as in Figure 14.5, showing an expanded view of the parasitic resonance

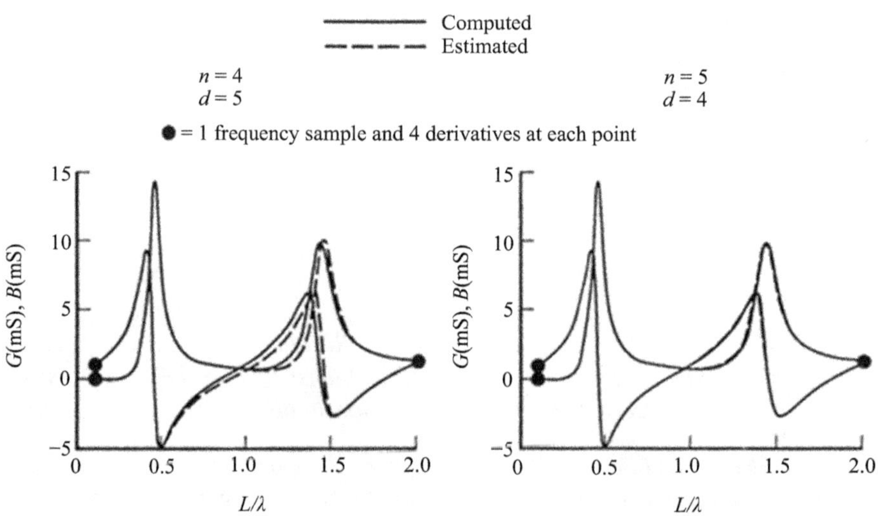

Figure 14.7 The NEC admittances and FDI estimates for a center-fed dipole antenna for two different FMs using the same sampled data at 2 frequencies and 4 derivatives at each

is an appropriate fit for a constant nonpole term. A FM having $n = d + 1$ is equivalent to a constant plus linear term in (14.14) for a nonpole term. This result is evidence that including a nonpole term or two in the spectral expression for a simple dipole antenna is appropriate.

14.7.2 Horizontal dipole antenna close to a ground plane

Results for a horizontal antenna near a perfectly conducting ground plane are presented in Figure 14.8. This case was selected to test the FDI model for a response function having a higher Q resonance. The solid curve in each plot was computed with a sampling interval of 0.001 in L/λ. It can be seen that for a relative bandwidth of about 10^{-3} in Figure 14.8(a), the rational function model provides a good match to the actual results even when sampled as far as 250 resonance widths away.

The plot in Figure 14.8(b) demonstrates how adding additional data samples at the conjugate frequency for $L/\lambda = -0.25$ improves the FDI result. Note that this negative-frequency sample requires no further NEC computation since the admittance at a negative frequency is the conjugate of the positive-frequency value.

14.7.3 Unequal-arm vee-monopole antenna

As shown in Figure 14.9, this antenna has, in addition to the monopole resonance, a very sharp parasitic resonance due to a differential (transmission-line) current mode

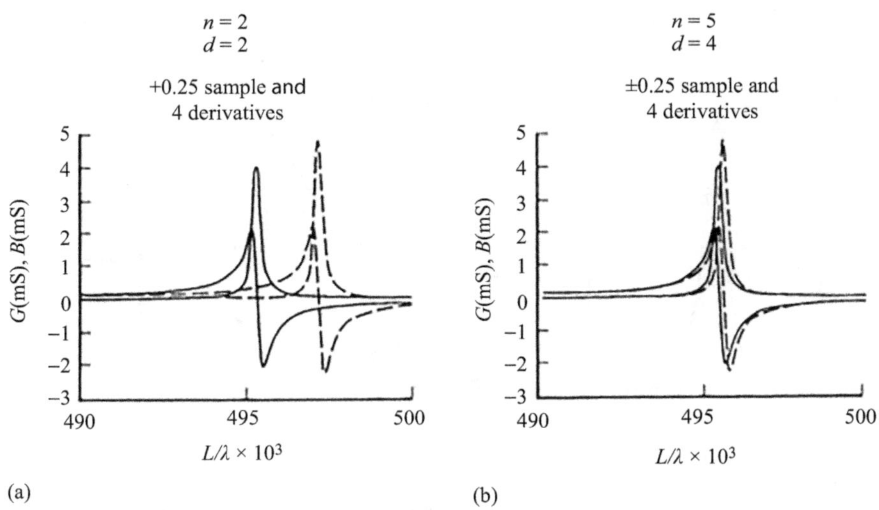

Figure 14.8 Input admittance of a dipole antenna with length L and thickness parameter Ω = 15 at height L/50 above a perfectly conducting ground. Directly computed values (solid lines) are compared with two FDI rational function models, dashed lines) with n = 2 and d = 2 for data samples at L/λ = + 0.25 (a) and n = 5 and d = 4 for data samples at L/λ = 0.25 and ±0.25 (b)

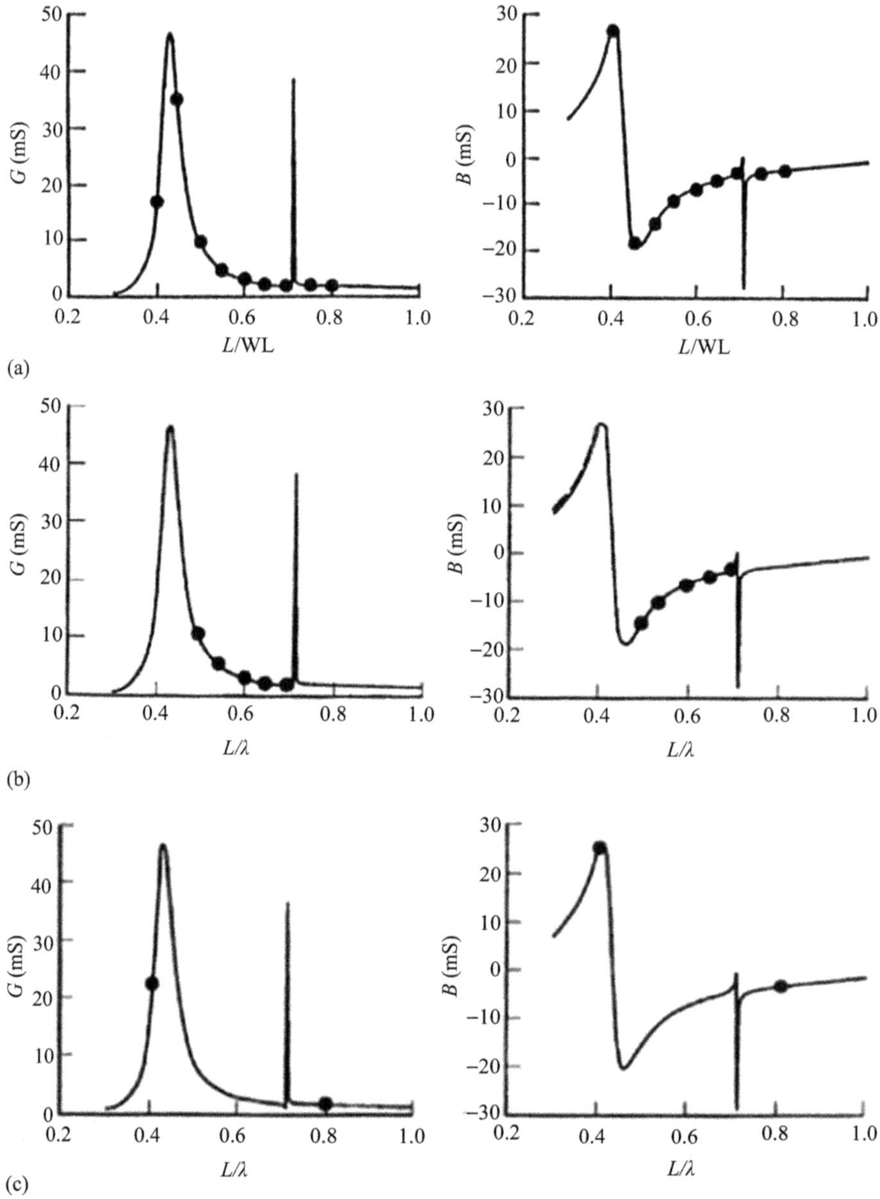

Figure 14.9 Input admittance of a forked monopole on a perfectly conducting ground. Directly computed values (solid lines) are compared with the FDI models (dashed lines) with n = 4 and d = 4, (a), n = 5 and d = 4 (b) with 1 derivative at each data point and (c) n = 5 and d = 4, with four derivatives at each data point.

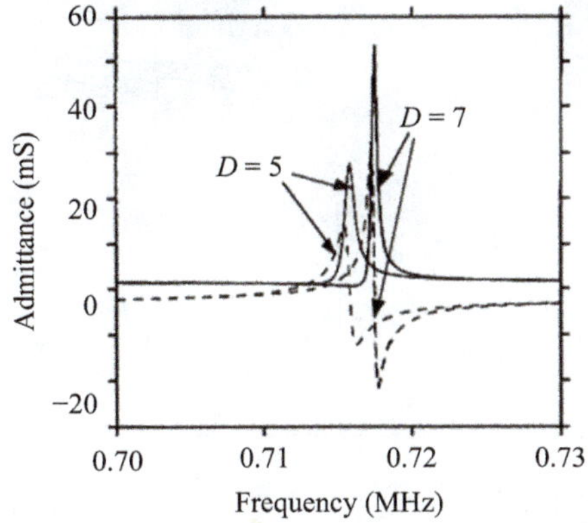

Figure 14.10 *The input admittance (solid line is conductance, dashed line is susceptance) of a forked monopole antenna in the vicinity of the parasitic resonance, where a differential-mode current can exist on the two unequal-length arms of the dipole [Burke (1992)]*

in the arms. The parasitic resonance corresponds to a pole very close to the real axis with a small residue, while the monopole resonance has a pole more distant from the real axis with a larger residue. This situation presents a severe challenge to the non-derivative model (a) and FDI models (b) and (c) derived by fusing 1 and 4 derivatives at each data point, respectively. The NEC-sampled and MBPE results are graphically indistinguishable.

Results for the forked monopole antenna over an expanded frequency range around the parasitic resonance are presented in Figure 14.10. Although the resonance is quite accurately located (to within 1% or so in frequency), there is some variation in the admittance values provided by two FMs, one using $D = 5$, and the other $D = 7$, function samples. The 7-sample model is the more accurate, as it is found to agree within a few percent with 21 additional GM samples spaced 10^{-4} L/λ apart beginning at 0.717 MHz.

14.7.4 Scattering from a slotted cylinder

A scattering result using frequency-derivative sampling is shown in Figure 14.11 [11]. These results were obtained using 1 frequency sample and 4 derivatives at the 5 locations depicted in the plot (sample points added to original graph). The result in Figure 14.11 is referred to as Cauchy's method in [11] but this is the same FM and procedure used throughout in connection with MBPE. The very sharp resonances obtained using a moment-method model and MBPE overlay essentially exactly.

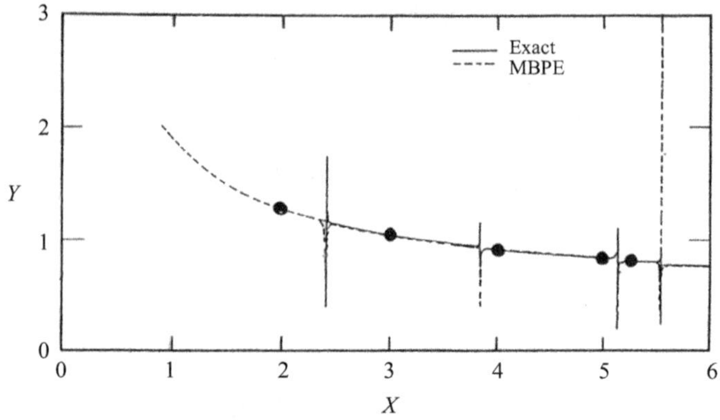

Figure 14.11 Scattered far field in V/m of a slit, PEC conducting cylinder using Cauchy's technique (MBPE) showing the 5 sampled frequencies for an FM using n = 10 and d = 11 [11]

14.8 An admittance-matrix function

The impedance and admittance matrices that arise in the moment method may be interpreted as N-port-network approximations to the problem being modeled. When only a single port is excited, as is the case for most antenna applications, then the current vector that results gives the self and mutual admittances for that port. Exciting the other $N-1$ ports one at a time similarly provides the $N(N-1)$ remaining admittances. Since typical problems rarely require all N^2 admittances, the inverse matrix is usually not computed and the solution is developed by LU decomposition, or iteration. Whatever excitation is used and however the solution is developed, the procedures outlined above can provide a rational-function approximation for the computed response.

A logical generalization of this somewhat limited approach is to represent some, or all, of the admittance-matrix coefficients in rational-function form. This is most easily done by either computing the inverse of the impedance matrix if all N^2 matrix coefficients are desired, or alternatively, computing current vectors caused by a sequence of single-port excitations. Whichever approach is used, observe that the admittance matrix can be written in the general form

$$\vec{\vec{Y}}(s) = \begin{bmatrix} X & \cdots & X & y_{1a}(s) & \cdots & y_{1b}(s) & \cdots \\ X & \cdots & X & y_{2a}(s) & \cdots & y_{2b}(s) & \cdots \\ X & \cdots & X & y_{3a}(s) & \cdots & y_{2b}(s) & \cdots \\ \vdots & \vdots & \vdots & \vdots & \vdots & \vdots & \vdots \\ X & \cdots & X & y_{Na}(s) & \cdots & y_{Nb}(s) & \cdots \end{bmatrix} \tag{14.15a}$$

where the X's denote the admittances that have not been computed, and the $y_{ij}(s)$ are those whose rational-function forms are modeled. Upon observing that the system poles are analytically the same whatever system observable has been used to obtain them, (14.15a) can be rewritten as

$$\vec{\vec{Y}}(s) = \frac{1}{D(s)}\begin{bmatrix} X & \cdots & X & n_{1a}(s) & \cdots & n_{1b}(s) & \cdots \\ X & \cdots & X & n_{2a}(s) & \cdots & n_{2b}(s) & \cdots \\ X & \cdots & X & n_{3a}(s) & \cdots & n_{2b}(s) & \cdots \\ \vdots & \vdots & \vdots & \vdots & \vdots & \vdots & \vdots \\ X & \cdots & X & n_{Na}(s) & \cdots & n_{Nb}(s) & \cdots \end{bmatrix} \tag{14.16b}$$

where $1/D(s)$ is the common denominator polynomial, and the $n_{ij}(s)$ are the numerator polynomials for each computed admittance. These would usually be found as column vectors since for single-port excitation, solution for the spatial current yields a column of admittances.

The validity of (14.16b) derives from the fact that the mode amplitudes are the only parts of the response that are excitation dependent, while the poles themselves are determined solely by the properties of the object for which a frequency response $F(s)$ has been obtained. Further note that if $Y(s)$ and $D(s)$ are known, then the corresponding $N(s)$ is obtained from their product. Thus although the response (or its frequency derivatives) must be found for all admittances whose rational-function model is desired, this would not require repeated solution via the matrix equations (14.5) or (14.7). Finally, observe that given part or the entire admittance matrix in the form of (14.14b), computing transient responses can be made much more efficient and convenient.

14.9 Conclusions

Two main ideas have been introduced in this chapter:

1) Use of MBPE based on rational-function approximations when obtaining electromagnetic transfer functions to reduce the number of frequencies at which samples are required, and
2) A sampling approach that employs frequency derivatives of the response using a new analytical technique based on differentiating the moment-method impedance equation, and which provides the derivative information in a time proportional to N^2 as contrasted with the N^3 dependency of solving the original problem.

Application of the frequency-derivative approach has been demonstrated for problems having well-defined resonances such as a dipole antenna in free space, and for more challenging problems having much narrower resonances such as a horizontal antenna brought close to a perfectly conducting ground plane and a "forked" monopole whose arms are slightly different in length. It should be clear that MBPE could develop into an important tool for not only increasing the efficiency of model applications but also in yielding more physically useful representations of those models.

References

[1] G. J. Burke, and E. K. Miller, "Modeling Antennas Near to and Penetrating a Lossy Interface," *IEEE Transactions on Antennas and Propagation*, AP-**32**, pp. 1040–1049, 1984.

[2] E. K. Miller, "Model-Based Parameter-Estimation Applications in Electromagnetics," *presented at NATO Advanced Study Institute on Modeling and Measurement in Electromagnetic Analysis and Synthesis Problems*, Il Ciocco, Italy, published by Sijthoff and Noordhoff, pp. 205–256.

[3] C. E. Shannon, "Communication in the Presence of Noise," *Proceedings of the IEEE*, Vol. 86, No. 2, pp. 447–457, 1998.

[4] R. Prony, "Essai Experimental et Analytique sur les Lois de la Dilatabilite de Fluides Elastiques et sur Celles del la Force Expansive de la Vapeur de L'alkool, a Differentes Temperatures," *J. l'Ecole Polytech. (Paris)*, Vol. 1, pp. 24–76, 1795.

[5] F. Hildebrand, *Introduction to Numerical Analysis*, McGraw-Hill, New York, 1956.

[6] S. M. Kay, and S. L. Marple, "Spectrum Analysis – A Modern Perspective," *IEEE Proceedings*, Vol. 69, No. 11, 1380–1419, 1981.

[7] G. J. Burke, and E. K. Miller, "Use of Frquency-Derivative Information to Reconstruct an Electromagnetic Transfer Function," *The Fourth Annual Meeting of the Applied Computational Electromagnetics Society*, Monterey, CA, 1988.

[8] J. N. Brittingham, E. K. Miller, and J. L. Willows, "Pole Extraction from Real-Frequency Information," *Proceedings of the IEEE*, Vol. 68, pp. 263–273, 1980.

[9] D. S. Jones, *Methods in Electromagnetic Wave Propagation*, *Engineering Science Series*, Oxford University Press, Oxford, 1979.

[10] E. K. Miller, G. M. Pjerrou, B. J. Maxum, *et al.*, *Final Technical Report on the Log-Periodic Scattering Array Program (U)*, MBAssociates, San Ramon, CA 94583, MB-R-70/105, 1970.

[11] K. T. Kottapalli, T. K. Sarkar, R. Adve, Y. Hua, E. K. Miller, and G. J. Burke, "Accurate Computation of Wideband Response of Electromagnetic Systems Utilizing Narrowband Information," *Computer Physics Communication*, Vol. 39, No. 4, pp. 682–687, 1991.

[12] G. J. Burke, *Enhancements and Limitations of the Code NEC for Modeling Electrically Small Antennas*, Lawrence Livermore National Laboratory, UCID-20970, 1987.

Appendix A
Further examination of poles in the spectral and time domains

Numerical modeling in electromagnetics (EM), whether based on integral, differential, or other formulations, inevitably involves a discretized, sampled-field approximation of the continuous physical reality that is being modeled. These descriptions exhibit a commonality with signal processing and model-based parameter estimation in terms of sampling requirements, information content and system, and the role of difference-equation linear predictors. Discussed here are some aspects of computational EM from a parameter-estimation viewpoint, emphasizing the variety of ways in which exponential-series and pole-series models can arise.

A.1 Introduction

A.1.1 Radar scattering as an encoding process

When an object (target) is illuminated by an EM field, the energy it scatters to a given observation point depends upon the target's geometry and material characteristics, the conditions of illumination (angle of incidence, polarization, frequency(ies), etc.) and the medium through which the field propagates. It is intuitively obvious, and demonstrated by analysis and measurement, that the scattered field (which will also contain, in general, interference and/or noise energy) has impressed or encoded on it information about the target that might somehow be extracted for purposes of: (1) detection (i.e., is there a target?); (2) classification (does the target belong to a class of interest?); (3) identification (if in the class of interest, which one of that class it is?); and (4) imaging or inversion (can the target's geometry and/or material properties be reconstructed?). An interesting view of this encoding process is described by Fiddy and Ross [1]) and Fiddy *et al.* [2].

The particular problem of deducing the geometry and material characteristics of a radar target from its scattered fields is challenging. Generally speaking, the amount of data which is needed, and the amount of processing it requires, can be expected to grow commensurately with the specificity and confidence desired in the answer being sought. The inverse problem is by far the most difficult of those applications listed as (1)–(4) above. Fortunately, many problems of practical importance do not require the most general answer. For example, airborne radar targets naturally fall into one or more of a few sets, that is, friend or foe, missile or aircraft, etc. In such circumstances, target classification and identification (C/I) can be based less on a rigorous inverse approach and more on whether their radar signatures match prestored information about the targets of potential interest.

But the quality and amount of target-scattered energy that are available place limits on how much information an observer can extract solely from measurement. Rarely, however, is an observer dependent wholly on the measurement alone to accomplish target identification. There is great value in whatever *a priori* information is available (e.g., atmospheric noise data, knowledge that airborne targets are planes, missiles, or helicopters) and/or modify the measurement itself (change frequencies, polarization, pulse shape, etc.). It is important that the observer processes the available data in such a way so as to utilize all available information, and however possible, adapt the controllable measurements to the target's characteristics. Exploiting such knowledge can greatly enhance the likely success of the associated signal processing.

A.1.2 Target characterization

Target characteristics may be recognized to belong to (at least) two domains. One, the target domain, is that in which the target has either directly measurable properties (such as mass, volume, shape) independent of its EM response, or properties which are derivable from those measurements (moment of inertia and higher-order moments, harmonic expansion of the surface profile, etc.). Either can be used to uniquely describe the target within some degree of accuracy. The other, or data domain, is that in which the target exists only indirectly, in terms of its effect upon a measured observable (a scattered EM field, for example), and from which various properties may be derived. When these data-domain properties can be uniquely related to target-domain, properties, then it can be said that an inverse problem has been solved. For many applications, however, classification or identification alone may suffice or even be all that is possible to achieve, in which case the data-domain properties may be most directly relevant. In the former application, the problem is more absolute in nature (i.e., determine the target), whereas in the latter, it is more comparative or relative (i.e., differentiate between targets). In either case, the way in which target-domain and data-domain properties interrelate is crucial to a successful outcome.

Another key to characterizing a target is the number of independent properties that are needed to accomplish either goal. This number, which we call the target rank, can depend on the characterization being used and the availability of any prior information. For example, if all targets are known to be penetrable, homogeneous spheres, then the only basic properties of concern are radius and composition. We might also expect that a desirable property of the characterization is that it be parsimonious, i.e., that with minimum rank it includes all independent properties over some set of measurements. This statement implies that the characterization in either the target domain or data domain is probably not unique, and that alternative property sets are available.

A.1.3 Information content and feature sets

Besides the issues of which target or data properties may be best, and what their associated ranks may be, there is the closely related question of the information content of the measurement. This might be loosely defined as the total number of bits required to represent the maximum number of independent properties (basic or derived) obtainable from the data. Implicit in this definition is a tradeoff between the number of properties and their complexity, as we speculate that the total information represented by the data is fixed. *A priori* knowledge, however, can greatly influence how the data are best used, changing the problem from one of system identification (determining a model for the data) to one of parameter estimation (given a model, estimating the model parameters). In other words, by reducing the number of unknowns via *a priori* knowledge (a model), the data can yield either more complete or more accurate information about the model parameters (the properties).

In radar terminology, the EM properties of a target are sometimes called its features, the ad hoc nature of which is demonstrated by the features that have at one time or other been used, some examples being: the maximum value of target cross section at a given frequency; the aspect-averaged cross section over a specified frequency band; or the ratio of cross sections for two orthogonal polarizations of the incident field. The success of such an approach will depend in part on the degree to which the features span the space of target-radar and target-geometry characteristics, individually and as a set. When the features chosen are target-radar characteristics or EM-observables data, we observe that it is the data domain in which C/I is being attempted. However, when the features are derived from the radar signature to yield target-geometry characteristics, then C/I is being pursued in the target domain. Clearly, target-geometry features would be more desirable, everything else being equal, because these are features closer to describing the target in ways that are recognizable to human observers. However, it is more straightforward to work with target-radar features that can unfortunately vary due to the aspect, frequency and polarization dependence of the target's scattering properties. It would be beneficial to have a feature set that is a fundamental property of the target and is independent of how the target is excited.

A.1.4 Complex resonances and other singularities

An example of such a feature set is the complex source-free resonances, or poles, popularized in recent years in connection with the singularity expansion method [3,4]. These frequency-domain poles are dependent only on the target geometry (and impedance, if it is imperfectly conducting). Their oscillatory or imaginary parts are established by the lengths(s) of propagation path(s) on the object, while the damping or real parts are associated with energy loss (radiative and dissipative) over that path. From a practical viewpoint, the poles seem to be unique properties of a target and, therefore, provide an eminently attractive mechanism for target identification. In some sense, too, they seem to represent a kind of minimum-parameter target description that has interesting implications for the inverse problem from an information-theoretic viewpoint. Finally, they may provide useful insight concerning other feature sets that have been or might be used for target identification.

Considering the use of frequency-domain poles for target identification presumes the availability of wideband or impulsive scattering data for a target. An alternate approach is to make the measurement as a function of angle in a bistatic or monostatic scattering mode at a fixed frequency. In this case, a different kind of pole (or more properly a singularity) arises due to equivalent or actual point sources which give the scattered field. Such sources can correspond to scattering centers and points of radiation from the target, and thereby might also serve as a feature set. To the extent that they are related to the target geometry, these space singularities could provide a more direct link between the target and data domains than do the complex-frequency poles. For example, the maximum physical extent of the target might be obtained directly from extrema of the space-singularity distribution, but could only be inferred from the wavelength of the smallest-value pole.

Space singularities can also be associated with measurements made as a function of position (e.g., along a line or circular arc) or as a function of frequency [5]. In the former situation, the singularities correspond to the direction angles of the plane waves that make up the total field or to other physical observables [6]. The result is that a spatial array can image a distribution of sources located on the far-field sphere. An analogous situation arises in measuring the field of a set of point sources as a function of frequency, but at a fixed point in space, where the singularities also correspond to the source positions. By making measurements from three orthogonal (or linearly independent) directions, the possibility arises of imaging a three-dimensional source distribution. Another example is provided by measuring the far-field of a linear array of point sources wherein the space singularities correspond to the source locations.

Alternate approaches are described by Fiddy and Ross [1] using complex zeros as information sources of the field and by Inagaki and Garbacz (1982) in terms of the eigenfunctions of operators with respect to discrete and radiating systems. In the latter case, it has been demonstrated that the width of the eigenfunction spectrum of the field decreases upon moving away from the source until a steady state is reached in the far field, due to the non-radiating, evanescent near fields.

An interesting property of these representations is their close connection with widely used signal-processing techniques. For example, a linear-predictor filter implicitly incorporates a series-exponential model, as can be demonstrated by applying Prony's method to a set of uniformly spaced (in x) samples of $f(x)$. Prony's method and related techniques provide a way to estimate values of the amplitudes (residues) and exponents (poles) and residues from the sampled data. Thus, there is the possibility of not only obtaining parameters which may be useful EM feature sets but also that of processing EM data for target identification.

A.2 Comparison of direct and inverse modeling

It is generally recognized that inverse problems are more difficult to solve than are direct problems because the direct problem can usually be reduced to the solution of a linear system, whereas inverse problems are intrinsically transcendental and nonlinear. These issues are discussed briefly below.

A.2.1 Numerical approximations to operator relationships

Linear problems in EM, acoustic and similar applications can be expressed in operator form as [7,8]

$$L(\mathbf{r},\mathbf{r}')f(\mathbf{r}') = g(\mathbf{r}) \tag{A.1}$$

where

g(**r**) is a field
f(**r**) is a source
L(**r**,**r**′) is an integral operator determined by the problem parameters and governing equations.

Equation (A.1) is a general statement relating sources and fields in the frequency domain. An analogous treatment can also be developed for the time domain [8]. When the source position vector **r**′ and field vector **r** lie on a locus of points B over where some known relations hold due to boundary conditions (e.g., vanishing of tangential electric field on a perfect electric conductor, or vanishing of the normal particle velocity on a rigid body), then (A.1) becomes an integral equation for the source, that is,

$$L_B(\mathbf{r},\mathbf{r}')f(\mathbf{r}') = g(\mathbf{r}); \mathbf{r},\mathbf{r}' \varepsilon B \tag{A.2}$$

with L_B denoting the operator L with the position vectors constrained to B.

A numerical representation of (A.2) can be derived using the moment method [7], leading to a linear system of the form

$$\mathrm{B_{ij}x_j = y_i} \tag{A.3}$$

where B_{ij} is a matrix of N^2 interaction coefficients relating the vectors of N sampled-field values y_i and the N sampled-source values x_j over the boundary B, and repeated indices are summed over N terms. We use the term "sampled" in a generalized sense because x_j and y_i represent actual spatial samples only for certain numerical treatments. The relationships in (A.3) can be inverted to obtain

$$\mathrm{x_i = B_{ij}{}^{-1} y_j} \tag{A.4}$$

where B_{ij}^{-1} is a matrix of inverse interaction coefficients (assumed nonsingular) and

$$\mathrm{B_{ij}{}^{-1} B_{ij} = I_{ij}}$$

with I_{ij} the identity matrix.

The matrices B_{ij} and B_{ij}^{-1} have straightforward physical interpretations, representing a constrained source-field, sampling relationship for a given problem.

In EM, they are referred to most often as the impedance and admittance matrices, respectively.

A numerical representation of (A.1) can also be expressed in terms of sampled values as

$$z_i = A_{ij}x_j = A_{ij}B_{jk}^{-1}y_k \tag{A.5}$$

where z_i is a sampled field on some locus of points in space other than B_{ij}. For far-field samples of z_j, the matrix A_{ij} is especially simple and can be expressed in terms of plane-wave projections onto B, but in general, A_{ij} involves samples of the operator L.

A.2.2 Direct and inverse solutions using sampled operators

The preceding discussion was presented to set the stage for clarifying the difference between direct and inverse problems. A common approach to both of these problems is to use an input–output model that leads to a generalized transfer function in which two basic problem types become apparent. For the analysis or direct problem, the input is known and the transfer function is derivable from the problem specification, with the output or response to be determined. For the case of the synthesis or inverse problem, two problem classes may be identified. The "easier" synthesis problem involves finding the input given the output and the transfer function, an example of which is that of determining the source voltages that produce an observed pattern for a known antenna array. The "harder" synthesis problem itself separates into two problems. One entails finding a source distribution that produces a given far field. The other and still more difficult is that of finding the object geometry that produces an observed scattered field from a known exciting field. The latter problem is the most difficult of the three synthesis problems to solve because it is intrinsically transcendental and nonlinear. Furthermore, such problems are subject to uniqueness constraints that can impose difficulties and uncertainties in developing their solutions.

In essence, both the direct and inverse problems can involve an Nth-order, sampled, approximated operator, and differ mainly in terms of how the operator (or its equivalent) is obtained. In the direct problem, the N^2 coefficients of this operator are derived from the known governing equations and problem parameters. In the inverse problem, an operator or its equivalent must also be obtained for a rigorous solution, where the absence of known problem parameters must be compensated by known input and output quantities.

Two approaches might be considered for the latter problem. One would be to obtain the N^2 coefficients directly from the known input and output quantities, which requires in principle solution of up to an N^2th-order linear system. The other would be to exploit the known governing equations as well as to reduce the number of coefficients requiring solution directly from the data.

If an inverse solution is being sought only to predict the target's response to a new incident field, then an operator matrix is all that is needed. Furthermore, this matrix need not be unique since the tangential field distribution on any closed surface which contains the target could be used to define it. However, the target geometry and electrical properties could be derived only from that matrix which represents the true boundary-value problem. This presumes that the object geometry can be related unambiguously to the coefficients of B_{ij}. The unknowns actually being sought would be the geometrical parameters needed to specify the Nth-order decomposition of the object's geometry, from which the coefficients in B_{ij} could be computed, possibly using an iterative approach. An example using a time-domain integral equation is provided by Bennett [9] with an earlier example of approximating transient responses given by Kennaugh and Moffett [10]. While the direct problem is a linear one when the governing equations are linear, the inverse problem is not because the object's boundary and/or its constitutive parameters are unknown, which is why inverse problems (using our definition) are so much more difficult.

It should be noted that this is the most general form of the inverse problem, and that it entails reconstructing the object's geometry and constitutive parameters. A simpler inverse problem is to estimate model or physical parameters from observational data to decide whether the object (process) belongs to a class of interest (the classification problem) and, if so, which one of that class it is most likely to be (the identification problem). These latter problems are amenable to resonance or pole-zero parameterization.

A.2.3 Implications of sampling

Inherent in all modeling of field problems is the idea of sampling. Sampling is involved in representing sources and fields whether the problem geometry permits solution via classical separation of variables or requires numerical techniques. For numerical solutions, sampling functions can exist over all of B, or be defined on subdomains of B. In either case, some lower limit on the number of samples is required to achieve a given degree of convergence in the solution.

The number of samples needed to represent a solution for a given problem represents a measure of the problem dimensionality, referred to previously as rank or the number of degrees of freedom. Problem rank is a fundamental property for both direct and inverse problems, in terms of solution difficulty and information content. It appears that the geometrical (target-domain) rank and EM or acoustic (data-domain) rank of a problem are not necessarily the same, a situation that requires consideration in treating inverse problems.

A.2.4 A unified approach

The above discussion emphasized that direct and inverse problems involve aspects of not only field theory but also information theory and signal processing. The literature in these separate areas is extensive. By comparison, relatively little evidence is available to show that a unified approach is likely to be considered when treating inverse problems. Unless these closely related aspects of direct and inverse problems are addressed in a unified way, progress toward achieving practical solutions is likely to be slower than might otherwise be possible.

An area where information-theory concepts are being pursued in inverse problems is in optical imaging. Representative examples in this area can be found in the work of Winthrop [11], Ross *et al.* [12], Gori and Ronshi [13], and Bertero and Pike [14]. The central theme of these and related works is that the information content of a field is a finite quantity that is relatable to a field's rank or number of degrees of freedom, one example being an eigen value measure. This concept is one of paramount importance in any inverse treatment, since it evidently establishes the number of independent parameters that can be extracted from the data, whether these are eigenvectors, complex-frequency poles, or other representations.

Object poles represent another way by which data rank can be quantified and object identification or even reconstruction might be attempted. These possibilities have not been very actively pursued, partly because estimating poles from noisy data is not easily or reliably accomplished, and the accompanying question of how poles can be used for identification has no clear answer. Results of an initial study of some of these questions are presented by Miller and Lager [15]).

One disadvantage of pole estimation is that only the source-free part of an object's response can be used in a straightforward way for this purpose. Otherwise, the model used and subsequent pole estimation can become much more complex because of the need to model the driven response as well. Unfortunately, this means that information contained in the early time, forced response might not be exploitable using only a pole-based scheme. The early time ramp response, however, can be used to estimate an object's cross-sectional shape and area [10]. By combining both ideas into a unified treatment, more effective utilization of the entire object response might be achieved as discussed by Miller *et al.* [16].

One approach to finding the poles of objects from their EM response in either the time domain or frequency domain using Prony's method or straightforward variations thereof is discussed next.

A.3 Processing pole-based representations

In many physical problems, especially those which involve wave phenomena such as in EM, acoustics, and seismology, a behavior results that can be described by systems of linear, partial differential equations. Solutions to such problems often can be expressed simply in exponential form (series or integrals), the occurrence and use are explored more fully below.

These exponential forms are of interest not only because they provide solutions but also because the parameters they contain, the exponents in particular, possess physical information basic to a given problem. In transient phenomena for example, the exponents (or poles) contain resonant frequencies and damping constants. Many other kinds of exponential forms which contain other kinds of physically relevant information also exist. Moreover, the poles (and associated zeroes or residues) can be extracted from a wide variety of data [5].

Exponential-form representations not only include parameters relevant to a problem's physics but are also valuable because they:

1. provide a way to condense the data needed to describe a problem's physical behavior;
2. offer insight into the physical phenomena involved;
3. provide a measure of the information content of the data; and
4. suggest a conceptual framework for solving the inverse problem.

Some example applications are given below in Miller [5]. Before discussing these specific issues however, some general background material is included in Section A.3.

A.3.1 Obtaining poles

Many kinds of mathematical transformations are used in solving physical problems. One of the more useful of these, and one which is ideally suited to the exponential forms of interest here, is the Laplace transform. For the generic waveform-domain exponential series

$$f(X) = \sum_{\alpha=1}^{W} R_\alpha \exp(s_\alpha x) \tag{A.6}$$

a Laplace transform leads to the generic spectral-domain pole series

$$F(X) = \sum_{\alpha=1}^{W} R_\alpha/(s_\alpha - X) \tag{A.7}$$

where the terms waveform and spectral domains are the most commonly encountered physical examples given by the forms in (A.6) and (A.7) respectively. We note that knowledge of the $W\ s_\alpha{}'s$ (the poles) and the $W\ R_\alpha{}'s$ (the residues) provides complete information both for $f(x)$ and $F(X)$. Thus, rather than needing many sampled values to reconstruct a waveform or spectrum, we need only store the two sets of numbers (generally complex) s_α and R_α. The result can be a greatly reduced number of bits required for storing such data while also providing an extrapolation function for $f(x)$ or interpolation function for $F(X)$, continuous in the independent variable.

Practical implementation of (A.6) or (A.7) requires that s_α and R_α be derivable from whatever analytical, computed, or measured data or information is available. A procedure for doing this from waveform data was developed in 1795 by Prony [Hildebrand (1965)]. Prony's method in various modified and extended forms may be recognized in many signal-processing methods in present use [17]. As outlined below, it can be used as originally developed to process waveform data, and in a slightly different, but closely analogous version, to process spectral data. We derive first the waveform-domain version of Prony's method, followed by its spectral-domain counterpart.

A.3.2 Waveform-domain Prony

Chapter 4 describes the final steps involved in implementing Prony's method. Included here for completeness are details of the mathematical derivation leading to the final linear-predictor equation on which Prony's method is based. Recall that determination of the poles from (A.6) using the Prony technique involves sampling

$f(x)$ at D data points $x_i = id$, $i = 0, 1, \ldots, D-1$ over a window of width $T = d(D-1)$. We thus rewrite (A.6) as

$$\mathrm{f_i} = \mathrm{f(X)} = \sum_{\alpha=1}^{P} \widehat{R}_\alpha \widehat{X}_\alpha^{i}; i = 0, 1, \ldots, D-1 \tag{A.8}$$

where

$$\widehat{X}_\alpha = e^{\hat{s}_\alpha \delta},$$

P is the number of poles being sought and the caret denotes a parameter-value estimate.

Equation (A.8) can be written as

$$\begin{aligned}
&\widehat{R}_1 + \widehat{R}_2 + \cdots + \widehat{R}_P = f_0,\\
&\widehat{R}_1\widehat{X}_1 + \widehat{R}_2\widehat{X}_2 + \cdots + \widehat{R}_P\widehat{X}_P = f_1,\\
&\widehat{R}_1\widehat{X}_1^{2} + \widehat{R}_2\widehat{X}_2^{2} + \cdots + \widehat{R}_P\widehat{X}_P^{2} = f_2,\\
&\widehat{R}_1\widehat{X}_1^{D-1} + \widehat{R}_2\widehat{X}_2^{D-1} + \cdots + \widehat{R}_P\widehat{X}_P^{D-1} = f_{D-1},
\end{aligned} \tag{A.9}$$

where each data sample can be seen to generate an equation involving higher-order polynomials in the exponential terms as a consequence of using uniformly spaced data samples. The requirement that a polynomial form be generated from uniform sampling of the waveform-domain signal model arises because if the $\exp(s_\alpha d)$ represent the roots of a Pth-order polynomial, that is,

$$\begin{aligned}
\sum_{\alpha=0}^{P} \widehat{a}_\alpha X^\alpha &= (X - \widehat{X}_1)(X - \widehat{X}_2)\cdots(X - \widehat{X}_P)\\
&= \prod_{\alpha=1}^{P} (X - \widehat{X}_\alpha); a_P = 1
\end{aligned} \tag{A.10}$$

it is then possible to reduce (A.9) to a linear-predictor form.

This can be done by first multiplying the equation series (A.9) respectively in turn by

$$\widehat{a}_0, \widehat{a}_1, \ldots, \widehat{a}_P$$

and adding the first $P + 1$ of these equations together. The result is that a new equation can be found, which upon using (A.10) can be reduced to

$$\widehat{a}_0 f_0 + \widehat{a}_1 f_1 \cdots, f_P = 0$$

If this operation is repeated another $D-P-1$ times by successively beginning with the second equation of series (A.9), ..., up to the $D-P-1$ equation, the following $D-P$ linear equations will result:

$$\vec{\vec{M}}_t\, \vec{a} = \vec{f}, \tag{A.11a}$$

where

$$\vec{\vec{M}}_t = \begin{bmatrix} f_0 & f_1 & \cdots & f_{P-1} \\ f_1 & f_2 & \cdots & f_P \\ \vdots & \vdots & \vdots & \vdots \\ f_{D-P-1} & f_{D-P} & \cdots & f_{D-2} \end{bmatrix}, \tag{A.11b}$$

$$\vec{f} = \begin{bmatrix} f_P \\ f_{P+1} \\ \vdots \\ f_{D-1} \end{bmatrix} \tag{A.11c}$$

and the vector of unknowns represents the estimated coefficients of the polynomial in (A.10) (sometimes called the characteristic equation). Observe that (A.11a) has the form of a linear predictor or difference equation, which occurs commonly in various signal processing procedures, and that the data matrix of (A.11b) is of Toeplitz form. The linear system in (A.11) provides the basis for obtaining the *P* coefficients in the polynomial in (A.10), from whose roots the *P* pole values are found as

$$\widehat{s}_\alpha = (1/\delta)\mathrm{In}(\widehat{X}_\alpha) \tag{A.12}$$

If $D = 2P$, the system is square of order *P* and can be solved directly for the *P* real a′s. For $D>2P$, the method of least squares can be used. In either case, the residues can be found by various means, such as fitting the first *P* equations of series (A.9) or by applying the least-squares technique to the entire set.

It is important to recognize that for the method to work, the data must be available in uniform steps of the exponent, so that (A.6) can be written as in (A.8). This limitation will be encountered in various situations discussed in Miller [5]. Note also that the technique just outlined can be used for data whose exponential terms are products of two or more exponents depending on different variables, where

$$f(x, y) = \sum_{\alpha=1}^{W_x} \sum_{\alpha=1}^{W_x} R_{\alpha\beta} \exp(s_\alpha^{(x)}) \exp(s_\beta^{(y)}) \tag{A.13}$$

We observe that a constraint or auxiliary equation is needed to obtain an inhomogeneous system of equations. It is conventional to impose condition (A.10) which leads to the linear-predictor form seen in (A.11a). Other forms could also be used [Moffatt and Mains (1975)],

$$\sum_{\alpha=0}^{P} \widehat{a}_\alpha{}^2 = 1$$

A.3.3 Spectral-domain Prony

The treatment of (A.7) to derive the pole values proceeds in an analogous fashion. First, rewrite (A.7) as

$$F(X) = \prod_{\alpha=1}^{P} R_\alpha \prod_{\beta=1}^{P}{}^{\alpha}(X - \widehat{s}_\beta) / \prod_{\gamma=1}^{P}(X - \widehat{s}_\gamma) \tag{A.14}$$

where the superscript a on the first product signifies omission of the $b = a$ term and again the carat denotes an estimated parameter. Upon recognizing that

$$\prod_{\gamma=1}^{P}(X - \widehat{s}_\gamma) = \sum_{\gamma=1}^{P} \widehat{a}_\gamma X^\gamma; a_P = 1 \tag{A.15a}$$

$$\prod_{\beta=1}^{P}{}^{\alpha}(X - \widehat{s}_\beta) = \sum_{\beta=1}^{P-1} \widehat{b}_\beta{}^{(\alpha)} X^\beta; b_{P-1}^{(\alpha)} = 1 \tag{A.15b}$$

Equation (A.14) can be expressed as

$$\sum_{\alpha=0}^{P-1} \left[\widehat{a}_\alpha X^\alpha F(X) - \tilde{b}_\alpha X_\alpha\right] = -F(X)X^P \tag{A.16a}$$

where

$$\tilde{b}_\beta = \sum_{\alpha=0}^{P} \widehat{R}_\alpha \widehat{b}_\beta{}^{(\alpha)}$$

and it should be noted that characteristic-equation coefficients are real for spectra where $F(X) = F^*(X^*)$. This fact can be used to reduce the number of real unknowns in (A.16a) from $4P$ to $2P$, and is assumed to be the case here.

If $D>2P$ samples are available for $F_i = F(X=X_i)$, where $i = 0, 1, \ldots, D-1$, (A.16a) provides the basis for obtaining the $2P$ real coefficients from

$$\vec{\vec{M}}_s \vec{\widehat{c}} = -F \tag{A.16b}$$

where

$$\vec{\vec{M}}_s = \begin{bmatrix} F_0 & X_0F_0 & \cdots & X_0^{P-1}F_0 & -1 & -X_0 & \cdots & -X_0^{P-1} \\ F_1 & X_1F_1 & \cdots & X_1^{P-1}F_1 & -1 & -X_1 & \cdots & -X_1^{P-1} \\ \vdots & \vdots & \vdots & \vdots & \vdots & \vdots & \vdots & \vdots \\ F_{D-1} & X_{D-1}F_{D-1} & \cdots & X_{D-1}^{P-1}F_{D-1} & -1 & -X_{D-1} & \cdots & -X_{D-1}^{P-1} \end{bmatrix}, \tag{A.16c}$$

$$\widehat{c} = \begin{bmatrix} \widehat{a}_0 \\ \widehat{a}_1 \\ \vdots \\ \widehat{a}_{P-1} \\ \tilde{b}_0 \\ \tilde{b}_1 \\ \vdots \\ \tilde{b}_{P-1} \end{bmatrix}, \tag{A.16d}$$

$$F = \begin{bmatrix} X_0^P F_0 \\ X_1^P F_1 \\ \vdots \\ \vdots \\ \vdots \\ \vdots \\ X_{D-2}^P F_{D-2} \\ X_{D-1}^P F_{D-1} \end{bmatrix}, \tag{A.16e}$$

The P pole frequencies are then obtained as the roots of a Pth-order homogeneous polynomial from the characteristic equation (A.15a). Similarly, the residues can be found using (A.15b) or, more directly, by returning to (A.7) and matching P points or using all D points and using a pseudo-inverse to obtain a least-squares solution.

A.3.4 Discussion

Note that Prony's method essentially linearizes the problem (except for finding the roots of a Pth-order polynomial) of estimating the s_a and R_a in (A.6) or (A.7) by separating their solutions. The above steps (in time-domain Prony) comprise Prony's original method of 1795 as described by Hildebrand [18]. It is worth noting that (A.11) is a difference equation whose form in the Z-domain is that of a pole-zero model.

If more data points (a total of D') were available than the minimum $2W$ values required to compute the W poles of $f(t)$, this would clearly result in an over-determined system because the number of equations would exceed the number of unknowns. Two approaches to handling this situation might be considered. In the first, we could continue to use $D' = 2P'$, that is, compute as many poles as permitted by the number of data samples available. This number of poles would, of course, be greater than the number actually contributing to $f(t)$ and the data matrix in (A.11) should then be singular unless, as is invariably the case, the data are noisy or of otherwise limited accuracy. This approach has been followed in much of the early work in EM that employed Prony's method.

Alternatively, we could use the original value of P and reduce the number of equations from P' to P by multiplying (A.11) by the transpose of the data matrix to obtain a least-squares solution. This results in

$$\vec{\vec{M}}'_t \,\hat{c} = -\vec{f}', \tag{A.17a}$$

where

$$\vec{\vec{M}}'_t = \begin{bmatrix} C_{00} & C_{01} & \cdots & C_{0,P-1} \\ C_{10} & C_{11} & \cdots & C_{1,P-1} \\ \vdots & \vdots & \vdots & \vdots \\ C_{P-1,0} & C_{P-1,1} & \cdots & C_{P-1,P-1} \end{bmatrix} \tag{A.17b}$$

$$\vec{f}' = \begin{bmatrix} C_0 \\ C_1 \\ \vdots \\ C_{P-1} \end{bmatrix} \tag{A.17c}$$

$$C_{ij} = \sum_{k=0}^{P-1} f_{i+k} f_{j+k}, \tag{A.17d}$$

and

$$C_i = \sum_{k=0}^{P-1} f_{i+k} f_{P+k} \tag{A.17e}$$

The two approaches could also be combined by choosing any value of P' between P and $D'/2$.

The matrix of (A.17b) has the form of an autocovariance matrix of data samples which is a common starting point for digital signal processing of transient waveforms. This form results directly if the minimization of the square of the error between the model and the data is realized by varying the model parameters. Prony's method does not consider this error explicitly, however, and as originally formulated, it is essentially a point-matching procedure. The process of developing the new data matrix in (A.17b) through multiplying the original matrix by its transpose also minimizes the squared error; this is sometimes called regularization or is also referred to as generating the pseudo-inverse.

There are additional ways to handle an overdetermined system. One of the more useful in our experience is the so-called moving-window technique which involves processing the data using several sequential time windows that "move" along the waveform by $D = nd$ (n an integer ≥ 1). In this way, the data are not used all at one time and several sets of estimates are obtained for the poles, a particularly helpful procedure when the data is noisy, as Prony's method can be very susceptible to noise [19]. If the noise is random, these individual windows will yield independent estimates for the pole sets, and authentic poles should cluster around an average value while spurious poles due to noise and curve fitting (caused by $P > W$) will be scattered rather randomly in the s-plane. Superimposing all pole-set estimates on one graphical plot can then help to identify the valid poles. An interesting alternate procedure [20] is to process the data in both forward and reverse directions with respect to time. The result is that poles actually present in the waveform are reflected about the $j\omega$ axis while noise and curve-fitting poles remain predominantly in the negative half-plane. It is worth mentioning that Kay and Marple [17] found that, when compared with a number of other more recently developed spectral-estimation procedures for handling noisy, time-series data, a variant of Prony's method yielded the best estimate for a noise-contaminated waveform having discrete and broadband components.

References

[1] M. A. Fiddy, and G. Ross (1979), "Analytic Fourier Optics: The Encoding of Information by Complex Zeros," *Optica Acta*, Vol. 26, No. 9, pp. 1139–1146.

[2] M. A. Fiddy, G. Ross, M. Nieto-Vesperinas, and A. M. J. Huiser, "Encoding of Information in Inverse Optical Problems," *Optica Acta*, Vol. 29, No. 1, pp. 23–40, 1982.

[3] C. E. Baum, "Electromagnetic Transient Interaction with Objects with Emphasis on Finite Sized Objects and Some Aspects of Transient Pulse Production," in *Digest of Spring URSI Meeting*, Washington, DC, 1972.

[4] C. E. Baum, "The Singularity Expansion Method," Chapter 3 in *Transient Electromagnetic Fields*, Springer-Verlag, New York, 1976.

[5] E. K. Miller, "Model-Based Parameter-Estimation Applications in Electromagnetics," in *Electromagnetic Modeling and Measurements for Analysis and Synthesis Problems*, edited by B. de Neumann, Kluwer Academic Publishers, NATO ASI Series E: Applied Sciences Vol. 199, pp. 205–256, 1987, 1991.

[6] M. Bleszynski, *Private Communication*, 1987.

[7] R. F. Harrington, *Field Computation by Moment Methods*, Macmillan Co., New York, 1968.

[8] A. J. Poggio, and E. K. Miller, "Integral Equation Solutions of Three-Dimensional Scattering Problems," in *Computer Techniques for Electromagnetics*, Pergamon Press, pp. 159–264, 1973.

[9] C. L. Bennett, "Time Domain Inverse Scattering," *IEEE Trans. on Antennas and Propagation*, Vol. 29, No. 3, pp. 213–219, 1981.

[10] E. M. Kennaugh, and D. L. Moffatt, "Transient and Impulse Response Approximations," *Proceedings of the IEEE*, Vol. 69, pp. 893–901, 1965.

[11] J. T Winthrop, "Propagation of Structural Information in Optical Wave Fields," *Journal of the Optical Society of America*, Vol. 61, No. 1, pp. 15–30, 1971.

[12] G. Ross, M. A. Fiddy, M. Nieto-Vesperinas, and I. Manolitsakis, "The Propagation and Encoding of Information in the Scattered Field by Complex Zeros," *Optica Acta*, Vol. 26, No. 2, pp. 229–238, 1978.

[13] F. Gori, and L. Ronshi, "Degrees of Freedom for Scatterers with Circular Cross Section," *Journal of the Optical Society of America*, Vol. 71, No. 3, 1981.

[14] M. Bertero, and E. R. Pike, "Resolution in Diffraction-Limited Imaging, a Singular Value Analysis. I. The Case of Coherent Illumination," *Optica Acta*, Vol. 29, No. 6, pp. 727–746, 1982.

[15] E. K. Miller, and D. L. Lager, "Inversion of One-Dimensional Scattering Data Using Prony's Method," *Radio Science*, Vol. 17, No. 1, pp. 211–217, 1982.

[16] E. K. Miller, G. A. Clark, G. D. Poe, B. D. Cook, and J. A. Jackson, "Combining Pole and Ramp-Based Techniques for Target Identification," *NATO Advisory Group for Aerospace Research & Development Meeting on Target Signatures*, London, 1984.

[17] S. M. Kay, and S. L. Marple, "Spectrum Analysis – A Modern Perspective," *IEEE Proceedings*, Vol. 69, No. 11, pp. 1380–1419, 1981.

[18] F. Hildebrand, *Introduction to Numerical Analysis*, McGraw-Hill, New York, 1956.

[19] D. G. Dudley, "Fitting Noisy Data with a Complex Exponential Series," Lawrence Livermore Laboratory Rept. UCRL-52242, 1977.

[20] M. L. Van Blaricum, J. R. Auton, and T. L. Larry, "Target Identification via Complex Natural Resonance Extraction from Radar Signatures," *NATO AGARD Electromagnetic Wave Propagation Panel Symposium on Target Signatures*, London, 1984.

Appendix B

Symbols and notation

A]	vector of coefficients of the characteristic equation
ATR	automatic target recognition
AWE	asymptotic waveform evaluation
$B(Y)$	accuracy value expressed in bits of parameter Y
$B_r(Y_i)$	bit value for parameter Y_i from run r
CEM	computational electromagnetics
CNA	classical numerical analysis
d	element spacing in a linear array
δ	time step
D	number of waveform-domain samples over window of time duration T required to compute $N_W = D/2$ poles
D′	a number of WD samples such that $D' > D$ needed to estimate data accuracy
D_α	directivity of the αth source in a discrete source array
$D(f_i)$	the number of correct digits in input sample f_i
$\overline{D}(f)$	average number of correct digits per data sample
$\overline{D}(\tilde{s})$	average number of correct digits per pole
DC	Dolph–Chebyshev
$\overline{D}(\tilde{R})$	average number of correct digits per residue
DoF	degrees of freedom of data or pattern
DSA	discrete-source array
DSP	digital signal processing
$D(X,P)$	Denominator polynomial of order P
λ and $\overline{m}$	eigenvalues and eigenvectors of a matrix $\overline{M}$
EE	estimation error
EM	electromagnetics
EMP	electromagnetic pulse
FD	frequency domain

FDEM	frequency-domain electromagnetics
FDPM	frequency-domain Prony's method
FM(s)	fitting model(s)
FME	fitting-model error
$f(t)$	waveform
f_i	ith sample of waveform f at time it_p
f_m	the highest frequency component of $f(t)$
F]	vector of data samples
FDI	frequency-derivative interpolation
$F_{i,U}$ and $F_{i,L}$	Upper and lower frequencies of FM_i
FPMs	first-principles models
GM(s)	generating model(s)
I_{in}	total information content of the input data
I_{out}	output information content of extracted parameters
IE	integral equation
L	object size in wavelengths in plane of far-field pattern.
LLNL	Lawrence Livermore National Laboratory
[M]	matrix of data samples
M	exponentiation factor for linear-array pattern
MBPE	model-based parameter estimation
ME	Maxwell's equations
$M_i(n_i,d_i)$	rational-function fitting model of index number i and numerator and denominator polynomial orders n_i and d_i
$MM_{i,j}(f)$	minimum match of overlapping FMs or FM and GM usually in digits so that a minimum 2-digit $MM_{i,j}(f)$would be a 1% difference
MOR	model-order reduction
np	non-pole term
N_P	number of poles being calculated
N_W	number of poles in the actual data
$N(X, P-1)$	numerator polynomial of order $P-1$ in FM
OC	operation count
P_{XXX}	pattern factor of an linear array of type XXX
$P_{DSA}(\theta)$	pattern of a discrete source approximation
PEC	perfect electric conductor
PM	Prony's method
PFM	Prony fitting model
R	rank of data
R_α	complex residue

RCS	radar cross section
RM	Romberg's method
s	complex frequency
S	number of currents on surface of object whose pattern is being obtained
$S_\alpha, \alpha = 1, \ldots, N$	source strengths of a discrete-source approximation
SD	spectral domain
s_α	complex pole
S_α	source in linear array
SCF	sinusoidal current filament
SEM	singularity expansion method
SID	system identification
S_P	the peak-signal-to-peak-noise ratio
SVD	singular-value decomposition
TDEM	time-domain electromagnetics
TDPM	time-domain Prony's method
T	time duration of waveform-domain data window given by $t_p(D - 1)$
t_p	time spacing of waveform-domain data samples
t_w	time spacing of waveform-domain data windows
TDEM	time-domain electromagnetics
W	number of data windows or total angle observation window
W	width of an exponentiated linear array
WD	waveform domain
UCF	uniform current filament
VF	vector fitting
x	generic waveform-domain variable
X	generic spectral-domain variable
X_α	Roots of waveform-domain characteristic equation
$z_\alpha = z_{\alpha,real} + iz_{\alpha,imag}$	complex source location in a linear array
$\langle B(Y) \rangle$	average bit value of parameter Y
2-D	two dimensional
3-D	three dimensional
$\Delta \cos \theta$	sampling interval for Prony pattern analysis and synthesis
$f^{(i)}$	waveform i'th derivative sample
$\hat{X}$	estimated quantity
$B(Y)$	bits of agreement for quantity Y
$\overline{D}(\overline{P})$	average number of correct digits per parameter extracted
Δt_N	Nyquist rate sample spacing

Appendix C
A sampling of representative MBPE and related references in various disciplines

Acoustics (25)

L. Ainsleigh, J. D. George, and V. K. Jain, "Ma)Imum Likelihood Parameter Estimation For Acoustic Transducer Calibration," *Report prepared for Naval Research Laboratory Underwater Sound reference Detachment*, 1988.

J. S. Hall, and J. E. Michaelsa, *A Model-Based Approach to Dispersion and Parameter Estimation for Ultrasonic Guided Waves*, *Acoustical Society of America*, 2010. https://doi.org/10.1121/1.3273894.

Y. Lu*, and J. Saniile, "Model-Based Parameter Estimation for Defect Characterization in Ultrasonic NDE Applications," *EEE 6th International Symposium on Electromagnetic Compatibility and Electromagnetic Ecology*, 2005, 2015.

Sanda Lefteriu, Marcos Suza Lenzi, Harden Benot, Michel Tourmour, and Wim Desmet, "Fast Frequency Sweep Method for Indirect Boundary Element Models Arising in Acoustics," *Engineering Analysis with Boundary Elements*, Vol. 69, No. 2, pp. 32–45, 2016. https://doi.org/10.1016/j.enganabound.

N. Xiang, "Model-Based Bayesian Analysis in Acoustics—A Tutorial," *The Journal of the Acoustical Society of America*, Vol. 148, pp. 1101–1120, 2020.

Y. Zhang, X. Zhu, and J. Gao, "Parameter Estimation of Acoustic Wave Equations Using Hidden Physics Models," *IEEE Transactions on Geoscience and Remote Sensing*, Vol. 58, pp. 4629–4639, 2020.

Giacomo Moretti, and Gianluca Rizzello, "Model-Based Parameter Analysis of Dielectric Elastomer Loudspeakers," *Conference Paper, July, IEEE/ASME International Conference on Advanced Intelligent Mechatronics (AIM)*, 2022. https://doi.org/10.1109/AIM52237.2022.9863404.

Aerodynamics (6)

J. Sanwale, and D. Singh, "Aerodynamic Parameters Estimation Using Radial Basis Function Neural Partial Differentiation Method," *Defence Science Journal*, Vol. 58, pp. 241–250, 2018. https://doi.org/10.14429/dsj.68.11843.

H. Liu, L. Dai, M. You, S. Tan, and K. Liu, "Aerodynamic Modeling Based on Segmented Parameter Estimation in Frequency Domain," *Part of the book series: Lecture Notes in Electrical Engineering* ((LNEE, volume 861), 2021).

H. Zhe, K. Yinan, Y. Weigang, and C. Gang, *Aircraft Parameter Estimation Using a Stacked Long Short-Term Memory Network and Levenberg-Marquardt Method, Production and hosting by Elsevier Ltd. on behalf of Chinese Society of Aeronautics and Astronautics*, 2023.

H. Zhe, Y. Kong, Y. Weigang, and C. Gang, "Aircraft Parameter Estimation Using a Stacked Long Short-Term Memory Network and Levenberg-Marquardt Method," *Chinese Journal of Aeronautics*, Vol. 37, pp. 123–136, 2023.

Astronomy and Cosmology (15)

A. F. Heavens, R. Jimenez, and O. Lahav, "Massive Lossless Data Compression and Multiple Parameter Estimation from Galaxy Spectra," *Monthly Notices of the Royal Astronomical Society*, Vol. 317, pp. 965–972, 2000.

A. Liddle, P. Mukherjee, and D. Parkinson, "Model Selection in Cosmology," *A&G*, Vol. 47, Vol. 37, pp. 430–433, 2006.

K. Liao, Xi-Long Fan, Xuheng Ding, Biesiada M. and Zong-Hong Zhu, "Precision Cosmology from Future Lensed Gravitational Wave and Electromagnetic Signals," *Nature Communications*, Vol. 8, pp. 1148, 2017. https://doi.org/10.1038/s41467-017-01152-9.

J. R. Eskilt, K. Lee, D. J. Watts, *et al.*, "COSMOGLOBE: Towards End-to-End CMB Cosmological Parameter Estimation Without Likelihood Approximations," *Astronomy & Astrophysics*, Vol. 678, 2023.

W. Su, J. Yan, S. Sun, *et al.*, "Simulated Gravity Field Estimation for Deimos Based on Spacecraft Tracking Data," *Frontiers in Astronomy and Space Sciences*, Vol. 11, 2024.

H. Meskhidze, "Beyond Classification and Prediction: The Promise of Physics-Informed Machine Learning in Astronomy and Cosmology," Black Hole Initiative, Harvard University, 20 Garden Street, Cambridge, MA 02138, 2024, emeskhidze@g.harvard.edu.

Biological (20)

G. Lillacci, and M. Khammash, "Parameter Estimation and Model Selection in Computational Biology," *PLoS Computational Biology*, Vol. 6, No. 3, p. e1000696, 2010. https://doi.org/10.1371/journal.pcbi.1000696.

A. Gábor, and J. R. Banga, "Robust and Efficient Parameter Estimation in Dynamic Models of Biological Systems," *A BMC Systems Biology*, Vol. 9, p. 74, 2015. https://doi.org/10.1186/s12918-015-0219-2.

D. R. Penas, P. Gonzalez, J. A. Egea, R. Doallo, and J. R. Banga, "Parameter Estimation in Large-Scale Systems Biology Models: A Parallel and Self-Adaptive Cooperative Strategy," Vol. 18, p. 52, 2017. https://doi.org/10.1186/s12859-016-1452-4.

J. A. Pittl, and J. R. Banga, "Parameter Estimation in Models of Biological Oscillators: An Automated Regularised Estimation Approach," *Bioinformatics*, Vol. 20, p. 82, 2019.

N. A. Eshtewy, L. Scholz, and A. Kremling, "Parameter Estimation for a Kinetic Model of a Cellular System Using Model Order Reduction Method," in *Mathematical Biology: Modeling, Analysis, and Simulations*, 2nd Edition, Ricardo Lopez-Ruiz (eds.), 2023.

N. TolleyI, "Methods and Considerations for Estimating Parameters in Biophysically Detailed Neural Models with Simulation Based Inference," *PLoS Computational Biology*, Vol. 20, No. 2, p. e1011108, 2024. https://doi.org/10.1371/journal.pcbi.1011108.

Chemistry and Chemical Engineering (15)

G. Franceschini, and S. Macchietto, "Model-Based Design of Experiments for Parameter Precision: State of the Art," *Chemical Engineering Science*, Vol. 63, No. 19, pp. 4846–4872, 2008.

M. Federico Galvanin, and B. Fabrizio Bezzo, "Online Model-Based Redesign of Experiments for Parameter Estimation in Dynamic Systems," *Industrial & Engineering Chemistry Research*, Vol. 48, No. 9, 2009.

C. Song, "Parameter Estimation of the Pyrolysis Model for fir Based on Particle Swarm Algorithm," *IEEE Conference Publication, Second International Conference, IEEE Spectrum*, 2011.

L.G.M. de Souza, H. Haida, D. Thevenin, A. Seidel-Morgenstern, and G. Janiga, "Model Selection and Parameter Estimation for Chemical Reactions Using Global Model Structure," *Computers & Chemical Engineering*, Vol. 58, pp. 269–277, 2013.

D. C. López, T. Barz, S. Körkel, and G. Wozny, "Nonlinear Ill-Posed Problem Analysis in Model-Based Parameter Estimation and Experimental Design," *Computers & Chemical Engineering*, Vol. 77, pp. 24–42, 2015.

E. Che Mid, and V. Dua, "Model-Based Parameter Estimation for Fault Detection using Multiparametric Programming," Department of Chemical Engineering, Centre of Process System Engineering (CPSE), University College London, London, 2017.

D. Zhang, Wang K, Xu Z, *et al.*, "Generalized Parameter Estimation Method for Model-Based Real-Time Optimization," *Chemical Engineering Science*, Vol. 258, p. 117754, 2022.

A. N. Prudente, R. V. A. Santos, A. M. Ribeiro, K. V. Pontes, I. B. R. Nogueria, "An Integrated Approach for Parameter Estimation of a Propyl Propionate Synthesis Model," *Chemical Engineering Research and Design*, Vol. 202, pp. 317–326, 2024.

Economics (8)

J. M. Wooldridge, "Applications of Generalized Method of Moments Estimation," *The Journal of Economic Perspectives*, Vol. 15, No. 4, 2001.

H. Seya, M. Tomari, and S. Uno, "Parameter Estimation in Spatial Econometric Models with Non-Random Missing Data," *Applied Economics Letters*, Vol. 32, No. 6, pp. 1195–1200, 2020.

C. M. Jacopo Lenti, and G. De Francisci Morales, "Likelihood-Based Methods Improve Parameter Estimation in Opinion Dynamics Models," *WSDM'24: Proceedings of the 17th ACM International Conference on Web Search and Data Mining*, pp. 350–359, 2023.

Electrical Engineering (65)

S. Wang, V. Dinavahi, and J. Xiao, "Multi-Rate Real-Time Model-Based Parameter Estimation and State Identification for Induction Motors," *IET Electric Power Applications*, Vol. 7, No. 1, pp. 77–86, 2013. https://doi.org/10.1049/iet-epa.2012.0116

C. Dzienis, R. Matussek, and H. Schulte, "Parameter Estimation in Electrical Power Systems Using Prony's Method," IET Electric Power Applications," *Journal of Physics Conference Series*, Vol. 659, pp. 1–12, 2015.

R. Ahmed, Model-Based Parameter Identifcation of Healthy and Aged Li-ion Batteries for Electric Vehicle Applications, *SAE International Journal of Electrified Vehicles*, 2015. https://doi.org/10.4271/2015-01-0252, saealtpow.saejournals.org

M. I. Abdelwanis, A. Abaza, R. A. El-Sehiemy, M. Ibrahim, and H. Rezk, "Parameter Estimation of Electric Power Transformers Using Coyote Optimization Algorithm With Experimental Verification," *IEEE Access*, Vol. 8, pp. 50036–50044, 2020. https://doi.org/10.1109/ACCESS.2020.2978398.

P. D. Achlerkar, and B. Ketan Panigrahi, "Recursive Least Squares-Based Adaptive Parameter Estimation Scheme for Signal Transformation and Grid Synchronization," *IEEE Journal Of Emerging And Selected Topics In Power Electronics*, Vol. 9, No. 2, pp. 2427–2439, 2021.

A. N. Ali, E. C. Mid, N. A. Ahamad, and E. Ruslan, "Parameter Estimation of DC Motor using Multiparametric Programming," *Journal of Physics Conference Series*, Vol. 2312, p. 012035, 2022.

K. M. Charu, "An Efcient Data Sheet Based Parameter Estimation Technique of Solar PV," *Scientifc Reports*, Vol. 14, p. 6461, 2024.

J. Mei, X. Meng, X. Tang, *et al.*, "An Accurate Parameter Estimation Method of the Voltage Model for Proton Exchange Membrane Fuel Cells," *Energies*, Vol. 17, p. 2917, 2024. https://doi.org/10.3390/en17122917.

Electromagnetics (>600)

G. J. Burke, E. K. Miller, S. Chakrabarti, and K. Demarest, "Using Model-Based Parameter Estimation To Increase The Efficiency Of Computing Electromagnetic Transfer Functions," *IEEE Transactions On Magnetics*, Vol. 25, No. 4, 1989.

K. R. Demarest, "Model-Based Parameter Estimation In Eletromagnetic Computer Modeling," Office of Naval Research Arlington, Virginia 22217-5000, Contract N00014-86-K-0326, 1989.

S. Chakrabarti, R. Ravichander, E. K. Miller, J. R. Auton, and G. J. Burke, "Applications of Mqdel-Based Parameter Estimation in Electromagnetic Computations," 89CHU85-4/90/0000M121.00 6 1990 IEEE, 1990.

E. K. Miller, "Using Model-Based Parameter Estimation for Characterizing Broadband Electromagnetic Transfer Functions," in *Proceedings of the 1992 Asia-Pacific Microwave Conference*, Adelaide, 1992.

P. M. Baggenstoss, and R. Kumaresan, "On the Estimation of Rational Transfer Functions from Samples of the Power Spectrum," *IEEE Transactions on Signal Processing*, Vol. 41, No. 3, pp. 1431–1435, 1993.

M. D'Amore, and M. S. Sarto, "Time-Domainanalysis of Field-Excited Transmission Line Systems By Using Model-Based Parameter Estimation," *Proceedings of IEEE Symposium on Electromagnetic Compatibility*, 1994.

J. Tobias de Beer, and D. C. Baker, "An Examinaation of the Effect of Mechanical Deformation on the Input Impedance of HF LPDA's Using MBPE," Department of Electrical and Electronic Engineering, U. Of Pretoria, South Africa, 1995.

J. E. Pekarek, and T. Itoh, "Use of Frequency Derivatives in the Three-Dimensional Full-Wave Spectral-Domain Technique," *IEEE Transactions on Microwave Theory and Techniques*, 1996. https://doi.org/10.1109/22.554579.

E. K. Miller, "Investigating the Use of Model-Based Parameter Estimation for Electromagnetic Phase Recovery," in *13th Annual Review of Progress in Applied Computational Electromagnetics*, Naval Postgraduate School, Monterey, CA, pp. 418–419, 1997.

D. R. Plug, "Model Based Parameter Estimation as a Model Abstraction Technique," *1998 IEEE Information Technology Conference*, Information Environment for the Future, 1998 (Cat. No.98EX228).

E. K. Miller, "Using Windowed, Adaptive Sampling to Minimize the Number of Field Values Needed to Estimate Radiation And Scattering Patterns," *14th Annual Review of the Applied Computational Electromagnetics Society, Naval Postgraduate School*, Monterey, CA. pp. 958–963, 1998.

R. J. Allard, and D. H. Werner, "Simultaneous Spatlal/Spectral Interpolationof Antenna Radiation Palterns," *IEEE Antennas and Propagation Society International Symposium Digest*, Cat. No.99CH37010, 1999.

V. Fiumara, V. Galdi, V. Pierro, and I. M. Pinto, "A Model-Based Parameter Estimation Approach for Numerical Analysis of Single-Mode Optical Fibers," *Journal of Lightwave Technology*, Vol. 17, No. 4, pp. 684–689, 1999.

E. K. Miller, and T. K. Sarkar, "Model-Order Reduction in Electromagnetics Using Model-Based Parameter Estimation," in *Frontiers in Electromagnetics*, D. H. Werner and R. Mittra (eds.), 2000. https://doi.org/10.1109/9780470544686.ch9.

R. Abacherli, M. Mattes, E. Suter, and J. R. Mosig, "Combining the Genetic Algorithm Approach and the Model-Based Parameter Estimation into an Adaptive Frequency Sampling Algorithm," 2001.

31st European Microwave Conference, September 2001.

M. Kahrizi, S. Safieddin Safavi-Naeini, S. K. Chaudhuri, and R. Sabry, "Computer Diagnosis and Tuning of RF and Microwave Filters Using Model-Based Parameter Estimation," *IEEE Transactions on Circuits and Systems I: Fundamental Theory and Applications*, Vol. 19, No. 9, pp. 1263–1270, 2002.

R. J. Allard, and D. H. Werner, "The Model-Based Parameter Estimation of Antenna Radiation Patterns Using Windowed Interpolation and Spherical Harmonics," *IEEE Transactions on Antennas and Propagation*, Vol. 51, No. 8, pp. 1891–1906, 2003.

M. Schick, and F. M. Landstorfer, "Accelerating Computations with a MoM-Based Computer Program Using a Model-Based Parameter Estimation Algorithm," *ACES Journal*, Vol. 18, No. 4, pp. 48–53, 2003.

D. Deschrijver, T. DhaeneJan, and Broeckhove, "Adaptive Model Based Parameter Estimation, Based on Sparse Data and Frequency Derivatives," *Computational Science - ICCS 2004*, 4th International Conference, Kraków, Poland, June 6–9, Proceedings, Part II, 2004.

C. J. Reddy, "Application of MBPE Method to Frequency Domain Hybrid Techniques to Compute RCS of Electrically Large Objects," *Tech Science Press cmes*, Vol. 5, No. 5, pp. 455–462, 2005.

Y. Kim, and H. Ling, "On the Optimal Sampling Strategy for Model-Based Parameter Estimation Using Rational Functions," *IEEE Transactions on Antennas and Propagation*, Vol. 54, No. 2, pp. 762–765, 2006.

H. R. Karami, R. Moini, and S. H. H. Sadeghi, "A Model-Based Parameter Estimation of Current Distribution along Lightning Return Stroke Channel," *Proceedings, 18th International. Zurich Symposium on EMC*, Munich, 2007.

S. H. Zainud-Deen, S. I. El-Doda, K. H. Awadalla and H. A. Sharshar, "The Relation between Lumped-Element Circuit Models For Cylindrical Dielectric Resonator and Antenna Parameters Using MBPE," *Progress In Electromagnetics Research M*, Vol. 1, pp. 79–93, 2008.

P. Hagg, "Model Based Parameter Estimation for a Missile," Masters' Degree Project, KTH Electrical Engineering, Stockholm, Sweden July, 2009.

X. Wang, and D. H. Werner, "Improved Model-Based Parameter Estimation Approach for Accelerated Periodic Method of Moments Solutions With Application to the Analysis of Convoluted Frequency Selected Surfaces and Metamaterials," *IEEE Transactions on Antennas and Propagation*, Vol. 58, No. 1, pp. 122–131, 2010.

D. Wu, Member, IEEE, W. Yu, and Z. Yin, "Parameter Estimation of Rational Models Based on Artificial Bee Colony Algorithm," *Proceedings of 2011 International Conference on Modelling, Identification and Control*, Shanghai, China, June 26–29, 2011.

F. Kaburcuk, S. Arvas, E. Arvas, and J. K. Lee, "Wideband Information in MOM Obtained from Narrowband Data," 2012 *IEEE International Conference on Ultra-Wideband*, 2012.

E. Miller, “Adaptive Windowed Estimation of Electromagnetic Frequency Spectra and Radiation Patterns to a Specified Uncertainty,” *Conference: IEEE, Digital Signal Processing and Signal Processing Education Meeting (DSP/SPE)*, 2013. https://doi.org/ 10.1109/DSP-SPE.2013.6642564.

X. Wang, D. H. Werner, and J. P. Turpin, “Application of AIM and MBPE Techniques to Accelerate Modeling of 3-D Doubly Periodic Structures with Non-orthogonal Lattices Composed of Bianisotropic Media,” *Proceedings 2011 IEEE International Symposium on Antennas and Propagation*, Vol. 62, No. 8, pp. 2637–2460, 2014.

Z. Hongwe, “Calculation of Frequency Dependent Parameters of a Buried Conductor with Earth Return by Model Based Parameter Estimation Method,” Corpus ID: 111501380, 2015.

Y. Yang, Y. Ma, and L. Wang, “The Simultaneous Interpolation of Target Radar Cross Section in Both the Spatial and Frequency Domains by Means of Legendre Wavelets Model-Based Parameter Estimation,” *Hindawi Publishing Corporation International Journal of Aerospace Engineering*, Vol. 2015, Article ID 543787, 2015.

Ngoy Mutonkole, Elizabeth Rite Samuel, Dirk I. L. d Villiers and Tom Dhaene, “Parametric Modeling of Radiation Patterns and Scattering Parameters of Antennas,” *IEEE Transactions on Antennas and Propagation*, Vol. 64, No. 3, pp. 1023–1031, 2016.

F. Kaburcuk, “Fast Wideband Solutions Obtained Using Model Based Parameter Estimation with Method of Moments,” *Advanced Electromagnetics*, Vol. 6, No. 3, pp. 13–17, 2017.

W. Bin Kong, X. Fang Yang, F. Zhou, *et al.*, “Fast Wideband Electromagnetic Analysis Using the Interpolation Technique and Fast Generating Matrix Method,” *ACES Journal*, Vol. 33, No. 12, Vol. 69, No. 2, pp. 1360–1367, 2018.

S. Zhang, J. Yao, B. Li, H. Lu, and W. Tang, “Electric Field Reconstruction Using Block Spatial Model-Based Parameter Estimation,” *2019 8th Asia-Pacific Conference on Antennas and Propagation (APCAP)*, Incheon, Korea (South), pp. 375–376, 2019. https://doi.org/10.1109/APCAP47827.2019.9472042.

H. Li, J.-M. Jin, Fellow, D. R. Jachowski, and R. B. Hammond, “Fast Frequency Sweep Analysis of Passive Miniature RF Circuits Based on Analytic, Extension of Eigenvalues,” *IEEE Transactions on Microwave Theory and Techniques*, Vol. 9, pp. 4–14, 2020.

Z. Wang, and S. V. Hum, “A Broadband Model-Based Paremeter Estimation Method for Analyzing Multilayer Periodic Structures,” *IEEE Transactions on Antennas and Propagation*, Vol. 69, No. 9, pp. 5771–5780, 2012.

H.-B. Yuan, W.-T. Bao, C. Hyun Lee, B. F. Zinser, S. Campione, and J.-F. Lee, “A Method of Moments Wide Band Adaptive Rational Interpolation Method for High-Quality Factor Resonant Cavities,” *IEEE Transactions on Antennas and Propagation*, Vol. 70, No. 5, 2022.

L. Codecasa, G. Guido Gentili, and M. Oldoni, “Exploiting Port Response for Multimode Wideband Analysis of Lossy Devices,” *IEEE Transactions on Microwave Theory and Techniques*, Vol. 71, No. 4, 2023.

D. Peumans, S. De Keersmaeker, C. Busschots, Y. Rolain, Fellow, and F. Ferranti, "Adaptive Frequency Sampling Based on Local Rational Modeling for Microwave Electromagnetic Simulations," *IEEE Transactions on Microwave Theory and Techniques*, Vol. 72, No. 8, 2024.

Fluid Mechanics

E. A. de Barros, A. Pascoal, and E. de Sa, "AUV Dynamics: Modelling and Parameter Estimation Using Analytical, Semi-Empirical, and CFD Methods," *IFAC Proceedings*, Vol. 37, No. 10, pp. 369–376, 2004.

P. J. Smith, S. Dance, and N. K. Nichols, "A Hybrid Data Assimilation Scheme for Model Parameter Estimation: Application to Morphodynamic Modelling," *10th ICFD Conference Series on Numerical Methods for Fluid Dynamics, Computers & Fluids*, Vol. 46, No. 1, 2011

S. N. Lipp,E. E. Niedert, H. L. Cebull, *et al.*, "Computational Hemodynamic Modeling of Arterial Aneurysms: A Mini-Review," *Frontiers of Physiology*, Vol. 11, p. 454, 2020. https://doi.org/10.3389/fphys.2020.00454.

Z. Shen, and Y. Tang, "A Two-Stage Inflation Method in Parameter Estimation to Compensate for Constant Parameter Evolution in Community Earth System Model," *Acta Oceanologica Sinica*, Vol. 41, pp. 91–102, 2022.

S. R. Morab, A. Sharma, and J. S. Murallidharan, "Computational Hemodynamics in Human Vasculature: A Review on Role of Rheology, Multiphase Flow, and Fluid–Structure Interaction," *Journal of the Indian Institute of Science*, Vol. 104, pp. 13–38, 2024.

General

L. Spreeuwers, K. Schutte, and Z. Houkes, "Model-Based Recognition and Parameter Estimation of Buildings from Multi-View Aerial Imagery Using Multi-Segmentation," Laboratory for Measurement and Instrumentation, Dept. of Electrical Engineering University of Twente, P.O. Box 217, 7500 AE Enschede, Netherlands TNO Physics and Electronics Laboratory, Netherlands, 1997.

M. Kahrizi, S. Safavi-Naeini, and S. K. Chaudhuri, "Computer Diagnosis and Tuning of Microwave Filters Using Model-Based Parameter Estimation and Multi-level Optimization," 2000 IEEE MIT-S Digest, 2000.

Z. S. Abo-Hammour, O. M. K. Alsmadi, A. M. Al-Smadi, M. I. Zaqout, and M. S. Saraireh, "ARMA Model Order and Parameter Estimation Using Genetic Algorithms," *Mathematical and Computer Modelling of Dynamical Systems, Methods, Tools and Applications in Engineering and Related Sciences*, Vol. 18, No. 2, 2012.

I. Farajpour, and S. Atamturktur, "Error and Uncertainty Analysis of Inexact and Imprecise Computer Models," *Journal of Computing in Civil Engineering*, Vol. 24, No. 4, pp. 528–531, 2013.

J. Zhao, W. Wang, and C. Sheng, "Parameter Estimation and Optimization," In *Data Driven Prediction for Industrial Processes and Their Applications. Information Fusion and Data Science*. Springer, Cham, 2018. https://doi.org/10.1007/978-3-319-94051-9_7.

A. Alim, and D. Shukla, "A Study on Sampling-Based Parameter Estimation Techniques in Big Data Analytics Environment," in *Intelligent Computing Applications for Sustainable Real-World Systems,* M. Pandit, L. Srivastava, R. Venkata Rao, and J. Bansal (eds.), 2019.ICSISCET

K. Kumar, "Data-Driven Modeling and Parameter Estimation of Nonlinear Systems," *European Physical Journal B*, Vol. 96, p. 107, 2023. https://doi.org/10.1140/epjb/s10051-023-00574-3.

K. Kondo, A. Fukushima, T. Yoshida, and K. Matsue, "Equation-Based Modeling and Optimization-Based Parameter Estimation in Multimodal Virtual Sensing Platforms for Smart Building," *Building and Environment*, Vol. 243, p. 110620, 2023.

A. Li, and Y. Li, *Rational Empirical Interpolation Methods with Applications, School of Mathematical Sciences, Zhejiang University, Hangzhou, Zhejiang 310058*, China, 2024.

Information Content

William S. Bickel, "Information Content and Sensitivity of Matrix Elements To Structural Changes in Complex Scattering System," Physics Department, University of Arizona B Tucson, Arizona 85721, US Army Armament Research and Development Command, Aberdeen Proving Ground, Maryland 21010, 1983.

T. Günther, S. Friedel, and K. Spitze, "Estimation of Information Content and Efficiency for Different Data Sets and Inversion Schemes Using the Generalized Singular Value Decomposition," Workshop on Electromagnetic Depth Sounding, 3.-7.-10.2005, Konigstein, 2003.

P. Ormerod, "Extracting Information from Noisy Time Series Data," Volterra Consulting Ltd, Sheen Elms, 135c Sheen Lane, London SW14 8AE, 2004.

SERGEYLOYKA, "On the Relationship of Information Theory and Electromagnetism," *IEEE 6th International Symposium on Electromagnetic Compatibility and Electromagnetic Ecology*, 2005.

Q. Xu, "Measuring Information Content from Observations for Data Assimilation: Relative Entropy Versus Shannon Entropy Difference," Tellus, 59A, 198–209, Journal compilation, Blackwell Munksgaard, 2006.

M. Donald Migliore, "On Electromagnetics and Information Theory," *IEEE Transactions On Antennas and Propagation*, Vol. 56, No. 10, 2008.

Mechanical and Structural (40)

A. Wolf, and O. Gottlieb, "Estimation of Rigid-Body Motion via an Indirect Measurement of An Elastic Appendage," *Journal of Vibration and Acoustics*, 2007. https://doi.org/10.1115/DETC2007-35

G. Franceschini, and S. Macchietto, "Model-Based Design of Experiments for Parameter Precision: State of the Art," *Chemical Engineering Science*, Vol. 63, No. 19, 2008.

H. Georg Bock, and T. Carraro, *Model Based Parameter Estimation: Theory and Applications*, Springer Publishing Company, Incorporated, Vol. 4, pp. 171–190, 2015.

Hu, Sau-Lon J., Liang Su, and Hua Jun Li. "Estimating Frequency Response Functions by pole-Residue Operations," *IOMAC 2017 - 7th International Operational Modal Analysis Conference*, 2017, pp. 291–294. https://digitalcommons.uri.edu/oce_facpubs/251 (https://digitalcommons.uri.edu/oce_facpubs/251).

P. Tran Tin, M. Tran, L. Anh Vu*, N. Quang Dung, and T. Thanh Trang, "Real Interpolation Method for Transfer Function Approximation of Distributed Parameter System," *Telkomnika*, Vol. 17, No. 4, pp. 1941–1947, 2019.

X. Chen, Y. Hou, and P. Xi, "Parameter Estimation of the Structured Illumination Pattern Based on Principal Component Analysis (PCA): PCA-SIM," Light: Science & Applications. Official Journal of the CIOMP 2047–7538, 2023. https://doi.org/10.1038/s41377-022-01043-9.

X. Youa, X. Yana, J. Liua, S. Lia, and Y. Liue, "A Novel Model-Based Parameters Estimation Combining Local Optimization and Global Optimization of Nonlinear Ship Models with Physical Experiment Dataset," Preprint submitted to Elsevier, 2024.

Medical (30)

K. R. Fowler, G. A. Gray, and M. S. Oluf, "Modeling Heart Rate Regulation—Part II: Parameter Identification and Analysis," *Cardiovascular Engineering*, 2007. https://doi.org/10.1007/s10558-007-9048-2.

A. L. Lloyd, "Sensitivity of Model-Based Epidemiological Parameter Estimation to Model Assumptions," in *Mathematical and Statistical Estimation Approaches in Epidemiology*, G. Chowell *et al.* (eds.), C! Springer Science+Business Media B.V., 2009. https://doi.org/10.1007/978-90-481-2313-1 6.

A. Moustafa Beni-Hassan, M. Fatma El-Hefnawi, and M. Yasser Kadah, "Model-Based Parameter Estimation Applied on Electrocardiogram Signal," *Journal of Computational Biology and Bioinformatics Research*, Vol. 3, No. 2, pp. 25–28, 2011.

M. S. Olufsen, and J. T. Ottesen, "A Practical Approach to Parameter Estimation Applied to Model Predicting Heart Rate Regulation," *Journal of Mathematical Biology*, Vol. 67, pp. 39–68, 2012.

V. K. Nguyen, and E. A. Hernandez-Vargas, "Parameter Estimation in Mathematical Models of Viral Infections Using R," *Methods of Molecular Biology*, Vol. 1836, pp. 531–549, 2018. https://doi.org/10.1007/978-1-4939-8678-1_25.

D. Fattahi, and R. Sameni, "Theoretical Performance Bounds of Model-Based Electrocardiogram Parameter Estimation," *IEEE Transactions on Biomedical Engineering*, Vol. 70, No. YY Preprint, Vol. 70, pp. 3181–3192, 2021. DOI: 10.36227/techrxiv.15176139.v2.

E. Taghizadeh, and A. Mohammad-Djafar, "SEIR Modeling, Simulation, Parameter Estimation, and Their Application for COVID-19 Epidemic Prediction," *Physical Sciences and Forum*, Vol. 5, p. 18, 2022. https://doi.org/10.3390/psf2022005018.

A. Gerardo, B. Chowell, and R. Luo "Parameter Estimation and Forecasting with Quanti!ed Uncertainty for Ordinary Differential Equation Models Using QuantDi!Forecast: A MATLAB Toolbox and Tutorial," *Statistics in Medicine*, Vol. 43, No. 9, pp. 182–1848, 2024.

Physics

V. Fiumara, "A Model-Based Parameter Estimation Approach for Numerical Analysis of Single-Mode Optical Fibers," *Journal of Lightwave Technology*, Vol. 17, No. 4, 1999. (/jlt/issue.cfm?volume=17&issue=4).

C. Fazekas, and I. Varga, "Modeling and Parameter Estimation of a Nuclear Power Plant," *9th International PhD Workshop on Systems and Control*, October 2008, Izola, Slovenia, 2008.

Y. Liu, Z. Liu, and S. Zhang, "Ensemble-Based Parameter Estimation in a Coupled General Circulation Model," *Journal of Climate*, Vol. 27, No. 18, 2014. https://doi.org/10.1175/JCLI-D-13-00406.1.

M. Andersson, "Parametrization of Physics-Based Battery Models from Input–Output Data: A Review of Methodology and Current Research," *Journal of Power Sources*, Vol. 521, p. 230859, 2022.

A. L. Milder, A. S. Joglekar, W. Rozmus, and D. H. Froula, "Qualitative and Quantitative Enhancement of Parameter Estimation for Model-Based Diagnostics Using Automatic Differentiation with an Application to Inertial Fusion," *Machine Learning: Science and Technology*, Vol. 5, p. 015026, 2024.

Power and Devices (10)

F. F. Chen, X. Junting, D. G. Thomas, and A. B. Murphy, "Model-Based Parameter Optimization for Arc Welding Process Simulation," *Applied Mathematical Modelling*, Vol. 81, pp. 386–400, 2020.

A. Sel, B. Sel, and C. Kasnakoglu, "GLSDC Based Parameter Estimation Algorithm for a PMSM Model," Selected Papers from Eight European Conference on Renewable Energy Systems (ECRES2020), 2021. (/journal/energies/special_issues/).

H. Rezk, A. G. Olabi, S. Ferahtia, and E. Taha Sayed "Accurate Parameter Estimation Methodology Applied to Model Proton Exchange Membrane Fuel Cell," *Energy*, 2022. www.elsevier.com/locate/energy.

F. Leguebedj, and D. Boukhetala, "Estimation of Synchronous Machine Parameters by Stand Still Frequency Responses Testing," *ENP Engineering Science Journal*, Vol. 3, No. 2, 2023.

Signal Processing

K. C. Lo, and A. Purvis, "A New Approach for Estimating Spectra from Randomly Sampled Sequences," *Circuits Systems and Signal Process*, Vol. 16, pp. 375–386, 1997. https://doi.org/10.1007/BF01246719.

Dušan Agrež, "Periodic Signal Parameters' Estimations from Random Sampling Measurements under Noncoherent Sampling Condition," *21st IMEKO TC4 International Symposium and 19th International Workshop on ADC Modelling and Testing*, Budapest, Hungary, September 7–9, 2016.

C. S. Burrus, "Prony, Pad'e, and Linear Prediction for Interpolation and Approximation in the Time and Frequency Domain Design of IIR Digital Filters and in Parameter Identification," *arXiv: Signal Processing*, 2019. https://arxiv.org/abs/1909.05298.

H. Zhang, J. Harlim, and X. Li, "Linear Response Based Parameter Estimation in the Presence of Model Error," *Journal of Computational Physics*, Vol. 43, pp. 110–112, 2021.

K. Kuma, "Data-Driven Modeling and Parameter Estimation of Nonlinear Systems," *European Physical Journal B*, Vol. 96, p. 107, 2023. https://doi.org/10.1140/epjb/s10051-023-00574-3.

Thermodynamics (7)

Z. O'Neill, S. Narayanan, and R. Brahme, "Model-Based Thermal Load Estimation in Buildings," Fourth National Conference of IBPSA-USA, New York City, 2010.

D. C. López, "Model-Based Identifiable Parameter Determination Applied to a Simultaneous Saccharification and Fermentation Process Model for Bio-Ethanol Production," in *Biotechnology Progress*, 2013. https://doi.org/10.1021/btpr.1753.

Y. Chen, J. Castiglione, R. Astroza, and Y. Li, "Parameter Estimation of Resistor-Capacitor Models for Building Thermal Dynamics Using the Unscented Kalman Filter," *Journal of Building Engineering*, Vol. 34, p. 101639, 2021.

H. Berian Pratama, and K. Koike. "Thermodynamic Model-Based Specification of Optimal Geothermal Power Generation System for High-Temperature Liquid Dominated Systems Using Flash and Flashbinary Cycles," *Renewable Energy*, Vol. 220, p. 119634, 2024.

Index

www.ingramcontent.com/pod-product-compliance
Lightning Source LLC
Jackson TN
JSHW081449120326
99266JS00003B/520
* 9 7 8 1 8 3 7 2 4 5 3 7 6 *